BETWEEN THE PLOUGH AND
THE PICK

INFORMAL, ARTISANAL AND
SMALL-SCALE MINING IN
THE CONTEMPORARY WORLD

BETWEEN THE PLOUGH AND
THE PICK

INFORMAL, ARTISANAL AND SMALL-SCALE MINING IN THE CONTEMPORARY WORLD

EDITED BY
KUNTALA LAHIRI-DUTT

Australian
National
University

PRESS

ANU PRESS

Published by ANU Press
The Australian National University
Acton ACT 2601, Australia
Email: anupress@anu.edu.au
This title is also available online at press.anu.edu.au

A catalogue record for this book is available from the National Library of Australia

ISBN(s): 9781760461713 (print)
9781760461720 (eBook)

Cover design and layout by ANU Press. Cover photograph by P. Madhavan, taken in 2006 as part of the ANU-based Asia Pacific Learning Communities project funded by Communities and Small-Scale Mining.

Contents

Section Three: Conflicts and governance

List of figures and tables

Preface and acknowledgements

Kuntala Lahiri-Dutt

This book presents experiences and insights gathered by a number of academics and policymakers through years of engagement with the thematic topic of informal, artisanal and small-scale mining, also known as ASM, the unprecedented rush by the global poor for commodities or jobs in extractive operations that have been unleashed by an increasingly resource-hungry world. My personal interest in the phenomenon began with my research into environmental and social changes that took place in mining areas during the mid-1990s in India. Those years roughly marked the beginning of a 'mining boom' in the country as the economy began to adopt more liberal economic policies. My early investigations of the impacts of mining on local environments and communities brought me to examine how—in the face of a degraded local environmental base—the communities living around mining areas survive. Around the coal-mining belt in eastern India had grown one of the world's largest continuously urbanised tracts into which the cities and towns not only sucked in people from all around, but also absorbed the surrounding rural land—the built-up tract expanding and transforming the rural areas. The undulating hills of the area began to be mined for stones and other construction material. Not only did the environment degrade but, with the expansion of mining and urban centres, the local peasantry began to decay as traditional rural farmers moved on to other, non-farm occupations, often working as wage labourers for contractors. These contractors were charged with procuring stones, and thus removed the overburdens, extracted the material for the mining companies, transported them to storage depots, loaded them onto the trucks and hauled them across all the sinews of transport—from the major arterial highways to local village roads, and narrow tracks that wound through the few bits of forests that were still left. This fundamental transformation of the social landscape was accomplished within the span of less than two decades.

This book takes up the much-needed task of investigating and explaining similar transformations—triggered by a number of ill-understood factors—that are sweeping through the mineral-rich tracts of countries in Africa, Latin America and Asia. In particular, the book deals with and puts under the microscope the political–economic processes of unleashing the extractive giants that have also pushed or lured the local communities to join in, leading to an extractivism unfamiliar to the industrialised world and hence cannot be explained by the same models of processes pertaining to that world. The extractivism of the poor creates an untidy and chaotic world that is difficult to explain and accommodate in the discourses of mining: of the triple bottom line, win–win solutions, unionised labour, corporate social responsibility, social licence to operate, sustainability and health and safety. Therefore, to shift the prevalent understanding of such mining as destructive to the environment, to make sense of the array of theoretical, conceptual and practical issues thrown up by it, to place them within the historical context and to relate them to the contemporary political economy of resource extraction is an urgent priority. The contributions in this book deal with a range of such issues pertaining to this little-understood world, attempting to make sense of it in different ways. It is also important to 'get inside the belly of the beast' to understand it, and several contributions from renowned scholars and practitioners help us understand a whole range of issues related to this kind of mining. Consequent to the immense variety that exists within informal, artisanal and small-scale mining, one can only expect the diversity of voices, interpretations and explanations that can be detected in these contributions.

Personally, it has been a long journey in understanding the social, economic and political aspects of the transformations of space and territories within which informal, artisanal and small-scale mining takes place. In this journey, I have benefited from having the privilege of knowing colleagues and receiving support from a number of organisations operating at the local, national and global levels. From India, my research took me to Indonesia, where a number of colleagues, including my dearest former PhD student Nina Lestari, gave me company in my early days of visiting remote locations, to Lao PDR, where I was fortunate to know Chansouk Invouvanh who exposed me to the fascinating history of Phon Tiu mine, and to Mongolia, where Hishgee Dondov took me deep into the rangelands of the Middle Gobi to observe how the lives of nomadic herders are changing in the face of the onslaught of mining. All three helped me to appreciate the vastly different contexts, allowing me to expand my vision and deepen my understanding. Besides the three mentioned, I am grateful to a number of individuals at a personal level for sharing their insights, which have enriched my intellectual journey. There are far too many for me to name each one of them individually within this space. I will name only two persons: Dr David Williams, my husband and partner in crime, who walked the rough paths with me, and my younger son, Ovimanyu Dutt, who made the great sacrifice of spending days and nights alone so that his mother could sprint around. I am grateful to both.

Besides individuals, a number of organisations need to be acknowledged. In the early years, the Communities and Small-Scale Mining (CASM) group (now defunct) offered two lots of development grants to establish the ASM Asia-Pacific Learning Network in 2004 and 2005. This allowed me to reach out to ordinary people, to build partnerships with them in a context that was characterised by a lack of knowledge at that time. The network is still alive at www.asmasiapacific.com. In more recent years, in 2013, the Australian Research Council generously provided funding for an action research project, titled 'Beyond the Resource Curse: Charting a Path to Sustainable Livelihoods for Mineral-Dependent Communities' (Discovery Project DP130104396). This project examined—jointly with civil society partners based in India—micro-level mineral-based livelihoods, placing them in context of the local history of agrarian transition. As a metal detector manufacturing company came forward to invest in the understanding of informal and artisanal gold-mining, with them as partners and jointly with my colleague Dr Keith Barney, I received funding yet again from the Australian Research Council for a project titled 'Going for Gold: Safe Livelihoods for Informal Gold Miners in South and Southeast Asia' (Linkage Project LP130100942). Together, the projects attempted to shift scholarly attention from viewing informal miners only as environment destroyers, to the context of poverty and rural political economy within which the mining communities operate. I thank the Resources, Environment and Development Group in the Crawford School of Public Policy where these projects were based. They culminated in a major international conference in 2015 that was held at The Australian National University, which also provided additional funding. Selected papers from the conference have undergone phases of reviews, revision and copy-editing to form a part of this book. Finally, I thank the Asia Research Institute of the National University of Singapore for offering me a Senior Visiting Fellowship in 2014 to write up parts of the chapters that I co-authored.

Notes on contributors

Professor Saleem H. Ali

Saleem H. Ali holds the Blue and Gold Distinguished Professorship in Energy and the Environment at the University of Delaware, and is also a Senior Fellow at Columbia University's Center on Sustainable Investment and Georgetown University's Center for Australia, New Zealand and Pacific Studies. Professor Ali has held the Chair in Sustainable Resources Development at the University of Queensland's Sustainable Minerals Institute in Brisbane, Australia (where he retains professorial affiliation). Previously, he was Professor of Environmental Studies at the University of Vermont's Rubenstein School of Natural Resources, where he was founding director of the Institute for Environmental Diplomacy and Security. His books include *Treasures of the Earth: Need, Greed, and a Sustainable Future* (2009, Yale University Press); *Environmental Diplomacy: Negotiating More Effective Global Agreements* (with Lawrence Susskind; 2015, Oxford University Press); and *Mining, the Environment, and Indigenous Development Conflicts* (2003, University of Arizona Press). Corporate and government experience includes employment in General Electric's Technical Leadership Program, a Baker Foundation Fellowship at Harvard Business School and a Research Internship at the UK House of Commons. He was chosen as a Young Global Leader by the World Economic Forum in 2011 and received an Emerging Explorer award from the National Geographic Society in 2010 and joined the United Nations International Resource Panel in 2017. Saleem received his doctorate in environmental planning from MIT, a masters degree in environmental studies from Yale University and a bachelor's degree in chemistry from Tufts University (summa cum laude).

Dr Keith Barney

Keith Barney is a Lecturer in Resources, Environment and Development Program at the Crawford School of Public Policy, The Australian National University. He has conducted research on sustainable forestry and land management issues in Southeast Asia for the past 14 years, including fieldwork in Lao PDR, Thailand, Malaysia, Cambodia and Vietnam. His conceptual interests lie at the intersections between political ecology, economic geography and agrarian studies. Keith has also conducted policy-based research with a number of international public policy organisations, including Forest Trends, the Rights and Resources Initiative and the Center for International Forestry Research, on issues relating to Asian forest markets and sustainable trade, land tenure and resource rights, and the implications of China's resource demand for local communities and ecologies in Southeast Asia.

Dr Sara Beavis

Sara Beavis is a Senior Lecturer at the Fenner School of Environment and Society at The Australian National University, where she teaches core courses on water science at undergraduate and graduate levels and undertakes research on water and sediment interactions and water resources management. She has published numerous papers on water and sediment geochemistry, and the impacts of natural and anthropogenic processes on water quality and water security. Her current research includes examining the impacts of climate variability on water security in the Pacific, the implications of climate variability and change on water and sediment quality in inland and coastal riverine environments, and the transport and fluxes of heavy metals associated with artisanal mining (in eastern Indonesia).

Dr Carmel Bofinger

Carmel Bofinger holds the position of Associate Professor at the Minerals Industry Safety and Health Centre within the Sustainable Minerals Institute at the University of Queensland. Carmel has a Masters of Environment and Community Health, and her role involves education and training, consultancies, project work and research in health and safety in the mining and minerals processing industry. She has previously worked independently and for mining companies and government agencies as a specialist occupational health and safety (OHS) and risk management consultant. Carmel also has extensive experience in providing training and education in OHS in large- and small-scale mining in many countries including Ghana, Mongolia, Zambia and Madagascar.

Professor Deborah Fahy Bryceson

Deborah Bryceson is a Research Associate at the African Studies Centre and the International Gender Studies Centre/International Development Centre, Queen Elizabeth House, Oxford University. She is currently a Reader in the School of Geographical and Earth Sciences at the University of Glasgow. She holds bachelor and masters degrees in geography from the University of Dar es Salaam, and a DPhil (sociology) from Oxford University. Her long-standing interest in rural and urban areas has involved extensive research into the interaction of livelihood, mobility and settlement in East Africa, and elsewhere on the African continent. Her early work spanned the topics of African food security, staple food markets, agricultural policy, rural transport and gender divisions of labour. During the 1990s, she pioneered the comparative study of the deagrarianisation processes in Africa, focusing on rural income diversification and associated household and community responses. More recently, she has concentrated her research on urban economies, urban growth and mobility patterns. Her current interests embrace the comparative study of East Africa's coastal cities, livelihood frontiers in Tanzanian mining and trading settlements, and the economic and social impact of HIV/AIDS.

Dr Arnab Roy Chowdhury

Arnab Roy Chowdhury is a Postdoctoral Fellow at the Public Policy Department in the Higher School of Economics (HSE), Moscow. Arnab received his PhD in Sociology from the National University of Singapore (NUS) in 2014. Prior to joining HSE, he taught in the Indian Institute of Management, Calcutta, and was a Postdoctoral Researcher in the Department of Communications and New Media at NUS. He was previously a University Grants Commission – Junior Research Fellow in Jawaharlal Nehru University and Indian Institute of Technology, Delhi. He has published a number of papers in international journals such as *Progress in Development Studies*, *Development in Practice*, and *Social Movements Studies*. He is also currently working on converting his doctoral thesis into a monograph. His research and teaching interests include forced migration, extractive industries, social movements, state–society relations, environmental sociology and post-colonial and subaltern studies.

Professor David Cliff

David Cliff was appointed Professor of Occupational Health and Safety in Mining and Director of Minerals Industry Safety and Health Centre (MISHC) at the University of Queensland, Australia in 2011. His primary role is providing education, applied research and consulting in health and safety in the mining and minerals processing industry. He has been at MISHC over 15 years. As part of the International Mining for Development program at the University of Queensland, funded by the Department of Foreign Affairs and Trade, David gained extensive experience in providing training and education in OHS in mining to many countries, focusing on developing capacity within governments in developing countries. He has also undertaken similar work for the World Health Organization. He is currently supervising the PhD research of a number of students from developing countries looking at ways to improve the OHS of mining in their countries.

Stacie Constantian

Stacie Constantian, MPH, is a research coordinator at the Harvard Humanitarian Initiative. She has eight years of experience working and living in developing countries, including two years as a Peace Corps Volunteer in Ecuador. She has conducted research evaluations in Uganda, Ecuador, Guatemala and Egypt, specialising in health applications of mixed-methods research in post-conflict settings.

Dr Marjo de Theije

Marjo de Theije is an Associate Professor at the Department of Social and Cultural Anthropology of Vrije Universiteit (VU), Amsterdam, and part-time member of the Centro de Estudios y Documentación Latinoamericanos (CEDLA) research staff. She received her PhD in social sciences at Utrecht University in 1999. Currently, she is also the academic director of VU Brasil Academic Program at VU University, and coordinates several international research projects. In her current research, she focuses on the cultural, social, economic and environmental aspects of small-scale gold-mining in the Amazon region (Bolivia, Brazil, Colombia, Peru and Suriname). Uncontrolled polluting activities of small-scale gold-mining often threaten the livelihoods of indigenous peoples. Cross-border tensions arise when miners from one country invade another, or smuggle gold between countries. With the recent instability in the world economy driving up the price of gold, and with mining techniques becoming more mechanised, the scale of the impact is increasing. Few national governments know how to respond to these developments and evidence-based policy responses are urgently required. These are what this project aims to provide.

Professor Katherine C. Donahue

Katherine C. Donahue is Professor of Anthropology and Chair of the Social Science Department at Plymouth State University, Plymouth, New Hampshire, United States. Her research is centred on environmental and social justice. She has done fieldwork in France, Tanzania and the United States. Her publications include *Slave of Allah: Zacarias Moussaoui v. The USA* (2007, Pluto Press) and, with David C. Switzer, *Steaming to the North: The First Summer Cruise of the US Revenue Cutter Bear, Alaska and Chukotka, Siberia, 1886* (2014, University of Alaska Press).

Dr Nicholas Garrett

Nicholas Garrett is a Director at RCS Global. He is a leading ASM consultant who has directed or led over 50 projects globally for companies, development organisations, governments and non-government organisations, particularly in complex, conflict-affected and high-risk areas. He completed a PhD (cum laude) on ASM and conflict minerals in the Democratic Republic of Congo at the Freie Universität Berlin, and holds an MSc in International Development Management from the London School of Economics.

Dr Maureen Hassall

Maureen Hassall holds a PhD in cognitive systems engineering, an MBA, and bachelor's degrees in engineering and psychology. Her research focuses on utilitising leading-edge human factors approaches to improve industrial risk management in ways that can help entities achieve better health, safety, well-being and business outcomes. Maureen's work involves delivering applied research, training and consulting services direct to industry, and to undergraduate and postgraduate students at the University of Queensland. Her work is motivated by over 18 years of working in industry, addressing various human and system operational challenges from a range of roles, including specialist engineering, line management, organisational change and business performance improvement roles.

Professor Amalendu Jyotishi

Amalendu Jyotishi is Professor at Amrita School of Business (Bangalore Campus), Amrita Vishwa Vidyapeetham and the adjunct faculty of University at Buffalo (SUNY). He has a PhD in economics from the Institute for Social and Economic Change, Bangalore, where he worked on the ecological economic issues of swidden agricultural systems. He also received the VKRV Rao Memorial Best PhD Thesis Award from the Institute for Social and Economic Change. His research covers issues relating to natural resources and institutions, on the one hand, and innovation, entrepreneurship in information technology business, on the other, both from an institutional economics perspective. Recently, he has been working on the issues relating to fish for food security in city regions, iron smelting and deforestation, informal gold-mining and climate change vulnerability issues, as well as innovation and cultural issues relating to small business including information technology business. He has published his research in journals, books, edited volumes and working papers. He is one of the core research members of Asian Initiative on Legal Pluralism and was the coordinator of the group from 2012 to 2015. Amalendu has collaborated in research projects supported by organisations such as the Netherlands Organisation for Scientific Research (NWO), Swedish International Development Agency (SIDA), World Bank, International Water Management Institute (IWMI), Oxfam (Great Britain) Trust, Aga Khan Rural Support Program (India), South Asian Network for Development and Environmental Economics (SANDEE), Australian Research Council (ARC) and Sir Ratan Tata Trust.

Joycelyn Kelly

Jocelyn Kelly is the Director of the Harvard Humanitarian Initiative's Women in War program, where she designs and implements projects to examine issues related to gender, peace and security in fragile states. She has been conducting health-related research using qualitative and quantitative research methods for over eight years, both in national and international settings. Her international work has focused on understanding the health needs of vulnerable populations in Eastern and Central Africa.

Dr Kuntala Lahiri-Dutt

Kuntala Lahiri-Dutt is a Professor at the Crawford School of Public Policy, ANU College of Asia and the Pacific, The Australian National University. Kuntala has extensively researched the social and ecological politics and gender equity issues related to extractive industries since 1993–94, initially in India and later on in other Asian countries such as Indonesia, Lao PDR and Mongolia. In particular, she has written on gender and livelihood issues in both large, industrial mining and in informal, artisanal and small-scale mines and quarries on the displacement of peasantry and indigenous peoples, and on the transformations of land and livelihoods in mining areas. A related field of research comprises the social and gender equity issues that are expressed at different geographical scales in water resource management. She has published on the livelihoods of poor and immigrants living on ecological boundaries of land and water, on chars or river islands in her book *Dancing with the River: People and Life on the Chars of South Asia* (2013, Yale University Press; co-authored with Gopa Samanta). Other books include *In Search of a Homeland: Anglo-Indians and McCluskiegunge* (1990, Minerva, Kolkata); *Women Miners in Developing Countries: Pit Women and Others* (2006, Ashgate; co-edited with Martha Macintyre); *Fluid Bonds: Views on Gender and Water* (2006, Stree, Kolkata); *Water First: Issues and Challenges for Nations and Communities in South Asia* (2008, Sage; co-edited with Robert Wasson); *Gendering the Field: Towards Sustainable Livelihoods for Mining Communities* (2011, ANU E Press); *Doing Gender, Doing Geography: Emerging Research in India* (2011, Routledge; co-edited with Saraswati Raju); *The Coal Nation: Histories, Ecologies and Politics of Coal in India* (2014, Ashgate); and *Experiencing and Coping with Change: Women-Headed Households in the Eastern Gangetic Plains* (2014, Australian Centre for International Agricultural Research).

Gernelyn Logrosa

Gernelyn Logrosa is a PhD candidate at the Minerals Industry Safety and Health Centre, University of Queensland. She holds bachelor and masters degrees in metallurgical engineering at the University of the Philippines. Gernelyn has several years of experience working with artisanal and small-scale miners, operators and ASM cooperatives in the Philippines, through fieldwork and forums, as Research Specialist of the Cleaner Mining Technologies, Engineering Research, and Development for Technology of the Department of Science and Technology. Gernelyn has also worked as Chief Metallurgical Engineer for mining projects in the Philippines.

Lynda Lawson

Lynda Lawson is Manager (Knowledge Transfer) at the Centre for Social Responsibility in Mining at the University of Queensland, and has extensive experience designing and delivering training programs for industry and government in Europe, Australia, Asia-Pacific, Africa and South America. She speaks fluent French and has a particular interest in mining and development in Madagascar and Francophone Africa. She is a PhD student at the University of Queensland, and her research area is artisanal mining, the mining of gemstones in Africa, women miners and entrepreneurship.

Danellie Lynas

Danellie Lynas is a Research Fellow and PhD candidate within the Minerals Industry Safety and Health Centre at the University of Queensland. Recently, she has been working with the Minerals Commission of Ghana to capacity build within the inspectorate to enable development and delivery of OHS training specifically targeting formal artisanal and small-scale mining. Past roles have included projects related to improving safety in mining equipment design, new mining technologies, injury prevention strategies and health-related policy review at corporate level. To date, her PhD research data collection has been undertaken in Papua New Guinea and Ghana.

Professor Andrew McWilliam

Andrew McWilliam is a Professor of Anthropology at Western Sydney University. Andrew's research interests are Timor ethnography, minorities and governance in Indonesia and East Timor, customary land and resource tenures, forms of religious practice in eastern Indonesia, mining and development, applied anthropology in economic development and Australian Aboriginal customary land interests and cultural heritage.

Dr Daniele Moretti

Daniele Moretti is an experienced sustainability consultant with particular expertise in ASM in the Asia-Pacific. He has worked for research organisations, governments and companies, including both producers and manufacturers. His PhD in social anthropology (Brunel University) focused on ASM in Papua New Guinea. He also holds an MSc in human resource management from the London School of Economics. Before his work as a consultant, he was Postdoctoral Research Fellow at the University of Cambridge and a lecturer at Brunel University.

Dr Rachel Perks

Rachel Perks is a Senior Mining Specialist within the Oil, Gas and Mining Unit of the Energy and Extractives Global Practice of the World Bank. She is based in Washington, DC, and provides technical assistance on ASM to several mining lending projects within her unit. Prior to joining the World Bank, she worked and lived for 12 years in the Horn and Central Africa. The focus of her work has been on managing the transition from conflict to peace in countries where natural resources have played a catalytic role in conflict, and may equally play a catalytic role in state building. Her PhD was on the formalisation of artisanal mining in Rwanda.

Dr Phuong Pham

Phuong Pham, PhD, is an Assistant Professor at the Harvard Medical School and Harvard T.H. Chan School of Public Health, and Director of Evaluation and Implementation Science at the Harvard Humanitarian Initiative. She has over 15 years of experience in designing and implementing epidemiologic and evaluation research, technology solutions and educational programs in ongoing and post-conflict countries such as northern Uganda, Democratic Republic of Congo, Rwanda, Central African Republic, Iraq, Cambodia, Colombia and other areas affected by mass violence and humanitarian crisis. She co-founded peacebuildingdata.org (a portal of peace building, human rights and justice indicators) and KoBoToolbox (a suite of software for digital data collection and visualisation).

Professor Ton Salman

Ton Salman studied philosophy and anthropology at the University of Amsterdam and also did his PhD there in 1993, on shanty-town organisations in Chile under dictatorship. After working at several Dutch, Chilean and Ecuadorean universities, he is now established at the Vrije Universiteit in Amsterdam, in the position of Associate Professor and head of the Department of Social and Cultural Anthropology. His research interests include social movements, democratisation, citizenship, ethnicity and small-scale gold-mining. Among his recent publications are (2015, with Felix Carrillo and Carola Soruco) 'Small-scale mining cooperatives and the state in Bolivia: Their histories, memories and negotiation strategies', in *Extractive Industries and Society* 2(2), 360–67, and (2016) 'The intricacies of being able to work undisturbed: The organization of alluvial gold-mining in Bolivia', in *Society and Natural Resources* 29(9), 1124–38.

Professor Ranabir Samaddar

Ranabir Samaddar is the Director of the Calcutta Research Group, and belongs to the School of Critical Thinking. He has pioneered, along with others, peace studies programs in South Asia. He has worked extensively on issues of justice and rights in the context of conflicts in South Asia. The much-acclaimed *The Politics of Dialogue* (2004, Ashgate) was the culmination of his work on justice, rights and peace. His particular researches have been on migration and refugee studies, the theory and practices of dialogue, nationalism and post-colonial statehood in South Asia, and new regimes of technological restructuring and labour control. He has authored a three-volume study of Indian nationalism, *Whose Asia Is It Anyway?: Nation and The Region in South Asia* (1996), *The Marginal Nation: Transborder Migration from Bangladesh to West Bengal* (1999), and *A Biography of the Indian Nation, 1947–1997* (2001). His recent political writings published in the form of a two-volume account, *The Materiality of Politics* (2007, Anthem Press), and *Emergence of the Political Subject* (2009, Sage) have challenged some of the prevailing accounts of the birth of nationalism and the nation state, and have signalled a new turn in critical post-colonial thinking.

Professor Sashi Sivramkrishna

Sashi Sivramkrishna did his masters in economics from the University of Bombay, Mumbai, and went on to complete his PhD at Cornell University, USA. He is currently Professor of Economics at the School of Business Management, Narsee Monjee Institute for Management Studies, Bangalore. His areas of research include contemporary macroeconomics, as well as economic and environmental history. His recent work has been published in the *Journal of International Development, Journal of Human Development and Capabilities, International Journal of Agricultural Resources, Governance and Ecology, Environment and History, Global Environment, Journal of the Economic and Social History of the Orient,* and *Economic and Political Weekly.* His book, *In Search of Stability: Economics of Money, History of the Rupee* (2015, Routledge), traces the history of the rupee from 1542 to 1971. Sashi is also an ardent documentary filmmaker; his films have been screened at international film festivals including the Royal Anthropological Institute International Festival of Ethnographic Film (UK) and Days of Ethnographic Films (Russia), as well as on India's national television channel (Doordarshan) and the National Geographic Channel.

Professor Alexandra Urán

Alexandra Urán is a Professor at the Anthropology Department of the Universidad de Antioquia in Medellín, Colombia. She has a bachelor's degree in mining engineering and anthropology, a masters degree in environmental anthropology from Kent University, England, and a PhD in sociology and social sciences from Kassel Universität, Germany. She is a senior researcher at the National System of Social and Environmental Research. She studies issues associated with the relationship between exploitation of natural resources and governance, ethnography of the state and natural resources and socio-political conflicts in Colombia. Recently, her projects have been focused on the development of artisanal mining in indigenous communities, and Afro-descendants' land in the Chocó region of the Colombian Pacific and in the Colombian Amazon. She is also an active researcher of the GOMIAM International Network (a knowledge network on small-scale gold mining and social conflicts in the Amazon), with insight into mining-related conflicts by comparing various cases in the different Amazon countries.

1

Reframing the debate on informal mining

Kuntala Lahiri-Dutt

It is common to see slums comprising makeshift housing, innumerable squatters, pavement-dwellers and street vendors occupying space in urban centres of Global South countries. They contain people who have migrated from rural areas in search of better living, people who carry out most of their trade in cash, or work in small workshops that manufacture components for larger industrial units. Together, these countless (and often uncounted) people, businesses, workshops and their services and produce comprise a significant part of their country's economy. They add value to the services and goods that make up most of the economy of the countries that comprise 'most of the world' (Chatterjee 2004: 59). However, these people and their economic contributions remain unknown or poorly known, loosely governed and mostly unrecorded, at times illegitimate, and often at the mercy of the state's law enforcement. Very large numbers of people constitute this economy and make a living within it while negotiating with the state and its various arms, and they remain unconsidered by the discussions in the formal parts of the economy.

Millions of poor all over the mineral tracts of Global South countries labour in a range of mineral extractive practices; some operations are small-scale, some are licensed, others evade tax payments and some have been traditionally carried out for generations in contrast to more recent establishments. Practices occur across the spectrum of mining, from the most artisanal and individual opportunistic enterprises to licensed small-scale firms that hire labourers on contract. Final products often reach non-local or even foreign markets. These mineral practices and those involved make up the subject matter of this book. These multitudes are from diverse origins, but ones located between the plough and the pick, sometimes shifting between both. Their livelihood choices and the context within which they take them up are still poorly understood, the conceptual problems they create by engagement with global commodity markets are not yet adequately debated, and the serious consequences of their labour for their own well-being and for the local environment of the areas where they operate pose an urgent problem for scholarly articulation. Miners use mercury for gold extraction, care little for rehabilitation of denuded, deforested and disturbed land, and have been at times marginalised and ostracised as reckless environmental and social criminals. Neither these people, nor their practices, have been at the forefront of debates on resource extraction, nor have they been interrogated in light of recent theoretical advancements in understanding the political economy of livelihood diversification and transformations of rural society. These informal extractive practices are described as artisanal and small-scale mining (ASM), implying that it is a 'sector' or smaller part of the resource extractive economy, and a problematic one. This definition, with clear indication to its sectoral belonging, leaves unanswered questions about its relationship with rural transformation, with globalisation and development processes, with the advent of large-scale extractive operations and extractivist economic policies followed by states, and with migration and labour studies intensely scrutinised through Marxist analyses.

One aspect of the problem centres on the question of developing a deeper understanding of a fluid, marginal and extremely poor people, and their livelihood changes under diverse pressures. This book proposes resolution through deep engagement with the problem holistically and through a multidisciplinary approach.

The term 'informal mining' in the book's title adds some clarity by informing readers that the type of mining dealt with is part of the heavily populated informal economy, in which 'most of the people' in mining today are concentrated, and which is primarily found in countries that Chatterjee (2004) describes as 'most of the world'. Such mining practices, even when licensed or legal but on a smaller scale, are characterised by informality in labour and production structures, and the rules and norms that determine practices. Several contributors to this book have used the more widely accepted global acronym 'ASM', and I reiterate that no difference is implied by my preferred use of the term 'informal mining'. Some ASM scholars interpret 'informality' to mean 'illegal' mining activities (for example, Verbrugge 2015) or extra-legal, following De Soto's argument[1] (Siegel and Veiga 2009); however, a wider and more complex understanding of the reality of informality is possible. The term 'artisanal' conveys labour intensiveness, low technology and low capital investment common to mineral extraction processes and their essentially premodern labour organisation.[2] The ennobling connotation of the term 'artisan' can be a misnomer, however. The confusion is perhaps why the International Labour Organization (ILO) consistently used a scale-based nomenclature in its various publications (Jennings 1999).

The use of 'informal' also adds a political–economic dimension to ASM literature that is increasingly nuanced and enriched by a range of explanatory approaches. The designation not only implies its contrasting binary in the formal economy to enable us to think broadly about a pluralistic, hybrid economy of resource use,[3] but also draws attention to the condition of informality,[4] besides aspects of labour (such as division and exploitation of labour) and aspects of production (such as simple reproduction with inexpensive and simple instruments sometimes for subsistence) within the practices and communities (Porter et al. 2011). It draws attention to the global politics over resources within which ASM has grown—neoliberal economic policies based on distinction between public and private property rights, and unitary and stable ownership based on the 'fixing of values'—to allow us to more easily understand 'the constant negotiability of value and the unmapping of space' (Roy and

---

1   This argument suggests that people take 'the other path' because of procedural difficulties and delays, and advises a cutting down of bureaucratic formalities associated with obtaining a licence.
2   Artisans are usually known to be endowed with a tradition of the craft they practice (Littrel and Dickson 2010). Although some peasants might have a generational history of mining as their primary occupation, as in the case of the Banjara tribe in South Kalimantan, Indonesia, or some communities amongst the Bheels in Central India, today's peasant miners are, in most cases, first-generation miners moving away from agriculture.
3   More details of this way of thinking about the economy as a whole are found in various works of Hart (2010), Gibson-Graham (2006) and Amin (2009), who emphasise that all human action always has some measure of informality.
4   The idea of informality is generally associated with modes of human settlements and is more widely debated within the urban context where settlement, housing and trade operate outside of legal structures but in full vision, cohabiting and sharing the space with more formal structures. Porter et al. (2011) argue that these are the 'other' of the way spatial planning envisions ordered social spaces, and hence are often presented as a law enforcement or policy problem. However, the basic questions are related to property rights—the rights to say who can do what and where—and issues around sustainability.

Al Sayyad 2004: 5). Property rights characteristics can also be seen in a more nuanced manner. When viewed from a historical perspective, property rights can be beyond the simplistic binary of public and private, or formal and informal. Jyotishi et al. (2017) illustrate this by using the dimensions of 'excludability' and 'rivalry' to identify a matrix of institutions governing artisanal gold-mining activities in an Indian context. They show that institutions governing mining not only move from informal to the formal sphere, but can also move in a reverse direction. Again, even within formal there are shades and layers of informality when labour arrangements are considered. In an analysis of the multiple coal-mining economies in India, using detailed empirical data from the field, Lahiri-Dutt (2016) shows how the formal and informal overlap and intermingle with each other to produce 'diverse worlds' of coal mining in India. ASM has been changing the global reliable and familiar map of mineral extraction in recent decades, pushing resource and commodity frontiers (Hilson 2013) and establishing capitalist frontiers. From the analytical perspective, the political–economic framework is expansive enough to accommodate considerable social and economic diversities apparent in the experiences and practices of informal, artisanal and small-scale mining.

Drawing on the previous example, these mines and the miners are squatters on the contemporary extractive landscape and collectively comprise a radically different kind of mining commonly envisaged by the term 'mining' in the context of industrialised, large-scale extractive operations. These mines and quarries redefine what is generally seen as the extractive economy at global and national scales of interpretation. Heavily capitalised and industrialised operations that have been the dominant mode of mineral resource extraction since the industrial revolution in Europe and the New World replaced the chaotic, ruthless world that had been unleashed by rushes for mineral resources in Australia and North America. Yet, their hegemony within our thought processes make us forget that the poor—in their desperate attempt to survive and make a living in the globalised world, and suffering from dispossession and unforeseen social, economic and political pressures—are no longer completely outside the wider political–economic processes influencing resource extraction (Smith 2011). They also make us forget that mining has a long history as an ancient livelihood in many of these countries, where modern industrial production has not kept pace with the rapidity with which they were incorporated within the global economy.

It is well documented that labourers in artisanal or informal mineral extraction practices are often the poorest and most exploited labourers, often erstwhile peasants caught in global change, and performing insecure and dangerous tasks. Although located at the margins of the mainstream mining economy, their extractive practices produce enormous amounts of mineral resources, securing a livelihood (and often a path out of poverty) for some. Although labourers in informal ASM comprise the poorest, they are also inextricably engaged with, or bound to, global commodity values and supply chains. As cheap labour, they are intimately involved in production of new extractive territories and rural economies by complicating what is meant by resource access and control. As migrant rural labour, they reshape agrarian communities and landscapes. Collectively, these miners also redefine our understanding of political economies, political ecologies and resource geographies of commodity extraction. This social and economic milieu also holds significant implications for scholarly understandings of contemporary mineral-dependent livelihoods, agrarian transitions, informality and the social meanings of destitution and poverty.

Diversity and difficulties of explanation

Almost all mineral-rich tracts in poorer countries of the Asia-Pacific, Africa and Latin America now have millions of squatters, labouring and making a living through mining in various scales and capacities, and together throwing up important conceptual and practical challenges. The great diversity of labour and production systems, organisations and mining practices blur the binary distinction between the formal and informal as they are commonly understood. In the context of a particular mineralised tract, mineral extraction ranges from smaller-scale mines and quarries worked by casual labourers in a few large enterprises— which run mechanised operations on reasonably sized leases and compete with many smaller firms with lower capital investment and production— to innumerable producers working their own land, most likely without licence or lease, using family labour, including children, and producing tiny amounts. Similarly, minerals extracted range from most to least valuable for many purposes: precious and semi-precious gemstones and metals such as gold; other metal ores such as copper, zinc, manganese, fluorspar and tin and coltan; industrial minerals such as limestone and marble; rare earths; and construction materials, including kaolins, feldspars, clays, sand and gravel. Together, 15 to 20 per cent of all global minerals and metals, including up to 80 per cent of all sapphires and up to 30 per cent of gold, is estimated to be produced in this way (Sippl and Selin 2012).

The mineral extractive practices have elements of petty commodity production, localised production, circulation and appropriation. These practices range from individual opportunistic enterprises to licensed small-scale firms hiring contract labourers. Although the labour is primarily for subsistence, extracted materials may fulfil wider market needs beyond those of a household or village community, and cannot be described simply as production for use rather than for exchange. Such mining may involve capital investment in trade and usury, and generate rent in kind rather than in money. But it also exhibits characteristics of informality such as extraction of surplus from the direct producer, political decentralisation and a fusion of economic and political power at the point of production, as well as a localised structure of power wherein landlords exercise judicial or quasi-judicial powers in relation to the dependent direct producers. Attempting to make sense of the situation, the International Council on Mining and Metals (ICMM 2010) classified these mining practices into five overlapping categories: traditional practices carried out for generations in an area that may form part of traditional livelihoods; seasonal activities that complement other livelihoods, such as agriculture or livestock rearing; permanent cohabitation in areas connected with large- or medium-scale mining, such as miners working abandoned areas, tailings dams or downstream of larger operations; shock mining resulting from unexpected events such as drought, economic collapse, commodity price fluctuations, conflict, retrenchment from mining parastatals and unexpected closure of commercial mines; and influx or opportunistic rush or rapid in-migration to an area where minerals have recently been found. However, these categories primarily look at the origin and present neither the vast diversity of extractive practices within this kind of mining, nor link them to the *causes*, and in a sense reflect a difficulty in approaching the thematic field.

The other difficulty is official recognition; unlike large-scale, formal, industrial mining, the definition of ASM or SSM (small-scale mining) varies widely between countries, with no universally accepted and precise definition. Only a few countries recognise ASM in state laws; for example, the People's Mining Act of Indonesia (of 1967) recognises the tradition of mining by communities in the country. In most countries, for example in India, mines are classified by size, production and labour requirement, or minerals commodities are classified in a way as to lend themselves to either large- or small-scale mining.[5] It has remained a challenge to accurately estimate the numbers involved. For example, the figure of 35 million provided in 2005 by Communities and Small-Scale Mining (CASM) is substantially higher than the 13 million provided in 1999 by ILO. Since then, the world has experienced the main commodity boom in the late 2000s, and the implications of neoliberal economic policies on the poor have also been more fully understood. For example, in Ghana there are an estimated 1 million men and women miners operating without a licence (Tschakert 2009). The United Nations Economic Commission for Africa, in its 2011 report (UNECA 2011) on the role of minerals in the continent's development, noted the growing number of rural people in ASM. Consider also that estimates were often based on incomplete secondary data and extrapolation from limited surveys, and primarily considered extraction of high-value material commodities and precious stones such as gold and diamonds. Therefore, the material commodity under consideration partly influences estimates. The extraction of most industrial or construction commodities, such as coal or even stone, sand and gravel, are not included in this estimate. If extraction of these materials is considered, the numbers of extracting peasants would increase. Moreover, as ASM often complements or replaces other rural-based livelihoods such as farming in season, and many people can be simultaneously involved in both, a clear identification becomes difficult to offer.

Yet another challenge is the lack of standardised production regimes in ASM/SSM or informal mining, which presents an array of production and labour practices. These practices range from the most artisanal to reasonably capitalised extraction—from the sole entrepreneur trying out his luck with a pick and basket, to large-scale rushes where thousands descend upon an area to create horrendous landscapes; from community groups following traditional livelihoods to modern mechanised operations where labourers toil in bondage. The only distinctive feature is that these people make a difficult, rough and risky living, making use of what Ali (2009) described as the 'treasures of the earth'. The clientelist system resembles other 'illegal' uses of resource such as logging, which ranges from highly capitalised actors with significant funds and industrial-scale operations, to mid-size entrepreneurs, down to village or family-

---

5   For example, in India coal is classified as a 'major mineral', implying that it can only be mined by big players, thereby illegitimising any other form of coal mining.

based groups. For example, in south and east Kalimantan in Indonesia, informally organised operations by PETIs (*penambang tanpa injin*, literally translated, 'those who mine without a licence') use reasonably capital-intensive earth-moving equipment to cut coal seams and truck them to Banjarmasin port. The material is transferred onto large coal barges that supply ships outside Indonesian waters for export to China and South Korea. Yet, next to them, often occupying the same space and extracting the same resources are the 'gurandils'. These fleeting masses literally 'jump from cliff to cliff', and operate clandestinely by staying small and invisible to the state machinery of law enforcement, some selling to the local markets and others supplying the barges that carry coal from the trucks to the ships. Not far from the general region is a traditional diamond-mining area where indigenous Banjars have mined diamonds for generations. This is different from the Brazilian Amazonia where multitudes of itinerant miners—wildcat *garimpeiros*—clear the jungle to extract gold and other precious commodities. The Galampseys in Ghana, Barranquilas in Bolivia, and the Ninjas of Mongolia each have different contexts and characteristics. These diverse contexts are only partly the products of historical continuity in mining, and each have a distinctiveness best explained through local studies that highlight the importance of place in shaping the political–historical trajectories.

While exact numbers of these workers are unknown, we realise that the scale and extent of their presence is difficult to explain with theories developed in the context of post-industrial resource extraction and capital accumulation processes. There are stories of spectacular 'rushes' of gold or diamonds—which are often equated with those that took place in the New World—occurring throughout the mineral-rich Global South. But in analysing such a 'New Gold Rush' in Zimbabwe, Kamete (2008) found that the rush was clearly linked to the economic crash and created a situation in which 'five per cent of the official national population, are directly involved in gold panning. The numbers are increasing at a phenomenal rate' (ibid.: 40). The extent is apparent from the fact that, by the early 1990s, more than 100 metric tons of gold was produced annually, and gold-mining had become the second-most important economic activity in Amazonia (following the combined ranching/ agriculture sector) (Godfrey 1992: 460). Godfrey stated, 'At least half a million gold miners now operate in Amazonia' (ibid.). To this might be added the large number of merchants, restaurateurs, sex workers and others who service the miners. He noted, 'in certain areas, gold-mining has become the leading economic sector … [and] the dynamics of gold-mining … call into question several common assumptions about the so-called agrarian frontier in Amazonia' (ibid.). Clearly, in Brazil, mining rather than agriculture has been pushing this frontier.

This evidence points to highly mobile rural communities scrambling for minerals, engaging with the market to extract values, competing with multinational mining companies or complementing them in many ways in search of minerals. The miners cannot be easily contained within concession boundaries or sedentarised as they move around and dig everywhere (referred to in Lao PDR as *khut thouathip*) to frustrate the governments (Lahiri-Dutt et al. 2014). The transitory nature of miners who follow the material commodities, with their undercover, surreptitious operations, make it difficult not only to explain but also to administer in the conventional way. They point to a significant change in contemporary extractive landscapes that is directly linked to stagnating rural economies.

Glimpses into a paucity of data show that this unprecedented rush has remained not only a theoretical challenge, but also is unimagined and largely unaccounted. As much as 65 per cent of Peru's gold production in 2005–06 came from non-formal sources—that is, other than from large mining companies. Gold production from informal sources in 1992 contributed an estimated 76 per cent of mineral export earnings of Tanzania (Tesha 2000). Of the nearly 600,000 carats of diamonds officially exported from Sierra Leone in 2006, as much as 84 per cent originated from peasant extraction (Government of Sierra Leone 2011). In Indonesia, in addition to gold and diamonds, coal, tin and manganese are mined informally in vast quantities, the majority reaching the international market (Lestari 2013). In Mongolia, where mining is the fastest growing segment of the national economy, former herders endure harsh winters to extract gold to reclaim their moral rights to natural resources (Lahiri-Dutt and Dondov 2016). Their numbers eclipse those employed by formal mining industries, and they produce more gold than the country's corporate mines. Approximately 20 per cent of Mongolia's labour force is thought to be engaged in mining and mineral processing without licence (UNDP 2009).

In official documentation and in social science studies, miners and mine labourers are dissociated from rural peasantry. The peasantry is conventionally regarded as a traditional occupational and demographic category comprising sedentary farming communities. The quarry labourer, as Bryceson (2002) observes, is typically regarded as an opportunistic rural migrant scrambling for income. Such conceptualisation enables a rethink of ASM as part of the wider global political economy. This type of mining is rewriting the meanings of commodities and materials, the meanings of places and territories, and the meanings of labour and social relations in production in a way that defies old explanations. Something unheard of before is happening: field-based scholars and experts are noting and commenting on mining, not just as involving capital and capitalist enterprise, but also as rural labour moving out and around. These rural people have been 'opening up' once remote locations to link them with global capital and commodity networks. Neither fully a labourer nor a peasant anymore, these people inhabit the margins of the mainstream mining economy; these labourers attest to the state's failure to accommodate their livelihoods within existing legislative and formal institutional frameworks (Jønsson and Bryceson 2009). Elsewhere, these miners apply techniques that involve staying small, bribing local government officers and mining in remote areas to avoid law enforcement. In yet other places, they present pathetic figures dulled by exploitation— earning the lowest wages, exploited in feudal or semi-feudal production systems, unrecorded and uncared for. They are the true outcasts of modernity and modern capitalist mining expansion. Collectively, these innumerable workers, toiling in mineral-rich regions, produce enormous quantities of minerals that we are little aware of.

Of the millions who shift from rural areas to informal mining or ASM— either as independent entrepreneurs or working as groups or wage labourers—very few can rise out of poverty. These people are the new mining precariats, toiling as unstable and unregulated labourers in mines and quarries in areas that are often difficult to access and lack any services. Informal labour in the Global South highlights their diverse and multiple precarity as historically specific to the poorer countries (Breman 2013). For this reason, the lack of social insurances such as old-age pensions, health and unemployment benefits and lack of safety within the mines and quarries, assume significance. Reframing therefore involves greater attention to labour conditions within ASM, and interrogation of the state's apathy that allows such persistent conditions.

An issue often confronting those who suggest that ASM is 'poverty driven' (Jennings 1999) is the preponderance, in certain locations, of heavy mining machinery that requires sizeable amounts of capital investment. One end of informal mining is heavily mechanised with wage labour hired by external investors to operate the machinery. This emphasises mode of production, the integrated complex of two interdependent factors—productive forces and social relations of production. If we accept that mode of production is an articulated combination of relations and forces of production that enables capital accumulation—in turn financing the transformation of rudimentary extraction processes into mechanised operations with sophisticated work practices and revenue-sharing arrangements—then it is easier to locate informal mining or ASM within petty commodity production. With increasing mechanisation and capitalisation, production quantities rapidly increase, leading to the most crucial policy question related to the environment—'what is to be done' to address the rampant environmental degradation in open-access commons such as forests and grazing fields (Suzuki 2013).

Mining experts have offered explanations for the prevalence and contemporary growth of ASM. Geologists in remote regions have noted many people digging for minerals—sometimes valuable metals or gemstones. They explain such mining by invoking the nature of the 'reserve': the manner in which deposits of ore or minerals were laid by Mother Nature for humans to use as resource. Geologists argue that certain deposits were either too small, too 'ribbonlike', too scattered, too shallow or too remote for large-scale mechanised operations. The ease of extraction, for example of surficial placer deposits, facilitates local people to dig them up. Mining engineers noted that if some deposits are 'leftover' by large mechanical operations, they can be extracted on a small scale using low levels of technologies. In these formulations, the explanation therefore hinges on not only the properties of the materials, but also the 'scale'. Scale of the extraction is related to the 'level' of technology used: larger scale requires higher levels of capital. The hidden assumptions are not difficult to detect. The first assumption is that such mining is a scaled-down version of larger-scale mining, implying there is no essential difference between the two. This is a purely technically biased worldview that reduces human endeavours to definitions based on their size. The second assumption is that the existence of such mining is environmentally determined, implying they existed because that was the *only* way these minerals could be extracted. Although nature does primarily determine the availability of minerals, it is unreasonable to suggest that nature determines resource extraction. The third assumption is that small scale and low technology are correlated. This assumption has inherent problems as we now know that both are no longer correlated as they were; many small-scale technologies are much more widely available and accessible by those who could not previously afford it. Capital investment in small-scale mining is no longer and not always local, thereby enabling the use of a wider range of technologies—from metal detectors to find gold, through to excavators, pumps and crushers.

Environmental geologists (for example, Douglas and Lawson 2000) noted the extremely deleterious cumulative effect of small diggings spread around an area. The impact of a large number of small operations can be harmful for the overall environment: forests are cut and farming land destroyed, rivers are chocked with sediments, the water table falls, the air fills up with dust and smoke. However, some experts also offered a variation of the environmentally deterministic argument to suggest that environmental degradation, leading to severe livelihood stress, explains the growth of informal, artisanal mining. A widely used example is that of Mongolia, where several successive and extremely severe winters destroyed herding livelihoods, forcing the destitutes to take up informal mining in the early part of the new century (Fernández-Giménez et al. 2012: 836). Such environmental determinism is locally true, and it is not difficult to find examples of a degraded subsistence resource base pushing local environment-dependent communities into informal mining. This explanation, however, is only partly applicable. Most often, instead of one single cause, several factors create a situation where informal mining becomes a more attractive livelihood option for the poor.

Contrary to these observations, archaeologists and historians have presented irrefutable evidence that communities in many parts of the world have traditionally extracted minerals. Archaeologists who draw evidence from material artefacts have argued that human civilisation did not necessarily 'progress' in a linear manner from hunter-gatherers to farmers to industrial labourers. They show that sometimes mining has been the primary occupation; however, some communities have actually never been farmers as was imagined commonly in the linear explanation of advancement of human endeavour. At other times, the use of minerals or extractive activities (such as stone-cutting) was an important part of human cultures (Knapp and Pigott 1997). Social historians as well as historical archaeologists show that mining and agriculture often coexisted. For example, Phimister (1974: 445) shows that in west and south-central Africa, local inhabitants engaged in the production and trading of gold, primarily alluvial gold, for over 900 years, which is even earlier than mining from gold reefs. Reef mining started as the demand for gold expanded, allowing the Iron Age miners to develop more sophisticated prospecting and mining methods.[6] Early alluvial gold-mining was a seasonal activity,[7] undertaken 'only because of the poverty of other local resources' (Phimister 1974: 445). Although Africa was the primary source of gold for Europe for a long time, very little is known about it today. Werthmann (2007: 394) cites three reasons for this paucity of knowledge: the remarkable secrecy about the sources of gold, the many unsuccessful attempts and failures to strike new deposits, and, more importantly, the 'myth-making' around mining. Caballero (1996) reviews how gold-mining continued as part of the flow of traditional livelihoods amongst the indigenous communities in Benguet province in the Philippines. In Peru, scholars (Deustua 1994; DeWind 1977) have shown that the fine details of the form of the economy varied depending on the nature of the mineral: silver mining entailed artisanal production with little technology, while copper mining demanded more capital-intensive processes of smelting and refining in plants.

In their global report on ASM, Hentschel et al. (2002: 9–10) provided an historical review of the manner in which the international development community has treated the issues. They show that the primary approach in dealing with ASM has shifted over the decades: during the 1970s, the primary focus was on definitional issues; the 1980s reflected a preoccupation with technical issues and physical impacts;[8] and, during the early 1990s, it shifted towards integration of technical, environmental, legal, social and economic issues. Throughout the 1990s, there was particular focus on legalisation of ASM, although development planners had begun exploring the relationship between large mining corporations and ASM, focusing on more specific labour issues including gender and child labour during the mid to late 1990s. Finally, with the establishment of CASM and the Mines, Minerals and the Sustainable Development (MMSD) Project by the British non-government organisation International Institute of

---

6   Again, Phimister (1974: 446) thinks that reef-mining methods were owed to alluvial gold, which, in some places, was worked by shafts sunk into riverbanks.
7   Besides the need for resources to trade, geological availability and climatic and riverine factors determined alluvial gold-mining. The Angwa, Mazoe and Ruenya rivers were of importance, as were areas of Manyika (currently Mozambique), where rivers are perennial.
8   Of these, the impact of the use of mercury in gold processing—on the local environment and on populations using it—remains a concern. The Global Mercury Project, funded by the United Nations, scrutinises mercury use by artisanal gold-mining communities. It continues to focus on health and environmental impacts of mercury use by artisanal gold-mining communities. The scholarly and international policy attention on various aspects of mercury use in processing gold, while also yielding a large body of literature, was productive for social science knowledge on the extractive practices.

Environment and Development (IIED), community issues including those related to livelihoods came to the forefront. The MMSD sponsored reports related to mining, of which the global report on ASM (Hentschel et al. 2002) remains one of the most authoritative accounts. CASM, under the auspices of British Foreign Aid and the World Bank, was to act as a global networking and coordination facility. With the mission 'to reduce poverty by improving the environmental, social and economic performance of artisanal and small-scale mining in developing countries' (ibid.: n.p.), it was also crucial in linking research with international development and policy-making. This close proximity with mainstream policy-making widened the scope of attention on ASM to the poverty and desperation of the people involved, and rescued ASM from association with sensational accounts of 'resource curse' and 'resource wars'; however, it did not really lead to full articulation of an analytical framework within which to place such mining.[9] A similar shift in scholarly literature has more explicitly engaged with development: from looking at 'socio-economic impacts of ASM' (Hilson 2003) and physical impacts, to the exploration of what might be causing the rural poor to move from traditional forms of subsistence to mineral extractive practices, giving rise to a significant body of development literature (Buxton 2012, 2013). As consensus was reached that ASM is largely poverty-driven, programs and projects searched for alternative livelihoods and cooperatives for these communities, albeit with not very significant results (Levin and Turay 2008).

Anthropologists tend to explain ASM differently. Examining ASM through deep ethnographical engagement and attention on the local context, scholars (such as High 2008) contest the notion and suggest that the argument that it is spreading throughout the Global South due primarily to poverty is simplistic. High argues that, in Mongolia, such mining is linked to customary ideas about patriarchy, generosity and the obligation to share wealth. Moreover, High and Schlesinger (2010) emphasise not just the long history of mining but also the centrality of mining—particularly of gold—to Mongolian politics and economy traditionally. Rather than analysing the herders-cum-informal miners as either victims or greedy agents, High suggests that artisanal mining is linked to 'Mongolian ideas about patriarchy, generosity and specifically the obligation to share wealth'. Explaining the 'cultural logic' of informal

---

9   In a previous work, I have criticised the use of macroeconomic theories such as resource curse to explain ASM (see Lahiri-Dutt 2006).

mining, she shows that many 'ninja miners' she spoke with during her 20 months in the field often 'cast themselves in opposition to mining companies, which they regard as "bad for Mongolia", "they steal our wealth"' (High 2012: 257). The moral economy of resources becomes apparent through these interpretations (also see Lahiri-Dutt 2014 for the Indian context). Ethnographic insights into social and cultural features of artisanal mining communities (such as by Walsh 2003; Grätz 2002; Nelson 2016) also allow the reader to move beyond the greed, blood and violence, and the monetisation of everyday life, which have now become associated with these mining communities. More importantly, they illuminate the importance of the specific context: deep connections to rural pasts, attitudes towards fortune-making, risk minimisation cultures and friendship ties in coping with an uncertain way of life that presents as many everyday challenges as opportunities.

The contemporary political economy of globalisation is pushing millions out of agriculture to negotiate the material world, which is the world of commodities. Rural communities are organising new systems of production and diversifying their economies and livelihoods. Sometimes, they are giving up farming altogether—a process that is described as 'de-peasantisation'[10] and/or deagrarianisation. In this process, peasants leave agriculture in favour of urban jobs, or move within the agricultural wage labour circuit diversifying their livelihoods, not as a temporary measure but permanently. Scholars such as Bryceson (2002) and Bryceson and Jønsson (2010) articulate that deagrarianisation is prominent in Africa where ASM has flourished as a means of rural livelihoods diversification. Pressured from external forces, rural communities are negotiating the free market and the state to make the best of constraining and enabling opportunities, accessing knowledge, resources and vistas of rights (Bebbington 1996). Following this argument, one can argue that rural communities, ousted by extractivism of states and international agencies, are adapting to the political economy of mineral extraction, and derive some benefits from the forces that are undermining their well-being (Verbrugge 2016).

---

10 Araghi (2009: 138) notes that de-peasantisation is neither synonymous with proletarianisation, nor is it a completed or self-completing process leading to the death of the peasantry, and observes that peasantisation and de-peasantisation are not necessarily two mutually exclusive phenomena.

This book examines informal, artisanal and small-scale mining, to locate it within the broader debates on rural transformation and resource politics of these countries, and the gamut of theoretical, analytical and practical challenges it presents to scholars, practitioners and policymakers. These challenges range from defining the composition of informal or artisanal and small-scale mining, theoretically explaining it, contextualising contemporary concerns over environmental integrity and labour safety standards, to governing it in a situation where no alternative livelihood plans work (Siegel and Veiga 2010). This book's contributions take up these challenges to reinterpret informal, artisanal and small-scale mining. Irrespective of historical antecedents, their collective numbers have grown to such an extent that, although scattered over (and bound to) mineralised and often remote areas, they have come under the intense scrutiny of scholars and other professionals.

Structure of the book

Contributions from multiple perspectives and disciplines reflect the diversity of ways that informal, artisanal and small-scale mining can possibly be perceived. Holistically, they represent a universe fraught with confusion and ongoing debates, and each contributes to one of three broad areas. Consequently, they are arranged into three sections. The first section deals with the historical antecedents of ASM and articulates the ways values—of commodities and of moral judgements—are constructed. The second section highlights aspects of precarious labour in this kind of mining to underline the broader political–economic perspectives. The third section focuses on aspects of conflicts and governance that pose significant difficulties in dealing with these extractive practices.

Deborah Bryceson outlines how gold rushes dramatically heighten people's expectations of the future. In societies where gold is highly valued as a source of wealth accumulation and adornment, and bears enduring economic value, the discovery of gold becomes irresistibly enticing. Contemporary gold rushes in the Global South have provided millions with a remunerative occupation, and often self-determination and local-level democracy, in ways both similar and dissimilar to the 'gold fever' that spurred large numbers to gold concentrations in the United States, Australia and Canada during the late nineteenth century. Bryceson explores what kinds of people take up shovels and picks to seek their livelihood or fortunes, and suggests that a historical and comparative outlook indicates the paradoxical aspects of artisanal mining as a transformative occupational frontier.

Katherine Donahue deals with an aspect of commodity production and trade: the creation of value not through labour but artificially, through middlemen and brokers. The globalised world of capitalised commodity production comprises buyers and sellers, in Tanzania and in the United States, creating a new sense of 'value' that goes beyond the use value of a commodity being its power to satisfy a want and exchange value being the quantity of other goods and services. Here, we see how economic value is created in a globalised market system through valuation principles and notions of value that create a private good for only those who own the stones. Thus, the notion of value of the commodity lies in the preparedness of someone to give up something in order to acquire what has now been constructed as a valuable commodity. Donahue shows how traditional Maasai communities participate in a global web of commoditised product—tanzanite, a gemstone that has expanded its market in recent years through publicity. She analyses this phenomenon of value creation not only for its impact on the Tanzanian economy, but also shows how the tanzanite moves through a globalised exchange system in which value is not necessarily determined by the amount of labour embodied in the gemstone.

Arnab Roy Chowdhury and Kuntala Lahiri-Dutt study gemstone mining by the Khonds, a local indigenous group or tribe. This chapter on the politically and economically volatile Kalahandi area of India shows that the recent spurt in mining has been associated with an increase in global demands and the entrance of various opportunistic outsiders who collude with the local state, caste leaders and class elites, police and bureaucracy. They present a political–economic explanation of de-peasantisation of an indigenous community to show that the Khonds have, since the opening of the Indian market of gemstones to the world, moved into the mining of gemstones. The Khonds, however, benefited little from entering the globalised world of capital—whilst out-migration has slackened, many more indigenous people are now without land and have turned into wage workers labouring in mines. In fact, the value they have created has not benefited them, as the tribes have remained just as impoverished.

Saleem Ali argues that the time has come to consider the overall trajectory of artisanal and small-scale mining from a broader socio-ecological sustainability lens. International agencies and development donors can play significant roles to rescue it from falling into the interpretative simplistic binary of seeing ASM either as a problem of illegality or as an exemplar of entrepreneurial development. Although ASM has existed for millennia, is ingrained in many cultural traditions and has an immense livelihood potential, the ecological and health impacts remain significant. Ali suggests technical interventions to improve mineral outputs or alternative techniques for safer extraction will reduce environmental costs associated with such mining. Further, he suggests that a middle path that considers ASM as part of a hybrid livelihood strategy and a transitional opportunity for catalysing development is now needed.

Chapters in the second section of the book primarily focus on different aspects of labour processes. This section begins with the contribution of Ranabir Samaddar, who brings attention to the 'informal' condition of labour forms and its variety in terms of work, agreement, labour process and nature of the job performed. He argues that in the contemporary neoliberal economy, the extensive presence of informal labour conditions within the economy at various scales—global, national and within different sectors—signals as if there is a law ordaining the informal conditions of labour in the interest of capital in almost all spheres of economy. Based on Marxist analysis, he shows that these informal conditions can be obtained only in productive activities described broadly as 'extraction'. Moreover, labour deployed in the domain of extraction will be transient and migrant labour—in other words, multiplication of the forms of labour. These three features are interlinked, because the neoliberal economy is essentially an extractive economy, forming the backdrop to the return of primitive accumulation. The extensive emergence of labour in decentralised and informal extractive processes contributes to primitive accumulation. He proposes two sets of investigations: to investigate the wide variety of forms in which production and extraction feed into each other in the neoliberal economy, and the role transit labour plays in this dynamic; and to examine the specific nature of politics produced by the regime of extractive and transit labour.

Danellie Lynas presents the innumerable safety and health concerns of women miners. Gender is crucial in ASM communities in determining who does what, where and how. Men and women are differently involved in and differently affected by ASM practices, cultural practices and legislative practices. Consequently, in a traditionally patriarchal country such as Papua New Guinea, such a gendered division of labour determines economic and social power differently according to one's gender. Restricted by lack of ownership over mining land, with almost all registered mining leases, agreements and customary land held by men and transmitted patrilineally, women find themselves disadvantaged. Lynas argues that understanding how gender-specific health and safety concerns impact on the lives of the women will allow implementation of programs that directly address their needs, improve their quality of life and allow them to participate safely in extractive practices.

Lynda Lawson's chapter is about the lives of women involved in the coloured gemstone trade of East Africa. It tells the story of two groups of women: gemstone miners in rural south-west Madagascar, and women in Madagascar and East Africa who make their living further along the gemstone supply chain by adding value to gemstones through faceting, polishing, jewellery making and trade. Their life histories show how their lives have transitioned and reveal that while women in general are locked into subsistence activity, some have become skilled craftswomen and entrepreneurs.

Gernelyn Logrosa, Maureen Hassall, David Cliff and Carmel Bofinger discuss the health and safety of miners. They argue that the introduction of certain technology has the potential to alleviate or exacerbate these impacts depending on how the technology is accepted and used by miners over time. They show that using the ISO 31000:2009 standard as guidance has both notable strengths and limitations. Risk management approaches can be used to identify and assess the key factors on whether it will alleviate or exacerbate worker, family and community health and safety impacts.

In the final contribution to this section, Rachel Perks, Jocelyn Kelly, Stacie Constantian and Phuong Pham bring forth the perspectives of researcher–policy practitioner. They argue that access to finance—often seen as a perennial barrier for miners' well-being and formalisation—is the primary cause of miners suffering undercapitalisation, even when there is a legal title. The lack of financial means pushes miners to borrow, typically from middlemen buyers. The authors offer an example, based on primary research, to show how some mining networks in Rwanda are organised around off-take agreements or royalty-sharing arrangements with international trading houses, European manufacturers and local export houses. These alternative ways of financing are discussed against the backdrop of their application to the global industrial mining industry, to ask if such arrangements might be a step towards formalisation.

Contributions to the third section of the book are grouped around the challenges of conflicts and governance. Amalendu Jyotishi, Kuntala Lahiri-Dutt and Sashi Sivramkrishna chronicle gold-mining and informalisation processes in the Nilgiri–Wayanad region of Southern India using historical documents of the colonial period. The 1970s saw the settlement of Tamil repatriates from Sri Lanka in the region, adding a crucial dimension that helped the continuation and growth of informal gold-mining. By using a property rights framework, it explores changes in institutions and governance structures through four phases: early colonial (up to 1870s), corporate speculation phase (from 1870s to 1900), open-access phase (from 1900 to 1970s) and the current phase from the 1980s.

Marjo de Theije and Ton Salman present an analysis of conflicts surrounding artisanal and small-scale gold-mining in a number of countries—Bolivia, Colombia, French Guiana, Peru and Suriname— located in the Amazon region, with a focus on notions of territoriality, marginality and access to the gold in forested and remote areas. They argue that physical and organisational distance from the centralised state plays a role in conflicts between state representatives, local communities and miners. The marginality of the territory is also created by the very act of mining activities that constitute territorial organisations and claims. They show that different politics in Amazonian countries cause conflict, but also generate opportunities for different actors. Miners cross the borders easily and without even being noticed, and differential access to the mineral-rich grounds sometimes triggers conflicts between actors.

Alexandra Urán's contribution is placed within the intensification of a neo-extractivist model of industrial and mechanised mining in Colombia during the last decade—a strategy that has excluded small-scale miners as local agents and decision makers. In areas that are literally at the margins of the official government, SSM activities occur beyond the reach of official administrative control. In the same area, the armed group has created an alternative political setting to control rural areas. The strong presence of the FARC guerrilla and other illegal armed groups in these rural areas have increased procedures for managing and administrating natural resources separately from the official legal system, and outside of the government's control, and they have also fabricated an extended conflict in response to the political competition for power. The chapter evaluates how the radical left in the political context may be included and recognised as a political actor to change the rules on mining and the administration of mining resources.

Sara Beavis and Andrew McWilliam focus on conflicts caused by the environmental, health and social impacts of artisanal and small-scale mining in the Indonesian archipelago where ASM is poorly governed. It highlights gold-mining in Southeast Sulawesi where there has been widespread and sustained use of elemental mercury for processing gold. This has resulted in extensive, unregulated releases of mercury into rivers that support valuable downstream land and riverine activities including irrigation, *tambak* fish ponds and estuarine fishing. Although the major rush since the first discoveries of alluvial gold is over, and the active mining population has declined substantially as a result of declining gold yields and greater government control over licensing and access for authorised mining companies, hydraulic mining of the riverbanks and channels continues. The legacy of mercury use has been long-lasting and deleterious for the environment. Moreover, traditional livelihoods have been affected as downstream flows in two affected river catchments still exhibit regular, extreme levels of turbidity and sediment loads—a result of intensive mining of riverbanks and beds in the upper catchments in combination with heavy seasonal monsoon rains.

The next chapter by Daniele Moretti and Nicholas Garrett brings readers back into the domain of governance. They note that much of the literature on ASM in the Asia-Pacific region suggests that national mining laws and mining governance frameworks within the region do not adequately provide for ASM. Yet, they argue, there are a few detailed and up-to-date case studies of such governance systems. Their contribution takes up the discussion of ASM governance in the Lao PDR, where ASM has long been present but has lately become an 'emerging issue', drawing increasing attention from government and international donors to show how national legal frameworks translate into, or differ from, realities on the ground. Further, it outlines ASM governance, in its current form and in light of proposed changes to ASM-specific laws, with a view to highlighting its main implications from a development perspective.

The final chapter of this book is by Keith Barney, who outlines the elusive idea of formalisation of ASM. From regulation, ASM debates have moved onto regularisation and formalisation. Yet, the form and processes remain controversial, due to the social illegibility (the lack of data and social knowledge required for fostering state control), the resourcefulness of miners (in contesting state initiatives) and labour exploitation and presence of middlemen. Most formalisation plans are conceived as processes of integration, as means to incorporate informal activities into the sphere of state legality through some kind of group formation. However, some of these initiatives tend to be oversimplified recipes, a one-size-fits-all prescription based on standardised rules.[11] Barney's contribution, drawing on evidence from three locations in India, Indonesia and Lao PDR, analyses the variations in material, social, economic, political and technological 'assemblages' of informal mining. These variations comprise what he describes as 'assemblages', producing the local contexts in all their enormous variations. He places the question of formalisation of informal mining in relation to these assemblages that comprise local socio-political struggles for rights to livelihood by local miners and resource-dependent communities, and the state–institutional governance and capital relations that reproduce vulnerability and establish conditions of informality. He recommends that public policies supporting formalisation need to better account for the connections between local processes, state institutions, the scale and use of technology, ecological externalities and the functioning of informal gold markets and commodity chains. By underlining the political nature of both miners' and local communities' livelihoods, social mobilisations and collective actions, he brings the readers back into the political economic debates.

The book ends with a postscript, which provides a rough summary of important issues emerging from the chapters and indicates new directions for possible research in future.

References

Ali, S.H., 2009. *Treasures of the Earth: Need, Greed and a Sustainable Future*. New Haven: Yale University Press.

---

11 Damonte (2016: 972) has argued that decentralisation policy based on the incorporation of local actors into the regional apparatus would foster processes of state hybridisation and gradual integration with local knowledge and skills, rather than coercion or the creation of offices that simply relocated central government authority and capacities at the sub-national or local level.

Amin, A. (ed.), 2009. *The Social Economy: International Perspectives on Economic Solidarity*. Chicago: University of Chicago Press.

Araghi, F., 2009. 'The invisible hand and the visible foot: Peasants, dispossession and globalization.' In A.H. Akram-Lodhi and C. Kay (eds), *Peasants and Globalization: Political Economy, Rural Transformation and the Agrarian Question*. London: Routledge.

Bebbington, A., 1996. 'Movements, modernizations, and markets: Indigenous organizations and agrarian strategies in Ecuador.' In R. Peet and M. Watts (eds), *Liberation Ecologies*. London: Routledge.

Breman, J., 2013. *Outcast Labour in Asia: Circulation and Informalization of the Workforce at the Bottom of the Economy*. New Delhi: Oxford University Press.

Bryceson, D.F., 2002. 'Multiplex Livelihoods in Rural Africa: Recasting the Terms and Conditions of Gainful Employment.' *The Journal of Modern African Studies* 40(1): 1–28. doi.org/10.1017/S0022278X01003792

Bryceson, D.F. and J.B. Jønsson, 2010. 'Gold Digging Careers in Rural East Africa: Small-Scale Miners' Livelihood Choices.' *World Development* 38(3): 379–92. doi.org/10.1016/j.worlddev.2009.09.003

Buxton, A., 2012. 'MMSD+10, Reflecting on a Decade of Mining and Sustainable Development.' IIED Discussion Paper. London: International Institute for Environment and Development.

Buxton, A., 2013. *Responding to the Challenge of Artisanal and Small-Scale Mining: How Can Knowledge Networks Help?* London: International Institute for Environment and Development.

Caballero, E., 1996. *Gold from the Gods: Traditional Small-Scale Miners in the Philippines*. Quezon City, Manila: Giraffe Books.

Chatterjee, P., 2004. *The Politics of the Governed*. New York: Columbia University Press.

Damonte, G.H., 2016. 'The "Blind" State: Government Quest for Formalization and Conflict with Small-scale Miners in the Peruvian Amazon.' *Antipode* 48(4): 956–76. doi.org/10.1111/anti.12230

Deustua, J., 1994. 'Mining, Markets, Peasants, and Power in Nineteenth Century Peru.' *Latin American Research Review* 29(1): 29–54.

DeWind, A., 1977. 'Peasants Become Miners: The Evolution of Industrial Mining Systems in Peru.' Columbia University (PhD thesis).

Douglas, I. and N. Lawson, 2000. 'The Contribution of Small-Scale and Informal Mining to Disturbance of the Earth's Surface by Mineral Extraction.' *Mining and Energy Research Network Research Bulletin* 15: 153–61.

Fernández-Giménez, M., B. Batkhishig and B. Batbuyan, 2012. 'Cross-Boundary and Cross-level Dynamics Increase Vulnerability to Severe Winter Disasters (Dzud) in Mongolia.' *Global Environmental Change* 22(4): 836–51. doi.org/10.1016/j.gloenvcha.2012.07.001

Gibson-Graham, J.K., 2006. *A Postcapitalist Politics*. Minneapolis: University of Minnesota Press.

Godfrey, B.J., 1992. 'Migration to the Gold-mining Frontier in Brazilian Amazonia.' *Geographical Review* 82(4): 458–69. doi. org/10.2307/215202

Government of Sierra Leone, 2011. 'An Overview of the Sierra Leone Minerals Sector.' Viewed at embassyofsierraleone.net/about-sierra-leone/mining/mineral-sector-overview (site discontinued)

Grätz, T., 2002. 'Gold Mining Communities in Northern Benin as Semi-autonomous Social Fields.' Working Paper no. 36. Halle/Saale: Max Planck Institute for Social Anthropology.

Hart, K., 2010. 'Informal economy.' In K. Hart, J.L. Laville and A.D. Cattani (eds), *The Human Economy*. Cambridge, UK: Polity Press.

Hentschel, T., F. Hruschka and M. Priester, 2002. *Global Report on Artisanal & Small-Scale Mining*. Minerals Mining and Sustainable Development, no. 70. International Institute for Environment and Development, World Business Council for Sustainable Development.

High, M., 2008. 'Wealth and Envy in the Mongolian Gold Mines.' *Cambridge Anthropology* 27(3): 1–18.

High, M., 2012. 'The Cultural Logics of Illegality: Living Outside the Law in the Mongolian Gold Mines.' In J.B. Dierkes (ed.), *Change in Democratic Mongolia: Social Relations, Health, Mobile Pastoralism, and Mining*. Leiden: E.J. Brill. doi.org/10.1163/9789004231474_013

High, M. and J. Schlesinger, 2010. 'Rulers and Rascals: The Politics of Gold Mining in Mongolian Qing History.' *Central Asian Survey* 29(3): 289–304. doi.org/10.1080/02634937.2010.518008

Hilson, G.H., 2003. 'Gold Mining as Subsistence: Ghana's Small-Scale Miners Left Behind.' *Cultural Survival* 27(1): n.p.

Hilson, G., 2013. '"Creating" Rural Informality: The Case of Artisanal Mining in Sub-Saharan Africa.' *SAIS Review of International Affairs* 33(1): 51–64. doi.org/10.1353/sais.2013.0014

International Council on Mining and Metals (ICMM), 2010. *Working Together—How Large-scale Miners can Engage with Artisanal and Small-scale Miners*. International Council on Mining and Metals (ICMM), Communities and Small-Scale Mining (CASM) and International Finance Corporation (IFC) Oil, Gas and Mining Sustainable Community Development Fund.

Jennings, N., 1999. 'Social and Labour Issues in Small-Scale Mines.' Report for discussion at the Tripartite Meeting on Social and Labour Issues in Small-Scale Mines. Geneva: International Labour Organization.

Jønsson, J.B. and D.F. Bryceson, 2009. 'Rushing for Gold: Mobility and Small-Scale Mining in East Africa.' *Development and Change* 40(2): 249–79. doi.org/10.1111/j.1467-7660.2009.01514.x

Jyotishi, A., S. Sivramkrishna, K. Lahiri-Dutt, 2017. 'Gold Mining Institutions in Nilgiri–Wayanad: A Historical–Institutional Perspective.' *Economic and Political Weekly* 52(28).

Kamete, A.Y., 2008. 'When Livelihoods Take a Battering: Mapping the "New Gold Rush" in Zimbabwe's Angwa-Pote Basin.' *Transformation: Critical Perspectives on Southern Africa* 65: 36–67. doi.org/10.1353/trn.2008.0009

Knapp, A.B. and V. Pigott, 1997. 'The Archaeology and Anthropology of Mining: Social Approaches to an Industrial Past.' *Current Anthropology* 38(2): 300–4. doi.org/10.1086/204613

Lahiri-Dutt, K., 2006. '"May God Give Us Chaos, so That We Can Plunder": A Critique of "Resource Curse" and Conflict Theories.' *Development* 49(3): 14–21. doi.org/10.1057/palgrave.development.1100268

Lahiri-Dutt, K., 2014. 'Between Legitimacy and Illegality: Informal Coal Mining at the Limits of Justice.' In K. Lahiri-Dutt (ed.), *The Coal Nation: Histories, Politics and Ecologies of Coal in India*. Aldershot: Ashgate.

Lahiri-Dutt, K., 2016. 'The Diverse Worlds of Coal in India: Energising the Nation, Energising Livelihoods.' *Energy Policy* 99: 203–13.

Lahiri-Dutt, K., K. Alexander and C. Insouvanh, 2014. 'Informal Mining in Livelihood Diversification: Mineral Dependence and Rural Communities in Lao PDR.' *South East Asia Research* 22(1): 103–22. doi.org/10.5367/sear.2014.0194

Lahiri-Dutt, K. and H. Dondov, 2016. 'Informal Mining in Mongolia: Livelihood Change and Continuity in the Rangelands.' *Local Environment* 22(1): 126–36. doi.org/10.1080/13549839.2016.1176012

Lestari, N.I., 2013. 'Mineral Governance, Conflicts and Rights: Case Studies on the Informal Mining of Gold, Tin and Coal in Indonesia.' *Bulletin of Indonesian Economic Studies* 49(2): 239–40. doi.org/10.108 0/00074918.2013.809847

Levin, E. and A.B. Turay, 2008. 'Artisanal Diamond Cooperatives in Sierra Leone: Success or Failure?' Ottawa: Partnership Africa Canada. Available at www.africaportal.org/dspace/articles/artisanal-diamond-cooperatives-sierra-leone-success-or-failure

Littrel, M.A. and M.A. Dickson, 2010. *Artisans and Fair Trade: Crafting Development*. Sterling, VA: Kumarian Press.

Nelson, H., 2016. *Black, White and Gold: Gold Mining in Papua New Guinea, 1878–1930*. Canberra: ANU Press.

Phimister, I.R., 1974. 'Alluvial Gold Mining and Trade in Nineteenth-century South Central Peru.' *The Journal of African History* 15(3): 445–56. doi.org/10.1017/S002185370001358X

Porter, L., M. Lombard, M. Huxley, A.K. Ingin, T. Islam, J. Briggs, D. Rukmana, R. Devlin and V. Watson, 2011. 'Informality, the Commons and the Paradoxes for Planning: Concepts and Debates for Informality and Planning.' *Planning Theory and Practice* 12(1): 115–53. doi.org/10.1080/14649357.2011.545626

Roy, A. and N. Al Sayyad, 2004. 'Prologue/dialogue. Urban informality: crossing borders.' In A. Roy and N. Al Sayyad (eds), *Urban Informality: Transnational Perspectives from the Middle East, Latin America and South Asia*. Oxford: Lexington Books.

Siegel, S. and M.M. Veiga, 2009. 'Artisanal and Small-Scale Mining as an Extralegal Economy: De Soto and the Redefinition of "formalization".' *Resources Policy* 34(1–2): 51–6. doi.org/10.1016/j.resourpol.2008.02.001

Siegel, S. and M.M. Veiga, 2010. 'The Myth of Alternative Livelihoods: Artisanal Mining, Gold and Poverty.' *International Journal of Environment and Pollution* 41(3): 272–88. doi.org/10.1504/IJEP.2010.033236

Sippl, K. and H. Selin, 2012. 'Global Policy for Local Livelihoods: Phasing Out Mercury in Artisanal and Small-Scale Gold Mining.' *Environment, Science and Policy for Sustainable Development* 54(3): 18–29. doi.org/10.1080/00139157.2012.673452

Smith, J.H., 2011. 'Tantalus in the Digital Age: Coltan Ore, Temporal Dispossession and "Movement" in the Eastern Democratic Republic of Congo.' *American Ethnologist* 38(1): 17–35. doi.org/10.1111/j.1548-1425.2010.01289.x

Suzuki, Y., 2013. 'Conflict between Mining Development and Nomadism in Mongolia.' In N. Yamamura, N. Fujita and A. Maekawa (eds), *The Mongolian Ecosystem Network: Environmental Issues Under Climate and Social Changes*. Tokyo: Ecological Research Monographs. doi.org/10.1007/978-4-431-54052-6_20

Tesha, A.L., 2000. 'Cooperation Between Small-Scale and Large-Scale Mining: Tanzania Experience.' Paper presented at the Growth and Diversification in Mineral Economies: Regional Workshop for Mineral Economies in Africa, Cape Town, South Africa, 7–9 November. Viewed at www.unctad.org/infocomm/diversification/cape/pdf/tesha.pdf (site discontinued)

Tschakert, P., 2009. 'Recognizing and Nurturing Artisanal Mining as a Viable Livelihood.' *Resources Policy* 34(1–2): 24–31. doi.org/10.1016/j.resourpol.2008.05.007

UNDP (United Nations Development Programme), 2009. 'Mongolia Human Development Report 2007.' Unpublished report to the UNDP. Ulaanbaatar: UNDP Mongolia.

United Nations Economic Commission for Africa (UNECA), 2011. 'Minerals and Africa's Development.' The International Study Group Report on Africa's Mineral Regimes.

Verbrugge, B., 2015. 'The Economic Logic of Persistent Informality: Artisanal and Small-Scale Mining in the Southern Philippines.' *Development and Change* 46(5): 1023–46. doi.org/10.1111/dech.12189

Verbrugge, B., 2016. 'Voices From Below: Artisanal and Small-Scale Mining as a Product and Catalyst of Rural Transformation.' *Journal of Rural Studies* 47: 108–16. doi.org/10.1016/j.jrurstud.2016.07.025

Walsh, A., 2003. '"Hot Money" and Daring Consumption in a Northern Malagasy Sapphire-mining Town.' *American Ethnologist* 30(2): 290–305. doi.org/10.1525/ae.2003.30.2.290

Werthmann, K., 2007. 'Gold Mining and Jula Influence in Pre-colonial Southern Burkina Faso.' *Journal of African History* 48(3): 395–414. doi.org/10.1017/S002185370700326X

Section One:
Historical antecedents and value-making

2

Artisanal gold-rush mining and frontier democracy: Juxtaposing experiences in America, Australia, Africa and Asia

Deborah Fahy Bryceson

Gold rushes dramatically heighten people's expectations of the future. Not all societies value gold, but in those where gold is viewed as a source of wealth accumulation, adornment and enduring economic value, news of the discovery of gold is irresistibly enticing for many. Most of those who succumb have no previous mining experience, and unwittingly enter an occupational transformation with unpredictable outcomes. The compulsive force of the gold rush phenomenon rests on mounting hype and hope of both local and migrant populations.

The nineteenth-century gold rushes of California and Australia gave rise to a rich and varied literature based on government record keeping, journalist accounts and miners' personal letters home, memoirs and diaries. By contrast, in our twenty-first century of instantaneous communication by cell phone and internet, gold rushes may generate media and academic interest after a lapse of time, but on-the-spot empirical accounts of events and activities of participants, be they miners, residents of the mining settlement or officials, are relatively negligible.[1]

---

1   There are notable exceptions; for example, the African case studies of Werthmann (2000, 2003, 2010) and Grätz (2002, 2010, 2013).

'Gold fever' is portrayed as a contagion affecting growing numbers of people who irrationally ignore the hardships, dangers and possibility of failure, keen to try their luck at catapulting into a more remunerative livelihood and better life. In effect, gold rushes are high-risk journeys into the unknown, with a strong possibility of disconnection between ends and means; the mirage that the individual sees on the future horizon as opposed to miners' meagre means to realise their goals given their lack of mining skill and the unpredictability of conditions at the mining site.

In their quest, what stands out is the individual's intent to succeed as a miner largely through learning while doing. The individual embarks on a process of becoming an artisanal miner through skill acquisition, economic exchange, psychological orientation and social positioning at the labour site, and gains a sense of proficiency and occupational professionalism through persistent work and collegial support (Bryceson 2010). An artisanal gold miner is self-propelled from the initial decision to migrate to a mineral site, arriving at the work site to begin an on-site apprenticeship and a mining career that entails material hardships and technical difficulties, and the inevitability of mineral depletion over time (Bryceson and Jønsson 2014). Before nineteenth-century mass artisanal gold-rush mining, gold was usually produced within tied labour relations of slavery, bonded labour or tribal tribute obligations that then formed the production base for long-distance commodity chains, involving a series of local middlemen and foreign merchants. Gold miners were locationally circumscribed and subject to tribal or other localised controls and exactions, with little awareness of the gold commodity chain that they supplied.

I argue in this chapter that gold rushes since the nineteenth century, involving individualised, highly mobile artisanal gold miners, may encourage the evolution of frontier democracy, depending on the extent to which hierarchically organised labour circumstances do not pose a constraint.

At the outset, some definitions are in order. 'Artisanal mining' is labour-intensive mineral extraction with limited capital investment using basic tools, manual devices or simple portable machinery (Bryceson and Jønsson 2014).[2] I use the term 'artisanal mining frontier democracy' to refer to miners' independent exercise of agency in the pursuit of their

2   It should be noted that hard-rock extraction worldwide tends to be performed by men, while women and sometimes children are more likely to be found panning, processing the hard rock, reworking tailings or working in the service sector of the mining settlement.

mining activities; free association amongst people from various localities, occupations and class backgrounds who are focused on the artisanal extraction of gold, with scope for collective self-governance and political activity in furtherance of their mining interests.[3]

There are three essential conditions of existence for artisanal frontier democracy. First, miners voluntarily enter and exit artisanal mining sites on a relatively egalitarian basis. Second, working together, they assert a strong collective identity as artisanal miners in the process of acquiring skills and evolving shared professional norms and ethics. Third, they mine in nation-states where they have scope to devise the *modus operandi* of their work relations and mineral-access rights consensually on a trial-and-error basis (Bryceson and Geenen 2016).

Increasing numbers of people worldwide appreciate gold's value as a consumption and investment good. Gold production has expanded over the past two decades of economic boom and bust, catalysing the appearance of large numbers of artisanal miners in long-established as well as new strike sites. Technological innovations have facilitated the spread of information about gold strikes. Many artisanal miners travel over long distances. What is significant in terms of the focus of this chapter is that some arrive at 'pure gold sites', metaphorically termed, where they have scope to be self-employed, rather than subject to waged, bonded or enslaved labour arrangements imposed by pre-existing property ownership and power structures. Untied, independent artisanal labourers, engaged in high stakes and hazardous work, are inclined to cooperate with one another for mutual benefit. Therein lies the foundational basis for frontier democracy.

This chapter chronologically considers the nineteenth-century experiences before turning to twentieth-century African and Asian gold-mining sites in Tanzania and Mongolia, where deagrarianisation has been a powerful catalyst to labour diversification (Bryceson 1999, 2002). I argue that gold rushes taking place globally during the last two centuries in varied cultures and historical contexts converge remarkably towards similar forms of economic and cultural organisation characterised by exceptionally hard work and hard play in challenging physical circumstances. But not all display the dynamic of frontier democracy. The fundamental question explored here is when, how and why frontier democracy arises in some

---

3   My use of the term 'frontier democracy' is centred on the notion of a labour, rather than land, frontier (Dumett 1998; Grätz 2013; Bryceson and Geenen 2016).

places and not others. The following sections, based on secondary literature, are intended as a preliminary exploration of where and when this tendency is likely to occur.

Nineteenth-century gold rush experiences

Forty-niner miners in California

The voluminous literature on the California gold rush includes a wealth of personal memoirs of nineteenth-century gold rush participants' migration and settlement (see, for instance, US Library of Congress n.d.; Street 1851; Brown 1894; Canfield 1906; Holliday 2002). It is apparent that there is a recording bias towards nationals with middle-class origins (Clay and Jones 2008).

The discovery of gold at Sutter's Creek in California in January 1848 catalysed an avalanche of local in-migration to the area. According to Vaught (2007), the person most instrumental in spreading news of the discovery was a San Francisco entrepreneur who advertised gold-panning equipment, thereby alerting the general public in the city and countryside beyond. By the end of the year, the news had reached Washington DC. President James K. Polk announced California's gold riches to the nation on 5 December 1848, prompting a large westward demographic shift of young men from eastern and mid-western states to the country's new California territory over the next four years. The local newspaper, *The San Francisco Californian*, charted the influx from its outset:

> It was the work of but a few weeks to bring almost the entire population of the territory together to pick up pieces of precious metal. In less than four months, the prospects and fate of Alta California had been revolutionised. Then, capital was in the hands of a few individuals engaged in trade and speculation; now labor has got the upper hand of capital, and the labouring men hold the great mass of wealth in the country—the gold. (*The San Francisco Californian*, 14 August 1848, cited in Rohrbough 1998: 7)

A couple of months later, a report in the London *Times* was sceptical about the rush and its effect:

> The discovery of gold in the Sacramento ... has produced a confusion of rank and a startling degree of equality ... we must hear more of this El Dorado before we bestow upon it ... our serious consideration. (*The Times*, early November 1848, cited in Hocking 2006: 21)

Discounting journalistic hyperbole, it is revealing that the California gold rush was immediately identified not only as a profitable opportunity, but one democratically open to able-bodied men.[4] An estimated 40,000 'forty-niner' migrants made their way to the goldfields in 1849 (Vaught 2007).

The gold strike occurred soon after the United States had wrested Alta California away from Mexico in the Mexican–American war. This had been an unpopular war, criticised for its expense and the conquest of territory considered of little value for the United States. Hence, President Polk was eager to broadcast news of the California gold strike. At the time, Alta California, as it was called then, was a territory not a state, and was not subject to US federal law. What makes the California gold rush so iconic as a frontier experience is that incoming multi-ethnic miners faced a seeming 'free-for-all' arising from the legal hiatus in the transition from Mexican to American political control. The existing Mexican settlers in Alta California had not been aware of the gold's existence. State regulation of mineral extraction was absent, with no mining licence system[5] or legal clarity about who the gold finds belonged to. California was quickly granted statehood in 1850, but the precedent had been set. The forty-niner miners retained relatively unimpeded access to mining.

Local migrants from San Francisco and the surrounding countryside numbered approximately 5,000 by the end of 1849. Thereafter, tens of thousands followed, coalescing to over 100,000 as migrants from western parts of the United States and others from the East Coast travelled overland or by boat around the Horn of South America or through the isthmus of Panama, disembarking and walking to the Pacific coast where they boarded a boat to California. Nicknamed 'argonauts', this surge of people populated the new American state in what was a territorially expanding nation. Over 90 per cent of the gold-rush migrants were male, under 30 and generally planned to return to their farms (Vaught 2007).

The first wave of migrants were mostly unskilled, focused on alluvial placer mining using simple technology, generally a shovel and pan. Rohrbough (1998: 12–13) observed, 'Open access to the gold in California seemed to represent the purest example of American economic democracy in

---

4   The stream of migrants was over 95 per cent male (Clay and Jones 2008).
5   Non-US nationals were charged a foreign diggers' tax, which included indigenous Hispanics, Europeans and later Chinese, giving American nationals a digging edge.

the middle of the nineteenth century'. Miners worked on the general principle of 'finders keepers'. Shinn (1884, cited in Rawls 1999) argued that the miners devised a code of conduct that governed the mining camps.[6] Individual panning gave way to use of a boxlike 'cradle' for sifting sediment that was optimally operated by four people.

As alluvial gold yields declined, digging commenced. A system of mine claims evolved consensually amongst miners, beginning with early placer rights based on discovery, rather than landed property rights. Miners were not allowed to hold more than one mining claim. Claims amounted to usufruct rights recognised and maintained by continuing labour input (McDowell 2013). The mounting numbers of the miners gave them not only a strong demographic presence, but also economic and political power.

Census and survey analysis during the California gold rush suggests that on average miners succeeded in gaining higher earnings relative to what they could have expected in their pre-rush residences. However, given the high cost of living driven by exorbitant food costs, their real earnings were disappointing. The average earnings of non-mining service sector providers, in fact, superseded those of miners (Clay and Jones 2008).

The duration of expansion of artisanal mining settlements was roughly six years. As surface gold deposits became depleted, environmentally destructive industrial hydraulic extraction methods were introduced to achieve large-scale hard-rock underground excavation, which required heavy investment in technology and technical training. Artisanal mining was increasingly sidelined. Having begun with placer mining open to large numbers of unskilled men seeking their fortune as individuals, larger-scale capitalist mining operations took over (Douglass 2002).

During the gold rush's heyday, social levelling was visibly evident in the simple work clothing people wore, which downplayed the fact that the migrant stream encompassed a wide range of lower- and middle-class people. The mining camp population was an economically and socially diverse assembly that necessarily had to operate on a trusting basis, both

6   Royce (2002) challenged this interpretation as over-romantic and oblivious to miners' inhumane behaviour towards Native Americans and non-white migrants.

to facilitate their work and also because carried large amounts of gold in their pockets due to the absence of banks. Conspicuous consumption, with the notable exception of alcohol, was generally avoided.

The gold rush climaxed by the mid-1850s, having involved an estimated 300,000 people. Cornford (1999) observed that it was one of the largest occupational migrations of labour in American history, with 4,000 migrant miners in 1848, growing to approximately 100,000 in 1852. By the end of the decade, the miners who remained were wage labourers rather than self-employed. The migration coincided with a period when smallholder agriculture was declining in the US. Most gold-rush miners returned to their home areas or moved on to new mining sites elsewhere, including Australia. Of those who stayed on, some made the transition to industrial mining, mostly as wage labourers, and in far fewer cases as capitalist investors, but the majority remaining in California started farming, notably wheat, or were absorbed into industrial and service activities, contributing to the state's non-mining development take-off (McCone and Orsi 1999).

Australia's imperial gold finds in Victoria

There were several parallels and linkages between the gold rushes in California and Victoria. Two generations after Europeans first settled in Australia, interest was kindled in the continent's mineral wealth. Relying on Blainey's (1964) detailed account of the Australian gold rush, its progression is outlined here.

Ironically, in the early 1850s, significant numbers of Australians had set sail for California to mine gold, not realising they were leaving a gold-rich country. James Esmond, an Australian who had returned from Californian diggings, struck gold in Ballarat in July 1851. News of Ballarat and other finds were eagerly snapped up by the national press, triggering a national gold-rush migration. Gold had also been discovered in New South Wales further to the northeast. Amidst these gold discoveries, New South Wales was subdivided, creating Victoria as a new, separate colony in which Ballarat was located.

Issuing of mining licences commenced in the latter part of September 1851 in Victoria. Deposits of gold and silver were the designated property of the Crown. Victoria's goldfields were considered the richest in the world at the time, prompting the government to immediately proclaim Crown rights, ruling that it was illegal to search for gold without a licence.[7] The government registered and regulated miners' activities through a licensing system that required a monthly payment of 30 shillings for a licence.

A succession of gold finds followed those of Ballarat, most notably in Bendigo. So many men made their way to the goldfields that labour shortages appeared in critical occupations—blacksmiths, sailors and civil servants. Nearby Melbourne experienced an acute shortage of men, while women gravitated to settlements in the hope of meeting and marrying a rich digger. In early 1852, news of the rich finds reached Europe, enticing adventure-seeking men to risk the hazards of a three-month sea journey from Europe to Australia. Tens of thousands arrived by boat. Foreigners began to outnumber Australians in the camps. The mining camps consisted of row after row of tents. Men from a wide array of occupations mined side by side. The most successful miners usually tended to be the physically strongest, rather than the best educated.

The cultural trappings of the virtually all-male population, earning above-average incomes, veered towards laborious weekdays and with pleasure-seeking, heavy drinking and liaising with prostitutes in ramshackle hotels on the weekend. Despite overcrowding and heavy drinking, the crime rate was not higher than in Australia generally, and a consensual moral economy existed at the diggings, where trust and morality were considered to exceed that prevailing elsewhere in the country.

In Victoria, miners were required to have and display their licence while working at all times. Miners who failed to produce a licence for police inspection faced a group of intimidating, armed constables and a fine of 40 shillings or seven days' imprisonment. The taxation system became a great source of irritation for miners. The presence of foreign miners—Irish, continental European and American—who had no political allegiance to the British Crown, contributed to the growing disgruntlement against licence fee enforcement. Miners greatly resented the legal imposition,

7   This ruling was immediately enforced, unlike in New South Wales where news of gold had attracted hundreds of men to panning and digging activity faster than the government could assert control over it. There were not enough police and soldiers to enforce the Crown's property rights, and the Governor of Australia informed London that it would be unrealistic and hazardous to prevent the incoming miners from their gold-rush search. A new Commissioner of Crown Lands for the gold district was appointed to set up the licensing system, settle disputes and provide justice in petty courts. However, as the gold rush spread to various scattered sites, miners often succeeded in avoiding fee payment in the more far-flung areas (Blainey 1964).

particularly when they were asked to pay while down in a pit, necessitating their ascent, or when the tattered or water-damaged state of their licence was challenged (Keesing 1967). As the goldfields became depleted, the number of miners failing to find gold rose. One observer estimated that no more than 20 per cent of miners had made finds. Miners' monthly licence payments became difficult (Chandler 1990, cited in Hocking 2006: 126).

Political activism surged when the new Governor of Victoria, Sir Charles Hotham, was appointed. Soon after his arrival he reported to the Colonial Secretary:

> no amount of military force at the disposal of Her Majesty's Government, can coerce the diggers, as the gold fields may be likened to a network of rabbit burrows. For miles, the holes adjoin each other, each is a fortification … nowhere can four men move abreast, so that the soldier is powerless against the digger, who well armed, and sheltering himself by the earth thrown up about him, can easily pick off his opponent—by tact and management must these men be governed; amenable to reason, they are deaf to force … (Hill 2010: 170)[8]

Only a month later, a couple of public disorder incidents prompted miners to amass and register their protest. In the heat of the moment, the crowd marched to the Eureka Hotel and set it alight. The arrest of three suspects for the burning of the hotel and the imposition of another government licensed 'man-hunt' led to a further mass meeting, and the demand for the sacking of the local police chief. Military reinforcements were sent to Ballarat, while political tensions escalated. At another mass meeting, miners established the Ballarat Reform League and elected digger leaders to develop a charter. In addition to calling for the cessation of the gold licence fees and the reform of Crown land laws and goldfields administration, the charter called for universal suffrage, the abolition of property ownership as a qualification for becoming a member of parliament, short-term parliaments and ballot voting. These were considered radical demands that sowed the seeds for Australian independence (Hill 2010).

The showdown between miners and government troops took place at a wooden stockade, which 1,500 miners had collectively built as a precaution against attack. On the night of 3 December 1854,

---

8 This was part of a communication from Hotham to Bar, 18 September 1854, Dispatch no. 112, VPRS 1085/P unit 8.

approximately 150 miners were surrounded by 300 troops in a surprise attack. Firing into the stockade, 22 miners were killed, many more were injured and over 100 were arrested, as opposed to the seven soldiers who lost their lives.

In the immediate aftermath of the Eureka rebellion, Governor Hotham met with heavy public criticism. He set up a commission of inquiry; however, before the commission handed down its findings, 13 miners were charged with high treason. The trials began in February 1855 and, one after another, the accused were acquitted. The commission of inquiry ruled in favour of the miners in recommending the abolishment of the miners' licence fee,[9] making land available for miners to purchase for house building and extending the franchise to enable diggers to be represented in parliament. Peter Lalor, a young digger who took a leading role in the Eureka Rebellion, was elected as a member of parliament for Ballarat in 1855. The miners' collective political action engendered a democratic victory oft-cited in Australian history books for establishing the foundational principles of governance in the struggle against British colonial rule.

Blainey (1964) cites official estimates that there were 30,000 adult miners in June 1852, ballooning to 100,000 in 1855 with availability of gold supply still at accessible levels, down to the water table at roughly 30 feet. The size of mining claims in Australia were smaller than in California, such that there was a great deal of movement between sites as old sites became depleted and word travelled about new sites. Arriving at a new strike site early gave diggers a big edge.

Despite its laissez-faire capitalist context, the enduring memory of the Australian gold rush is of an egalitarian spirit, freedom and mateship of the gold diggers, culminating in their demands for democratic reform in 1854 at Ballarat's Eureka Stockade (Goodman 1994: 8). Gold was seen as a 'democratic mineral' (Bate 1999: 40). Diggers came together to protest on the basis of moral economy rights, confronting the bureaucratic demands of imperial officialdom; in so doing, they were seen to be advancing democratic nationalist demands (Goodman 1994).

---

9   'The licence fee has been undermined by widespread evasion, which has become a practiced and skilful art, and that its effective enforcement produced violent results ... To carry out the law in its integrity ... required a constant exercise of authority ... scenes between the police and the miners, were a daily occurrence, where mutual irritation, abuse, and gross violence would ensue ... The present system of a licence fee [should] be given up.' Report from the Commission Appointed to Inquire into Conditions in the Goldfields, Victorian Parliamentary Papers, Legislative Council, Votes and Proceedings, A76/1854–55, Vol. II (cited in Hill 2010: 190).

Probing the nature of nineteenth-century artisanal mining democracy

Social levelling, identified with 'mateship' in Australia, and 'the great adventure of the common man' in California (McWilliams 1949, cited in Cornford 1999: 78) offered potential economic advantage to a broad spectrum of novice 'greenhorn' miners, many of whom could not have hoped to gain the same level of earnings in other occupational avenues in their home areas. However, not everyone who arrived in the goldfields achieved economic success or feelings of camaraderie. Many were frustrated by late arrival, lack of luck or ability to adapt to the new work setting.[10]

Above all, a sense of democracy was delimited to men of European descent. Racial and ethnic tensions intensified as the availability of gold dwindled. In the first years, men of whatever background dug side by side; however, as the gold supply dwindled and incomes stagnated or declined, rancour and discrimination set in, with hostility defrayed by individuals moving on to new strike sites (Paul 1969). The Californian artisanal mining democracy eroded as American citizens were distinguished from Europeans, Latin Americans, Chinese and Native Americans. Distrust amongst diggers was strongest against foreign-speaking miners, including the French, Chileans, Mexicans and Chinese, leading ultimately to expulsion of Mexicans and open hostility towards the Chinese (Rohrbough 1998).

By far the most marginalised participants in the gold rush were Native Americans. Before the gold rush, the territory was characterised by the native Californian population who engaged in a hunting and gathering mode of livelihood, and did not mine or value gold (Rawls 1999). In the first months of the rush, when the labour force was drawn primarily from Alta California's local population, more than half the gold diggers were Native Americans employed by Mexican rancho owners. Only a few of them dug or traded independently. However, as the old Hispanic paternalist order of agrarian semi-feudal estates disintegrated, the Native American presence faded. The incoming migrant mine population engulfed Native American land. Conflicts over land usage were an excuse to deploy US army troops and volunteer militia companies. By 1854, army

---

10   The historian Oscar Lewis (2007) estimated that less than 5 per cent of gold-rush miners left the goldfields with more money than when they arrived.

expenditure on these incursions reached almost US$1 million, with total removal and killing of the Native American population in many places (Goodman 1994). America's creed of 'manifest destiny' had taken hold at immense loss of livelihood and lives of the Native American population.[11]

So too, Australian artisanal mining democracy was severely delimiting. Aboriginals were marginalised. Gold was certainly not a democratic mineral for them, as they lost their land and were excluded from citizenship; thus, their grievance testimonies were not admissible in court. The Chinese miners, at times constituting as much as one-third of the diggers (Goodman 1994), were considered to be quiet and hardworking, but were subjected to derogatory treatment as foreigners, and ostracised from the camaraderie of mateship. General wariness towards the Chinese prevailed on the grounds that they were worryingly present in growing numbers and remained aloof from Christian conversion, at a time when a Christian moral discourse was omnipresent. In 1856, European diggers initiated an attack on Chinese diggers at Buckland where 750 of their tents were destroyed. Many Chinese miners returned to China, discouraged by the hostile threats of violence directed at them.

Thus, the democracy narrative of artisanal mining can only be recounted with the proviso that it was deeply flawed, not unlike Greek democracy's indifference to the rights of women and slaves. Artisanal mining democracy in California and Victoria represents an historical interlude in which artisanal miners became enmeshed in a frontier nation-building narrative. They took centre stage in a pioneering ethos associated with the countries' spatial and political frontier expansion. Moving into low-population density areas where the regional state's presence was new and the state regulatory apparatus was not well established, artisanal miners had scope for occupational self-determination.

Other countries experiencing artisanal gold rushes in the nineteenth century with more established state governance were less amenable to miners' democratic self-direction. These included the Ghana gold boom of the 1870s in the British Gold Coast, which saw an influx of European

---

11   At the start of the gold rush in 1849, the Californian Native American population was estimated at 100,000. By 1860, their numbers had declined to approximately 35,000. While some died of disease, Russell Thornton is unequivocal that the largest number and most deliberate killings in the United States took place in northern California during the gold rush decade when 'case after case was recorded of Indian villages being massacred by larger or more powerful groups of Non-Indians' (Thornton 1987: 109, 201).

expatriates intent on setting up mining companies for the pursuit of mechanised mining alongside a coastal Creole mercantile elite who were sub-leasing mining land for industrial mining from chiefs (Dumett 1998). This constellation of mechanised mining interests played a formative role in the gradual development of industrial mining in Ghana during the late nineteenth century and into the twentieth century, when Ghana became Africa's second-largest gold producer after South Africa.

Similarly, the gold rush of the late 1880s in the South African Rand virtually precluded open entry of artisanal miners. Cecil Rhodes and Alfred Beit had already established controlling interests over the Kimberly diamond fields. They were in a strong position to seize economic and political hegemony in gold production together with other key investors, and came to be known as South Africa's 'Randlords'. The comradeship of diggers was pre-empted by the Randlords' supremacy; the fact that the gold deposits were embedded in a mass of deep hard rock further contributed to artisanal miners' difficulty in getting a significant foothold (Meredith 2007). By 1888, speculative investors had established 450 gold-mining companies. Over time, the mining of deep hard rock seams, monopolised by such companies, was based on the employment of black wage labour.

Twentieth- and twenty-first-century frontiers in Africa and Asia

Moving a century forward to the twentieth and twenty-first centuries, gold certainly has not lost its lustre. The world price of gold began its ascent during the troubled years of the global oil crises of the mid-1970s, stabilising during the 1980s and early 1990s, only to scale new dizzying heights as the world plunged into economic recession in the latter 2000s (Figure 2.1).

The following two country case studies drawn from Africa and Asia illustrate how national governments by default or design may have benign or indifferent policies to the presence of artisanal miners, which afford artisanal miners scope for democratic occupational self-determination, at least in the early stages of the gold rush.

$/Troy ounce*

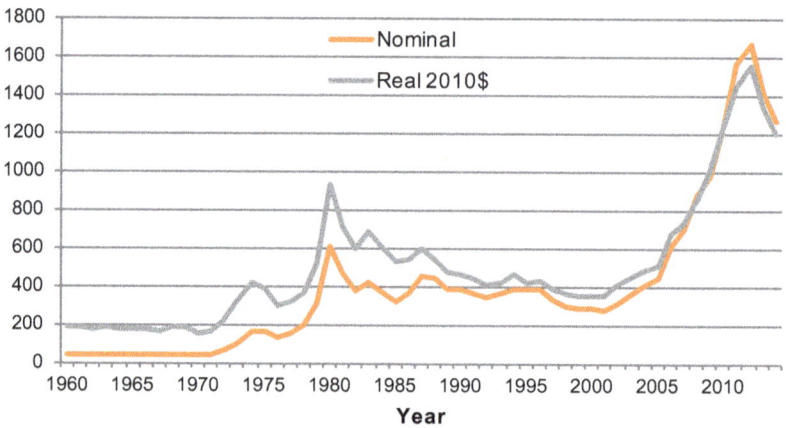

Figure 2.1: Trends in the gold price since 1960

Source: Author's graphical representation. Data source: Global Economic Monitor Commodities. Viewed at data.worldbank.org (20 June 2015).

Tanzania: Evolution of artisanal mining under the blind eye of the state

It is impossible to chart Tanzania's turn-of-the-century artisanal gold rush in terms of precise dates and names of strike sites, since the country's gold boom quietly coalesced from the early 1980s at a time of extreme duress. Following the 1979 international oil crisis, the country was economically crippled at the same time as the international price of gold was on the ascent (Figure 2.1). In need of foreign exchange, the Tanzanian Government passed the *Mining Act 1979*, allowing Tanzanian nationals to prospect for minerals and initiate mining activities not requiring investment in industrial machinery.

Tanzania's gold seams largely follow the bifurcation of the East African rift, forming a 'ring of gold' (Bryceson et al. 2012). Clearly farmers and ex-workers at the industrial gold mine had been aware of the rift's potential riches; however, prior to the 1979 Act they were not at liberty to engage in artisanal mining. Multiple gold discoveries started unobtrusively unfolding. With the increasing use of mobile phones in the twenty-first century, news of the gold finds spread. Gold rush fever gripped the mineral-rich regions (Jønsson and Bryceson 2009). Artisanal mining took off; yet, at the national level, there was very little recognition of the occupational tsunami underway. Certainly the press were not reporting gold strikes. At promising gold sites, in-migration of artisanal miners and service providers caused populations to balloon from a few hundred nearby villagers to 10,000 or more within three months. In 2011, the total population of artisanal miners throughout Tanzania was estimated at 685,000[12] out of a population of 40 million. Artisanal gold production spurred urban growth within the gold-mining regions (Bryceson et al. 2012).

In the mining camp, miners and residents represented a spectrum of occupational backgrounds, education levels and classes. However, the majority migrated from rural areas and entered a multi-ethnic, cosmopolitan experience divorced from the ascriptive age and gender hierarchies of the countryside. The rank and privilege of birth, education and class were replaced with the rough and tumble of market relations and work output. Tanzanian artisanal gold-mining, as elsewhere through time, was a relatively open-entry occupation that did not necessitate a large amount of starting capital. The functional division of labour involved a pyramid of 'claim owners', who were restricted in number and usually aloof from pit production, 'pit managers', who organised the work process on behalf of the owner, and 'diggers' (Bryceson and Jønsson 2010). The division of output between these three categories varied from place to place, but usually approximated roughly a third of the hard rock mined by diggers being allocated to each category (Jønsson and Fold 2012).

While the pit managers were the 'headmen' orchestrating pit work, there was nonetheless a spirit of camaraderie and teamwork. This rested on the material fact that everyone was exposed to the physical hazards of the pit, and necessarily cognisant of the primacy of cooperation to ensure survival. So too, they were all subject to the fluctuation of mining fortunes, such that diggers could rise to become pit managers, who in turn could experience a reversal and return to being diggers when their resources to finance digging operations declined (Jønsson and Bryceson 2009; Bryceson and Jønsson 2010). In the pit, the miner had to prove himself to his workmates, showing his skill, capability to work hard and trustworthiness. Thus, in choosing to join one work team as opposed to another, these considerations prevailed over those of ethnic ties.

---

12   Provisional estimate according to Dr Crispin Kinabo of the Geology Department, University of Dar es Salaam.

Claim owners by and large did not participate in gold production, and comprised only 1 per cent of the total population of artisanal miners. Yet, ironically by law, they were the only Tanzanian artisanal miners with legal mining rights. This raised a number of ambiguities and potential tensions between the state and artisanal miners. The main source of contention was the state's mediation between small- and large-scale foreign mining interests.

In a number of cases, beginning in the late 1990s, artisanal miners were told that their diggings encroached on mining company concessions (Kulindwa et al. 2003; Lange 2008).[13] Artisanal miners organised political protests against the state and foreign mining firms, but they were usually forced to leave the site. In the lead-up to the opening of Tanzania's biggest industrial gold mine at Bulyanhulu in Kahama, owned by African Barrick Gold, artisanal miners claimed that 50 of them had been buried alive and killed during eviction from the concession by Tanzanian police using graders to destroy their pits. Barrick and the police denied the allegations, while various international agencies, called to give independent assessments, were divided about what had happened.

More recently in 2013, Barrick Gold's North Mara mine faced violent contestation from artisanal miners, who waged a continuous daily protest campaign against the mining company, which they blamed for the death of five artisanal miners (Hodges and Biesheuvel 2013). With these pressures weighing on their business reputation, coupled with the decline in the global gold price, Barrick Gold began a process of divesting in African Barrick Gold in Tanzania (*Financial Times* 2014).

In the space of three decades, Tanzania has transformed from a primarily agricultural country to a mainly mining country, measured in foreign export earnings. During the 1980s, its artisanal miners pioneered gold exploration,[14] generating a regional economic boom in the country's 'ring of gold' (Bryceson et al. 2012). They benefited from the open-entry nature of artisanal mining and participated in occupational transformation that yielded more lucrative earnings than farming. In the process, they gained a new occupational identity as artisanal miners, freedom from the age and gender divisions of labour enforced by male gerontocracy

---

13 For example, this occurred with regards to Tom Mines in Mahenge, Barrick Mines in Bulyanhulu, North Mara Gold Mine in Tarime and East African Mines IAMGOLD in Buck Reef.
14 Artisanal miners inadvertently acted as gold prospectors for the large-scale mining companies that followed them.

in villages and, in some cases, challenged the growing hegemony of industrial mining. This took place despite the World Bank and national government's prioritisation of foreign direct investment of large-scale industrial mining companies.

Mongolia: Ambiguities of artisanal mining in a national mineral boom

Traditionally, Mongolia's economy was comprised primarily of nomadic herding households. However, rapid deagrarianisation over the last three decades associated with the collapse of the country's Soviet-influenced commandist economy, and catastrophic weather conditions between 1997 and 2002, resulted in drastic reductions in the size of households' livestock herds. Similar to Tanzania, facing heavy debt, the Mongolian Government adhered to the World Bank's insistence on economic liberalisation and encouragement of foreign direct investment, which triggered foreign mining companies' attraction to Mongolia's gold wealth.

Artisanal mining surged as well (Cane et al. 2015). Historically, Mongolian herders had been aware of gold-rich areas, but under both the Qing dynasty of the nineteenth century and later under a repressive Soviet-style government, there were heavy penalties imposed on people who dared to engage in artisanal gold production (High and Schlesinger 2010). Furthermore, Buddhist ethics traditionally ordained a sense of harmony with the environment, which posed a moral dilemma for herders-cum-miners accustomed to adapting rather than disrupting the natural landscape (High 2008). The 'gold rush', as it became known nationally, marked a drastic rupture with the traditional economy and culture of Mongolia. The first wave of artisanal miners in the mid-1990s was primarily composed of former industrial miners who had lost their jobs when the state mining industries folded after the collapse of the Soviet-influenced government, followed by a second wave of ex–state farm workers. The third and by far the largest wave of migrants was composed of herder families reacting to the loss of livestock and general deagrarianisation (Grayson et al. 2004).

Most of the literature to date on Mongolian artisanal mining relates to its environmental impact, but very little has been written about the social and political changes underway. Mette High's (2008, 2012) insightful study of herders-cum-miners in one of the main mining regions reveals what happened at the local level within communities and households.

In a country where people were accustomed to migrating in family units to graze their herds, migration to mining areas often involved the entire family, while others already located close to a mine site could mine part-time while continuing to tend their herd.

These tendencies contributed to widespread occupational shifts of herders, factory workers and even bureaucrats towards gold-mining, spurred by the promise of enhanced earnings from the high price of gold in the world market.[15] Hence, Mongolian artisanal mining became a site of immense labour absorptive capacity. In 2013, artisanal miners numbered over 100,000 out of a national population of just 3 million, representing 20 per cent of the rural workforce (Villegas 2013). The mining sector experienced astonishing growth, generating 75 per cent of national export revenue by 2008, with artisanal mining far superseding industrial employment and output in the late 2000s (High 2008).

The legal position of artisanal miners changed under pressure from the World Bank, various foreign and local non-government agencies, and rising resource nationalism on the part of the Mongolian public and state. During the 1990s, Mongolia opened the door to foreign investment, but by the 2000s there was increasing resentment about the presence of foreign mining companies, and the feeling that Mongolian nationals were not benefiting sufficiently from the country's mineral wealth. In 2002, a draft law was proposed whereby artisanal miners were required to be licensed and organised into cooperative groups; however, local governors had the right to arbitrarily terminate artisanal miners' agreements and licences. Rising concerns about environmental destruction, the use of banned mercury and growing resource nationalism on the part of the public who resented the presence of foreign mining companies caused the government to insist on Mongolian state equity participation up to 50 per cent in all industrial mining. While the formulation of temporary regulations allowed artisanal miners some room to manoeuvre, the legality of artisanal mining remained hazy.

Although artisanal mining is identified with distress migration away from herding, wealthy households and part-time herders are active participants. High's 2008 research in Uyanga traced the effects of a gold rush that began in 1999. She observed class and ethnic diversity amongst

---

15   High (2008) observes that gold miners are likely to earn four times the national average salary of approximately US$60 per month.

the miners as well as a tendency for social levelling that encompassed women. Interestingly, returns to individuals in the miners' work groups were divided equally at the end of each workday, regardless of age and sex. This egalitarianism starkly contrasted with the social hierarchy of patriarchal herding households where women's herding and domestic work was undervalued. Nor did they receive payment on par with men during Mongolia's socialist era on the collective farms and in the towns (Lahiri-Dutt and Dondov 2016). Above all, the existence of Mongolian artisanal mining was appreciated for its role in labour absorption and poverty alleviation. People from various walks of life gained higher than average livelihoods during the gold rush. Many herders transferred their economic dependency from herding to artisanal mining (Grayson et al. 2004). Meanwhile, their enhanced purchasing power supported the expansion of a labour-generating service sector in remote areas, with the potential for small town growth.

Twenty-first-century frontier democracy dissected

The two country cases of frontier democracy reviewed here—Tanzania and Mongolia—have a similar chronology, whereby peasants and herders facing hostile conditions, be it drought and famine or, alternatively, structural adjustment austerity, found gold-mining as a lifeline or lucrative alternative. In both countries, the golden opportunity generated widespread egalitarian tendencies that crossed class, ethnic and, in Mongolia, gender boundaries.

While nineteenth-century gold rushes were characterised by discovery and excavation of gold in remote, sparsely populated regions viewed as political frontiers in the eyes of the state, by the twenty-first century, populations had become denser and state territories were generally permanently fixed. Given ample supplies of local manual labour within the country, nation-states are now far less likely to publicly encourage international in-migration of miners, in contrast to what had happened during the Californian and Victorian gold rushes. However, foreign migrants from neighbouring states can cross borders unnoticed and usually pass as natives. Artisanal miners, migrating in large numbers from within their national borders to gold rush sites, are most common.

Like nineteenth-century artisanal gold rush sites, those of the late twentieth and twenty-first century represent a spectrum of outcomes, ranging from temporary easing of artisanal mining controls to repressive policies aimed at actively dissuading or eliminating artisanal miners. For example, gold-rush mining in West Java, triggered by the 1997 Asian financial crisis and the fall of the autocratic Suharto Government, led to an estimated 26,000 illegal artisanal miners flooding into Pongor; the composition of the migration flow reflected the legacy of insecurity from the Suharto period (Lestari 2007, 2011). Migrants from Banten Province or nearby cities aimed to occupy the most valuable sites, and formed rival gangs to achieve that end. Some of Indonesia's military generals became involved with medium-scale operations involving equipment, hiring artisanal miners or monopolising existing pits and renting them out. Artisanal miners' self-employed status was thereby compromised. Conflict between Antam mining company and the artisanal miners flared in December 1998, and intensified into the millennium. Clearly, the artisanal gold rush of the late 1990s and early 2000s in Indonesia was far from approximating collegial occupational transformation and a democratic frontier in labour relations.

Ghana's gold rush in the 2010s has included migrants from China, neighbouring African countries and a sprinkling of other nationalities of which Russians are noteworthy, but the majority of miners are Ghanaian nationals intent on earning a livelihood. Chinese miners started being arrested in 2013, this was followed by a presidential crackdown on the local illegal artisanal *galamsay* miners (Levin 2013). Several military operations have been staged with the objective of eradicating their digging activities. The media demonised the *galamsay*, but criminalising them did not deter their presence (Hilson and Hilson 2015). Steeped in gold history, Ghana's artisanal mining context was metaphorically a bubbling cauldron of melded vested interests of small-, medium- and large-scale gold producers, in situ local communities, chiefs and regional and national governments. The ensuing contestation and conflict acted to throttle artisanal miners' individual and collective democratic voice.

Conditions of existence: Gold rush frontier democracy then and now

I have explored the dynamics of frontier democracy with respect to the numerical size and demographic composition of miners; the nature of the labour process, notably organisational aspects of the work teams, their social and political identity and coherence as a collective labour force; the material and legal aspects of mineral access; the artisanal mining presence vis-à-vis the state; and artisanal miners' interaction with other external agents, notably the indigenous population and corporate mining interests.

The narrative emerging from this historical review is one of remarkably similar social transformations experienced by gold rush migrant populations spanning the nineteenth to the twenty-first century, despite widely differing cultural origins of the migrants. In the nineteenth century, European populations dominated the frontier artisanal labour expansion in colonial and post-colonial North America, Australia and New Zealand, whereas the twenty-first century, with its unprecedented international gold prices (Figure 2.2), witnessed the spread of innumerable artisanal gold sites across Asia, Africa and Latin America. Regardless of migrant gold rush miners' differences in beliefs pertaining to social hierarchical order, nature and resource utilisation, most miners were likely to embark on artisanal mining with a strong sense of expediency, adopting a lifestyle of hard work and hard play, infused with an infectious sense of expectation. Artisanal miners working side by side to find and extract gold shared a common identity and ethos. Their associational ties as miners were underpinned by an innovative can-do determination and camaraderie, a 'Eureka ethos', which sometimes transgressed local social boundaries of propriety.

$/ounce

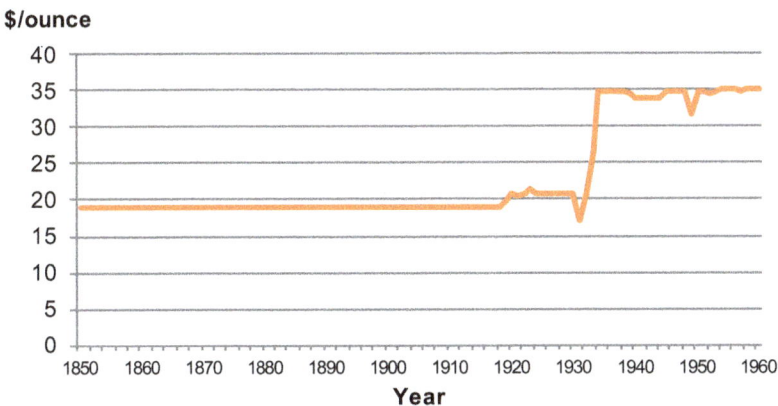

Figure 2.2: International gold prices, 1850–1960
Source: Author's graphical representation. Data source: usagold.com/reference/prices.

Curiously, during the mid-nineteenth century, public opinion and the media were highly favourable to the gold rush miners, often seeing them as pioneering role models, whereas in the twenty-first century, they tend to be ignored or viewed as a nuisance standing in the way of 'modern mining', as is the case in Tanzania and Mongolia. In countries with longer mineral production histories, more entrenched mining company interests and governments supporting large-scale mining, artisanal miners were usually disparaged as environmental polluters and illegal workers, subjected at times to criminalisation, as exemplified by Ghana. Yet many artisanal miners, facing a stiffening atmosphere of hostility, nonetheless found room for occupational manoeuvre in these settings, although the experience of frontier democracy largely eluded them.

A number of criteria associated with national and local governments preclude or undermine frontier democracy. First, contested local interests in land and mineral rights: given higher population densities than prevailed 150 years ago, local property interests of people with different modes of livelihood tend to be both complex and deeply entrenched. Second, mining policies are now skewed towards large-scale mining. Transnationalism of corporate industrial mining over the twentieth century has brought large-scale mining interests to the fore in far more places than ever before. The global surge in market liberalisation policies from the 1990s onwards engendered a proliferation of national and foreign-owned mining companies. Third, under global economic liberalisation, increasing mutuality of interests of multinational mining corporations and national governments has evolved (Moody 2007; Campbell 2009). All three criteria appear to be more prevalent in the twenty-first century. Such circumstances stack the odds heavily against artisanal miners' activities and the emergence of frontier democracy.

Despite these obstacles, manifestations of frontier democracy have emerged in various places in the twenty-first century, but through different forms of mutuality between the state and artisanal miners. The development of frontier democracy during the nineteenth century arose in a historical period of nation-state expansion into remote geographical areas where artisanal mining proved useful to state-building agendas of governments, albeit often at the expense of the aboriginal populations.

In the twenty-first century, with national borders worldwide relatively fixed, interests of artisanal gold miners and state agents are most likely to converge when government politicians seek to win votes from the mass numbers of artisanal miners. They may also converge in the process of individual government officials or commercial middlemen seeking pecuniary gain along the gold commodity chain between local sources of gold supply and the foreign export of gold. For example, in Tanzania it is rumoured that most artisanal unprocessed gold is smuggled to Dubai, evading royalty payment (Fold et al. 2014).[16] This kind of subterfuge in foreign exports would most likely benefit state and/or market elites in the capital city.

The duration of a gold rush tends to be short-lived, as miners using handheld artisanal implements reach a depth where they cannot proceed further without considerable capital investment in equipment. Four to six years is the norm before there is a substantive tail-off of the rush. The significance of temporality in relation to the unfolding non-renewable nature of mineral deposits, fluctuating rates of in-migration, state attempts to assert control and the availability of alternative options have altered labour patterns over the centuries. In the nineteenth-century Californian and Victorian gold rushes, it is believed that many if not most artisanal miners migrated back to their home areas as large-scale capitalised mining displaced artisanal mining. In both places, the gold rush catalysed economic diversification in services and industry, beckoning former miners, especially since many had witnessed the high returns to labour in services as the gold rush proceeded.

Perhaps the most significant and enduring difference between the nineteenth and twentieth centuries is the post–gold rush phase. In the twenty-first century, smallholder farmers find it difficult to compete with larger-scale producers and agro-industry. Many are not attracted to returning to their rural home areas, having grown accustomed to higher levels of income and a freer, non-traditional lifestyle. In the service sectors around the mining sites, formal jobs are very scarce, and the informal sector is already oversubscribed. Thus, artisanal miners are likely to try to eke out a subsistence reworking tailings—a precarious existence with low returns and high risks of falling foul of enforcement agencies of the mining companies or state. Fundamentally, artisanal miners' persistence is premised on the lack of better work options elsewhere, contrasting starkly with the nineteenth-century's industrial expansion and proliferation of urban employment that offered abundant alternative work options.

---

16 Given the sensitivity of the subject matter this is difficult to document, but Geenen (2015) has succeeded in providing a detailed account of the gold commodity chain for the eastern Democratic Republic of Congo.

Paradoxes of frontier democracy: From mirage, to material reality, to enduring legacy

In exploring frontier democracy with respect to artisanal gold rush miners during the nineteenth century as opposed to now, I have conceptualised frontier democracy as the realisation of economic opportunity, class levelling, trust and cooperative interaction amongst manual labourers arising from their occupational solidarity as migrant artisanal miners and, in some cases, their engagement in collective political protest against external agents seen to be standing in the way of the manifest destiny of their search for gold.

Dumett (1998: 8) observed that gold has given rise to 'legends, exaggerations, and delusions' leading to 'myths and fantasies' through recorded history. I would add, with respect to the era of nationalism-cum-globalisation spanning the mid-nineteenth century to the present, the production of artisanal gold has generated many positive economic and political labour trends followed by a clutch of paradoxical outcomes. Positively, gold-mining's accessibility to masses of people willing to work with simple handheld tools has provided employment to impressive numbers of artisanal miners; forged a strong occupational identity within a short period of time fostered by collectively experienced risktaking and hard work; and generated a democratic ethos in a number of places when and where the 'vent for democracy' is not preempted by the state, existing local vested interests in gold-mining or large-scale industrial mining.

These outcomes challenge the common belief that artisanal gold miners are necessarily overly acquisitive and irresponsible. True, the quest for El Dorado is an important motivational force underlying miners' migration over long distances, and persistence of hardship and uncertainty. But, too often, the emphasis on artisanal miners' end goal overlooks their mode of working, notably the collegial ties and occupational integrity of working together on a common occupational footing regardless of class and educational level, and their creativity in devising a modus operandi of shared work practices. It is not accidental that Fetherling (1988) called the phenomenon 'the Gold Crusades' to denote the intensity of purpose and common identity involved. Gold seekers were credited with transforming California from 'a half-tamed wilderness to a settled commonwealth' (Billington 1956: 218), while similarly gold-mining was viewed as altering Victoria from 'a minor pastoral settlement to the most celebrated British colony' (Serle 1963: 369). Despite moral criticism of artisanal miners by vocal segments of the population at that time, this pioneering narrative has prevailed, identifying artisanal miners with foundational democratic principles (Goodman 1994). The irony is that artisanal miners' contribution to economic and social development at present could be argued to be far more laudable than that of earlier gold rush miners, who participated in a highly 'delimited democracy' focused primarily on European settlers, at the expense of local people. Twenty-first-century artisanal mining has been more socially and ethnically inclusive, engaging larger numbers of the working population, mostly nationals.

It is lamentable that their contribution to egalitarian democracy has received negligible recognition, and is woefully under-recorded in terms of what they have achieved materially and culturally. Present-day artisanal miners should be encouraged to record and disseminate their stories of how they negotiated the organisation of artisanal mining, regardless of differences in class and education. Their testimony is vital to ensure appreciation of twenty-first-century artisanal mining's social significance. At the very least, artisanal mining has provided vital poverty alleviation through employment generation and on-the-job occupational training for millions in Africa and Asia. Additionally, the Eureka ethos and labour frontier democracy represent invigorating forces of change, invaluable for national rehabilitation and renewal in the present, as well as during nineteenth-century nation-building.

The final irony is that neoliberal casino capitalism has paradoxically generated the conditions for the recent boom in artisanal mining, as well as its ultimate demise. Capitalism's gold price spiral, its expanding unregulated labour markets and the ongoing process of deagrarianisation displacing millions from age-old agrarian work patterns has triggered the artisanal gold boom. Yet, these same neoliberal economic forces, combined with the natural depletion of gold at depths accessible to artisanal miners, are likely to spell the eventual eclipse of artisanal mining by large-scale global mining interests. Generally, national governments, beholden to international financial institutions' neoliberal agenda, play a central role in this process. Inevitably in the twenty-first century, small-scale producers, in whatever occupational sphere, face the Goliath of international capitalism. The democratic legacy of artisanal gold-rush mining of the late twentieth and early twenty-first centuries is too valuable for the future to be ignored, forgotten or derided.

References

Bate, W., 1999. *Victorian Gold Rushes*. Victoria: The Sovereign Hill Museum Associations.

Billington, R., 1956. *The Far Western Frontier 1830–1860*. New York: Harper.

Blainey, G., 1964. *The Rush that Never Ended: A History of Australian Mining*. Melbourne: Melbourne University Press.

Brown, J.S., 1894. *California Gold: An Authentic History of the First Find*. Oakland: Pacific Press Company.

Bryceson, D.F., 1999. 'African Rural Labour, Income Diversification and Livelihood Approaches: A Long-term Development Perspective.' *Review of African Political Economy* 80: 171–89. doi.org/10.1080/03056249908704377

Bryceson, D.F., 2002. 'Multiplex Livelihoods in Rural Africa: Recasting the Terms and Conditions of Gainful Employment.' *The Journal of Modern African Studies* 40(1): 1–28. doi.org/10.1017/S0022278X01003792

Bryceson, D.F., 2010. 'Africa at Work: Transforming Cccupational Identity and Morality.' In D.F. Bryceson (ed.), *How Africa Works: Occupational Change, Identity and Morality*. Rugby: Practical Action Publishing. doi.org/10.3362/9781780440248.001

Bryceson, D.F. and S. Geenen, 2016. 'Artisanal Frontier Mining of Gold in Africa: Labour Transformation in Tanzania and the Democratic Republic of Congo.' *African Affairs* 115(459): 296–317. doi.org/10.1093/afraf/adv073

Bryceson, D.F. and J.B. Jønsson, 2010. 'Gold Digging Careers in Rural Africa: Small-Scale Miners' Livelihood Choices.' *World Development* 38(3): 379–92. doi.org/10.1016/j.worlddev.2009.09.003

Bryceson, D.F. and J.B. Jønsson, 2014. 'Mineralizing Africa and Artisanal Mining's Democratizing Influence.' In D.F. Bryceson, E. Fisher, J.B. Jønsson and R. Mwaipopo (eds), *Mining and Social Transformation in Africa*. London: Routledge.

Bryceson, D.F., J.B. Jønsson, C. Kinabo and M. Shand, 2012. 'Unearthing Treasure and Trouble: Mining as an Impetus to Urbanisation in Tanzania.' In D.F. Bryceson and D. Mackinnon (eds), *Mining and Urbanisation in Africa: Population, Settlement and Welfare.* Special Issue of *Journal of Contemporary African Studies*: 631–49. doi.org/10.1080/02589001.2012.724866

Campbell, B. (ed.), 2009. *Mining in Africa: Regulation and Development.* London: Pluto Press.

Cane, I., A. Schleger, S. Ali, D. Kemp, N. McIntyre, P. McKenna, A. Lechner, B. Dalaibuyan, K. Lahiri-Dutt and N. Bulovic (eds), 2015. *Responsible Mining in Mongolia: Enhancing Positive Engagement.* Brisbane: Centre for Social Responsibility in Mining.

Canfield, C. de L. (ed.), 1906. *The Diary of a Forty-niner.* San Francisco: M. Shepard Co.

Clay, K. and R. Jones, 2008. 'Migrating to Riches? Evidence from the California Gold Rush.' *The Journal of Economic History* 68(4): 997–1027. doi.org/10.1017/S002205070800079X

Cornford, D., 1999. 'We All Live More Like Brutes Than Humans.' In J.J. Rawls, R.J. Orsi and M. Smith-Baranzini (eds), *A Golden State: Mining and Economic Development in Gold Rush California.* Berkeley: University of California.

Douglass, W.A., 2002. 'The Mining Camp as Community.' In E.W. Herbert, A.B. Knapp and V.C. Pigott (eds), *Social Approaches to an Industrial Past: The Archaeology and Anthropology of Mining.* London: Routledge.

Dumett, R.E., 1998. *El Dorado in West Africa: The Gold Mining Frontier, African Labor, and Colonial Capitalism in the Gold Coast, 1875–1900.* Oxford: James Currey.

Fetherling, D., 1988. *The Gold Crusades.* Toronto: University of Toronto Press.

Financial Times, 2014. 'African Barrick Gold "could be independent in 12 months".' 23 September.

Fold, N., J.B. Jønsson and P. Yankson, 2014. 'Buying into Formalisation? State Institutions and Interlocked Markets in Small-Scale Gold Mining.' *Futures* 62: 128–39. doi.org/10.1016/j.futures.2013.09.002

Geenen, S., 2015. *African Artisanal Mining from the Inside Out: Access, Norms and Power in Congo's Gold Sector.* London: Routledge.

Goodman, D., 1994. *Gold Seeking: Victoria and California in the 1850s.* Australia: Allen & Unwin.

Grätz, T., 2002. 'Gold Mining Communities in Northern Benin as Semi-autonomous Social Fields.' Working Paper no. 36. Max Plank Institute for Social Anthropology.

Grätz, T., 2010. 'Gold-mining and Risk Management: A Case Study of Northern Benin.' *Ethnos* 68(2): 192–208. doi. org/10.1080/0014184032000097740

Grätz, T., 2013. 'The '"Frontier" Revisited: Gold Mining Camps and Mining Communities in West Africa.' Working Paper no. 10. Zentrum Moderner Orient (ZMO).

Grayson, R., T. Delgertsoo, W. Murray, B. Tumenbayar, M. Batbayar, U. Tuul, D. Bayarbat and C. Erden-Baatar, 2004. 'The People's Gold Rush in Mongolia—the Rise of the "Ninja" Phenomenon.' *World Placer Journal* 4: 1–112.

High, M., 2008. 'Wealth and Envy in the Mongolian Gold Mines.' *Cambridge Anthropology* 27(3): 1–19.

High, M., 2012. 'The Cultural Logics of Illegality: Living outside the law in the Mongolian gold mines.' In J. Dierkes (ed.), *Change in Democratic Mongolia: Social Relations, Health, Mobile Pastoralism, and Mining.* Leiden: E.J. Brill. doi.org/10.1163/9789004231474_013

High, M. and J. Schlesinger, 2010. 'Rulers and Rascals: The Politics of Gold in Qing Mongolian History.' *Central Asian Survey* 29(3): 289–304. doi.org/10.1080/02634937.2010.518008

Hill, D., 2010. *The Gold Rush: The Fever that Forever Changed Australia.* Australia: William Heinemann.

Hilson, G. and A. Hilson, 2015. 'Entrepreneurship, Poverty and Sustainability: Critical Reflections on the Formalisation of Small-Scale Mining in Ghana.' Working Paper. Oxford & London: International Growth Centre.

Hocking, G., 2006. *Gold: A Pictorial History of the Australian Goldrush*. Victoria: The Five Mile Press.

Hodges, J. and T. Biesheuvel, 2013. 'African Barrick Gold Sued in U.K. by Tanzanians over Death.' *Bloomberg Business*, 30 July. Available at www.bloomberg.com/news/articles/2013-07-29/african-barrick-gold-sued-in-u-k-by-tanzanians-over-mine-deaths

Holliday, J.S., 2002. *The World Rushed In: The California Gold Rush Experience*. (Based on Swain's diary). Norman: University of Oklahoma Press.

Jønsson, J.B. and D.F. Bryceson, 2009. 'Rushing for Gold: Mobility and Small-Scale Mining in East Africa.' *Development and Change* 40(2): 249–79. doi.org/10.1111/j.1467-7660.2009.01514.x

Jønsson, J.B. and N. Fold, 2012. 'Dealing with Ambiguity: Policy and Practice Among Artisanal Gold Miners.' In D.F. Bryceson, E. Fisher, J.B. Jønsson and R. Mwaipopo (eds), *Mining and Social Transformation in Africa: Mineralizing and Democratizing Trends in Artisanal Production*. London: Routledge.

Keesing, N., 1967. *History of the Australian Gold Rushes: By Those Who were There*. Australia: Angus & Robertson Publishers.

Kulindwa, K., O. Mashindano, F. Shechambo and J. Sosovele, 2003. *Mining for Sustainable Development in Tanzania*. Dar es Salaam: Dar es Salaam University Press.

Lahiri-Dutt, K. and H. Dondov, 2016. 'Informal Mining in Mongolia: Livelihood Change and Continuity in the Rangelands.' *Local Environment* 22(1): 126–39. doi.org/10.1080/13549839.2016.1176 012

Lange, S., 2008. 'Land Tenure and Mining in Tanzania.' CMI Report no. 2. Bergen: Chr Michelsen Institute.

Lestari, N.I., 2007. 'Illegal Gold Mining in West Java—Can Antam's Community Development Programs Win Over Cynical Locals?' Artisanal and Small-Scale Mining in Asia-Pacific Case Study Series. Viewed at www.asmasiapacific.org (site discontinued)

Lestari, N.I., 2011. 'The Informal and Small-Scale Industry in Indonesia.' The Australian National University (PhD thesis).

Levin, D., 2013. 'Ghana's Crackdown on Chinese Gold Miners Hits One Rural Area Hard.' *New York Times*, 29 June.

Lewis, O., 2007 [1949]. *Sea Routes to the Gold Rush*. Oakley, CA: Oakley Press.

McCone, M. and R.J. Orsi, 1999. 'Preface.' In J.J. Rawls, R.J. Orsi and M. Smith-Baranzini (eds), *A Golden State: Mining and Economic Development in Gold Rush California*. Berkeley: University of California.

McDowell, A.G., 2013. 'From Commons to Claims: Property Rights in the California Gold Rush.' *Yale Journal of Law and the Humanities* 14(1): 1–72.

Meredith, M., 2007. *Diamonds, Gold and War: The Making of South Africa*. London: Pocket Books.

Moody, R., 2007. *Rocks & Hard Places: The Globalisation of Mining*. London: Zed Books.

Paul, R., 1969. *California Gold: The Beginning of Mining in the Far West*. Lincoln: University of Nebraska Press.

Rawls, J.J., 1999. 'Introduction.' In J.J. Rawls, R.J. Orsi and M. Smith-Baranzini (eds), *A Golden State: Mining and Economic Development in Gold Rush California*. Berkeley: University of California.

Rohrbough, M.J., 1998. *Days of Gold: The California Gold Rush and the American Nation*. Berkeley: University of California Press.

Royce, J., 2002 [1886]. *A Study of American Character*. Santa Clara: California Legacy Book.

Serle, G., 1963. *The Golden Age: A History of the Colony of Victoria 1851–1861*. Melbourne: Melbourne University Press.

Street, F. J., 1851. *California in 1850 Compared with What it was in 1849, and a Glimpse of its Future Destiny.* Cincinnati: R.E. Edwards & Co.

Thornton, R., 1987. *American Indian Holocaust and Survival: A Population History since 1492.* Norman: University of Oklahoma Press.

US Library of Congress, n.d. *California as I Saw It: First Person Narratives of California's Early Years, 1849–1900.* US Library of Congress digital collection. Washington DC. Available at www.memory.loc.gov

Vaught, D., 2007. *After the Gold Rush: Tarnished Dreams in the Sacramento Valley.* Baltimore: John Hopkins University Press.

Villegas, C., 2013. 'Ninja Miners and Rural Change in Mongolia.' Viewed at www.asm-pace-org/blog/item/itninja-miners-and-change-in-mongolia.html (site discontinued)

Werthmann, K., 2000. 'Gold Rush in West Africa: The Appropriation of "Natural Resources": Non-industrial Gold Mining in South-Western Burkina Faso.' *Sociologus* 50(1): 90–104.

Werthmann, K., 2003. 'The President of the Gold Diggers: Sources of Power in a Gold Mine in Burkina Faso.' *Ethnos* 68(1): 95–111. doi.org/10.1080/0014184032000060380

Werthmann, K., 2010. '"Following the Hills": Gold Mining Camps as Heterotopias.' In U. Freitag and A. von Oppen (eds), *Translocality: The Study of Globalising Processes from a Southern Perspective.* Leiden: E.J. Brill. doi.org/10.1163/ej.9789004181168.i-452.31

3

Tanzanite: Commodity fiction or commodity nightmare?

Katherine C. Donahue

Rafiki Nderema[1] leaned back in his chair, finishing off the last piece of goat. A plane landed nearby at the Kilimanjaro International Airport. We were talking about the tanzanite mines not far from the airport. Rafiki is a part-time dealer in tanzanite. 'It can be dangerous,' he said, 'so I hire a woman to take the bus to the mines and pick up the tanzanite. No one knows she is carrying it. Sometimes she changes buses, just to make sure.' When she is confident she is not being followed, she brings the raw gemstones to Arusha, Tanzania, where Rafiki lives, and where tourists usually begin their safaris to nearby national parks or to the Serengeti. Rafiki's wife prepares the tanzanite by cleaning and heating it. A relative helps them with the processing. The treated tanzanite is then marketed.

Tanzanite is available in many shops in Arusha, and it is frequently sold on the streets. Experienced buyers know what to look for, but cost-conscious tourists sometimes end up with lumps of dark glass. The tanzanite gemstone business is just one of several enterprises Rafiki undertakes to support his family. He works in the safari business for himself and for

---

1   The name is a pseudonym, as is that of a later source. Pseudonyms are used here because the gemstone industry can be dangerous. Hold-ups and shootings still occur in the region. Tanzania is not the only place where gem dealers are robbed; it happens in the US as well. One American jeweller with whom I spoke says that Colombian gangs in particular acquired a specialty in such thefts, staking out jewellery stores for possible marks.

other companies, as both a driver and a guide. He is a fine mechanic. He supports not only his wife and children, but also occasionally other family members. His businesses bring him into contact with tourists who may be interested in tanzanite. But to go into business in a bigger way is hard. Credit is difficult to get, and financing large purchases of anything is difficult. His enterprises are all small in scale, but Rafiki gets by. This is a very Tanzanian story.

The problem

> The commodity description of labor, land, and money is entirely fictitious ... Nevertheless, it is with the help of this fiction that the markets for labor, land, and money are organized. (Polanyi 1957: 72)

Karl Polanyi's discussion of the fiction of land, labour and money as commodities was neither the first nor the last analysis of the social meanings attached to things. Things exist in a world of goods (Douglas and Isherwood 1979), they have social lives (Appadurai 1986) and they are circulated and controlled in a web of socio-ecological relations (Donahue 2003). The fictitious nature of commodities, of their supply and of the demand for them, is exemplified in the story of a stone. This chapter discusses the recent commodification of tanzanite, a product sold globally as a semi-precious gemstone. Reported to be '1,000 times more rare than diamonds' (Nones 2005), tanzanite is found solely in one small region of Tanzania, East Africa. Popular with Americans, it has been widely marketed in the US and the Caribbean. Advertisements use the mystique of Mt Kilimanjaro and the Maasai to do so.

However, after 11 September 2001, allegations that al-Qaeda bought tanzanite to finance its operations led to close scrutiny, and almost the collapse, of the tanzanite market. Journalists Robert Block and Daniel Pearl, on the trail of al-Qaeda's financial dealings before 11 September 2001, reported in a November 2001 *Wall Street Journal* article on the links between al-Qaeda and the tanzanite trade (Block and Pearl 2001). I describe more fully here how the impact of that article, which appeared at the same time that Douglas Farah (2001) of the *Washington Post* reported on al-Qaeda's connections to the diamond trade in Sierra Leone and Liberia, led to an immediate decline in sales of tanzanite. The adverse publicity concerning 'conflict' or 'blood' diamonds since 2001 (see Farah 2001, 2004a; Campbell 2002), and especially since the 2006 release of the film

Blood Diamond, has awakened interest in the role of easily transportable and therefore easily smuggled gems in money laundering, in the purchase of weapons and other goods, and in the financing of insurgencies. However, since early 2002, the US and Tanzanian governments have argued that there is no evidence that al-Qaeda is currently involved in the tanzanite trade.

The lure of profit in Tanzania, where the average person currently lives on about US$920 a year (see World Bank 2014), led to the quick growth of Mererani, the main town near the tanzanite mines, with attendant problems of HIV/AIDS, prostitution and drug use. While the commodification of tanzanite has created profit for gemstone buyers and dealers, and skilful marketing has brought tanzanite to the attention of cruise ship passengers in the Caribbean at the same time that it has been featured as a desirable present for birthing mothers and film stars on the Hollywood red carpet, it has also led to unsafe mining practices, violence over disputed mining rights and pitched battles between small-scale miners and TanzaniteOne, a mining corporation with South African roots. Before the sale of TanzaniteOne to a local Tanzanian company in December 2014, armed security officers for TanzaniteOne patrolled the mining block controlled by this corporation, both above and below ground, in order to keep small-scale miners out. Several miners and a TanzaniteOne employee have died in these disputes. Flash-flooding of mines and faulty air pumps have also led to deaths in Mererani. Tanzanite has thus become both a commodity fiction and a commodity nightmare.

Approaching the field

Several colleagues at my university and I have taken students to Tanzania to research community-based conservation in two areas of the country. One is on the western slopes of Kilimanjaro, north of Arusha, between the villages of Tinga Tinga and Engare Nairobi. I first encountered tanzanite in the summer of 2002. Students in that year's field course had heard about tanzanite from publicity concerning the 1997 film *Titanic*. The students were not the only ones who thought the large blue stone, 'The Heart of the Ocean', worn by the actress Kate Winslet, was actually tanzanite. Internet advertisements also featured the connection. Babb (2013) noted that jewellers in Jaipur, India, believed the film was responsible for introducing tanzanite to the global market.

In January 2004, I visited two stores in Key West, Florida, a cruise ship destination, which prominently advertised the fact that they sold tanzanite. When I inquired, a store employee said that it was one of their fastest selling commodities. Later, back in Tanzania in the summer of 2004, I spoke at some length with a tanzanite dealer, who described the process he went through to buy and prepare it. Subsequently, I have interviewed jewellers in Florida, Massachusetts and New Hampshire in the United States, have read relevant court cases and searched accounts of gemstone commodity production. I attended and wrote about the trial of Zacarias Moussaoui, the so-called 20th hijacker of 9/11 (Donahue 2007), who was also named in a lawsuit connected to tanzanite, and I have pursued the numerous stories of funding sources for al-Qaeda.

What follows here is a discussion of the process by which a stone has become commodified. It has been assigned value despite, or perhaps because of, the fictions that surround this commodity. Other scholars, such as Kelsall (2004), Lange (2006), Schroeder (2010) and Helliesen (2012) have skilfully written about the impact of tanzanite mining, the designation of tanzanite as a conflict gem and on the businesses and people of Tanzania. The violence that attends tanzanite mining has been reported in Tanzanian newspapers (see, for instance, Philemon 2013). Numerous reports have been written about conditions in the mines by researchers for institutions such as the International Labour Organization, the US Department of Labor, United Nations Children's Fund (UNICEF) and the World Bank, some of which are discussed in this chapter. Here, I describe some of the social and political issues that have developed because of the tanzanite industry, but my focus is on the development of the market for tanzanite, the adroit ways in which it has been marketed and the reception of that marketing on the part of buyers in countries such as the US. The emphasis, therefore, is on the downstream ripples created by this particular commodity, ripples that spread to the United States and back to Tanzania. In the first section, I describe the discovery of tanzanite and its subsequent marketing to a worldwide clientele, but in particular to Americans by companies such as Tiffany and TanzaniteOne. Next, I describe the connection between the United States and the tanzanite trade, focusing in part on the allegations that tanzanite was a 'conflict gem'. In the third section, I describe the relationship between the large-scale mining company and the small-scale miners, and the conditions in which the miners work. In the fourth section, I explore the role played by local Tanzanians, some wealthy, some not, in attempting to control portions of the tanzanite trade.

The tanzanite story: Fact or fiction

Tanzanite had a difficult, dirty and dangerous beginning in the Maasai Steppe of Tanzania. Various stories credit its discovery to at least three people: an Indian tailor, Manuel D'Souza from Arusha, Tanzania; a Maasai cattle herder, Ali Juyawatu; and a Meru cattle herder, Jumanne Ngoma. In July 1967, according to one report, D'Souza was prospecting for rubies in the area east of Arusha when Maasai cattle herders showed him some blue stones. Legend, widely spread on the internet, has it that a lightning strike started a grass fire, as frequently happens in the dry season there. When the fire burned out, some beautifully coloured gemstones were found in the burnt grass. The stones were a rare form of zoisite, a calcium aluminum silicate. The semi-precious gemstone in its natural state is often a rather uninteresting purple/brown in colour. But when this particular form of zoisite is heated to 300–500 degrees centigrade, it turns a sapphire blue, or purple or red. In the gem trade, this trait is referred to as 'trichroic', as having three colours when the stone is turned to the light. To date, this form of zoisite is found only in the Manyara region, Simanjiro district of Tanzania. The mining area is in the Mererani hills southeast of Arusha, not far from the Kilimanjaro airport, and north of the Simanjiro plain. The Maasai people, primarily cattle herders, have lived there since their migration south from Kenya hundreds of years ago. Some have become brokers for tanzanite, but the miners themselves often come from elsewhere in Tanzania, leaving behind their own poverty to engage in more poverty, to dig and crawl through the more than 300 pit mines and accompanying tunnels to depths of 500 m. Some miners work 18-hour shifts. Some miners were reportedly only seven years old. Drugs, alcohol, sex, HIV/AIDS and crime are commonly reported there. In 2004, Daniel Yona, the minister of energy and minerals, declared during a visit to Mererani that 60 per cent of the inhabitants had HIV/AIDS (*Arusha Times* 2004a). This was most likely an exaggeration, however unwitting, but the figure did indicate the social perceptions of the area.

When buyers from Tiffany, the upscale New York–based jewellery company, first saw the stones in the 1960s, their interest was aroused. It was Henry B. Platt of Tiffany who reportedly gave the name 'tanzanite' to the stones, and who said that the discovery of the gemstone was 'the most exciting event of the century' (*Time* 1969). Another statement from Tiffany said that tanzanite was 'the most beautiful blue stone to be discovered in 2,000 years' (Tanzanite Foundation n.d.). Although the market did not expand through the 1970s, its popularity grew during the 1980s through good advertising. By the time of the release of the film *Titanic* in 1997, interest in tanzanite had expanded to the point that jewellery stores in the Caribbean, in Florida and even in Alaska had difficulty keeping it in stock. Not unlike the techniques used in marketing diamonds, tanzanite had acquired its own mystique and legend. 'From the foothills of Mt. Kilimanjaro', said the internet marketers on one website. 'Unlocking the riches of Mt. Kilimanjaro', said another (Nones 2005). One website claimed that the Maasai give tanzanite to women who have just borne a child. It displayed a photo of a Maasai woman wearing a tanzanite stone. While it is the case that many Maasai women wear blue clothing, some websites, such as that for Tanzanite Gallery (2007), a gem dealer, stated:

> Recent Maasai tradition tells of how pieces of tanzanite were given by Maasai Chiefs to their wives on the birth of a baby to bestow upon the child a healthy, safe and positive life. According to their customs, only women who have been blessed with the miracle of new life have the honour of wearing blue coloured beads and garbs.

In this vein, the Tanzanite Foundation, a non-profit group established by the TanzaniteOne mining company, was created. The website for the Tanzanite Foundation once stated that it:

> strives to develop the tanzanite industry by growing demand and creating value for stakeholders in the tanzanite value chain. By striving to standardize methods of practice and conduct, the Tanzanite Foundation aims to uphold an ethical route to market in accordance with the Tanzanite Tucson Protocols, and invests in meaningful and sustainable upliftment projects developed in harmony with the indigenous communities in Tanzania. (TanzaniteOne 2007)

The foundation, run by a public relations firm, launched a tanzanite marketing campaign in 2006 specifically aimed at the 'Push Present' market, meaning the market for presents to women who have just pushed through labour (*Colored Stone* 2005). According to one online dictionary, Karen Heller in the *Austin American-Statesman* on 31 March 1992 wrote of the Oscars:

> Let us say that Annette Bening has lost all that baby weight—Warren must have given her a ThighMaster as a push present—and looked understated yet ravishing. (Double-Tongued Dictionary 2007)

The website for the Tanzanite Foundation included in its image gallery a photo of a Maasai woman wearing blue, with a baby on her back and a large tanzanite necklace around her neck, hanging just above the baby's head (TanzaniteOne 2008). By 2007, the foundation had created awards for the designers of jewellery made as part of their 'Be Born to Tanzanite' campaign. They did move beyond marketing to build two schools, an orphanage, a clinic and the Maasai Ladies' Project, teaching Maasai women how to use wire to make jewellery using discarded tanzanite, and serving as an advocacy group overall. However, because of unrest in Mererani, by August 2014 the Tanzanite Foundation had closed its operations, as did TanzaniteOne.

An American story: The market for tanzanite

The market for tanzanite in the United States and Caribbean is a particularly American story. In the 1990s, it was something new and still somewhat affordable for Americans, and was considered a good investment if you knew what you were looking for. The production and marketing of tanzanite became problematic once reports surfaced that tanzanite was bought and smuggled out of Tanzania by al-Qaeda members. After the attacks of 11 September 2001, Robert Block and Daniel Pearl of the *Wall Street Journal* investigated the trade in tanzanite at its source; their report brought a halt to the importation of tanzanite to the United States. The US State Department, the Tanzanian Government and the gem industry reached an accord in February 2002—the 'Tucson Tanzanite Protocol'—that reopened imports to the US. After that, the market for tanzanite became steadier, but the production at the source became less so. Schroeder (2010) has described the impact of the move toward ensuring ethical mining practices on the Tanzanian Government and the stakeholders in the tanzanite mining industry. The argument has been made that disputes between small-scale miners and AFGEM (African Gem Resources)/TanzaniteOne, as well as poor infrastructure and low foreign direct investment have caused a drop in production. The American and Tanzanian governments worked to take steps toward the regularisation and control of the export of tanzanite, steps mirrored in other mining and export sectors, and new marketing strategies such as those by TanzaniteOne to support interest in the stone. Thus, the future of small-scale and large-scale miners of tanzanite was linked to American politics and concerns about terrorist acts on American soil.

The al-Qaeda connection

On 16 November 2001, Robert Block and Daniel Pearl's piece ('Underground Trade: Much-Smuggled Gem Called Tanzanite Helps Bin Laden Supporters') on al-Qaeda's involvement in buying and smuggling tanzanite appeared in the *Wall Street Journal* (Block and Pearl 2001). After the bombings of the US embassies in Kenya and Tanzania in 1998, and particularly after 11 September 2001, the United States searched for any sources of money that might support al-Qaeda, and proceeded to shut down a number of banks and aid associations. Al-Qaeda apparently sought out other means of moving money after 1998, including buying diamonds in Liberia and Sierra Leone (see Farah 2001, 2004a, 2004b; Campbell 2002). Block and Pearl reported that al-Qaeda members were also buying tanzanite in Tanzania, which the buyers then smuggled out of Tanzania into Kenya, and thence to buyers in Dubai and elsewhere. Block and Pearl identified an imam of a mosque in Mererani who was sympathetic to Osama bin Laden, and a tanzanite dealer who worked with that imam. Daniel Pearl was murdered in Pakistan in January 2002, while still on the trail of al-Qaeda's finances for the *Wall Street Journal*. As a result of Block and Pearl's article and other news stories, large gem retailers such as QVC, Birks, Peoples, Wal-Mart, Zale Corp. and Tiffany & Co. all stopped selling tanzanite, much to the concern of the Tanzanian authorities, for whom the US market for tanzanite represented the majority of its exports of that gemstone (Roskin 2002; Hübschle 2004). The US State Department investigated, and said it found little evidence connecting al-Qaeda with tanzanite. By 9 February 2002, the Tucson Tanzanite Protocols were announced in Tucson, Arizona, which is a major centre for the global gem industry. Edgar Maokola Majogo, the Tanzanian Minister of Energy and Minerals, representatives of international gem dealers' associations and Michael O'Keefe of the US State Department Office of East African Affairs were all present. The Tucson Tanzanite Protocols stated that the Government of Tanzania 'has taken and continues to take significant steps to safeguard this gemstone' by licensing miners and traders in Tanzania, and ensuring for all countries 'transparency and accountability in the supply chain'. Furthermore, warranties were to be required of exporters and 'all those in the downstream chain of commerce' (Tanzanite Gems 2002).

As a result of the reported links between al-Qaeda and tanzanite, a lawsuit was filed in the federal district court in New York City. In February 2002, lawyers for families of three victims of the 11 September terrorist attacks filed a wrongful death suit against STS Jewels, Inc., owned by

Sunil Agrawal, a tanzanite dealer, and the Tanzanite Mineral Dealers' Association (TAMIDA). The plaintiffs wanted an injunction banning STS Jewels from selling tanzanite and:

> forcing it to contribute all proceeds from any past tanzanite sales to a court-supervised 9/11 victims' relief fund. The suit also sought $1 billion in compensatory damages from other defendants, which included not only TAMIDA but also Osama bin Laden, the former Taliban government of Afghanistan, the Iraqi government, and the supposed '20th hijacker' of 9/11, Zacarias Moussaoui. (Tanzanite Gems 2002)

The suit was withdrawn with prejudice on 8 April 2002, meaning that the case could not be brought again. According to Sunil Agrawal, the head of STS Jewels: 'To a potential plaintiff, the lawsuit's withdrawal sends the message: there's nothing there … I feel vindicated' (Tanzanite Gems 2002; see also Markon 2002).

In March 2002, after the Tucson Protocols were announced, Tanzanian Minister Majogo accused AFGEM, the South African tanzanite mining company, of inviting Robert Block of the *Wall Street Journal* to Arusha and Mererani to write about the tanzanite trade. Minister Majogo accused AFGEM of intending to make it difficult for the local Tanzanian miners to sell their unbranded, undocumented gems on the market. AFGEM was at the time preparing to ensure that each gemstone sold from their mine was to be identified and guaranteed, and they reported that the controls they were initiating were designed to bring order to the tanzanite industry (Beard and Kondo 2002; Kondo 2002).

By early 2002, questions had been raised as to the veracity of Block and Pearl's article. According to one source, the imam of the mosque described by them was incorrectly identified. Other names and facts were said to be inaccurate, and the people identified in the article denied their sympathies with Osama bin Laden's cause (see, for example, Maharaj 2002; Hübschle 2004). Annette Hübschle, in a report on crime in southern Africa, suggested that Block and Pearl's article was a case in which 'journalists in need of a good news story may have blown an allegation out of proportion' (Hübschle 2004: 9). Michael O'Keefe of the US State Department, who was present at the announcement of the Tucson Tanzanite Protocol, said:

> I think they looked at the trade … saw it was chaotic, not totally controlled, with a certain percentage lost to smuggling, and they tied that to al-Qaeda … [But] our resources are a hell of a lot bigger; we have an intelligence community. (Beard and Kondo 2002)

The gem dealers were happy. By September 2002, the *Jewellers' Circular Keystone* reported that:

> U.S. State Department representative Michael O'Keefe has repeatedly told the jewelry industry that there is no evidence that any terrorist organization is funding operations through the sale of tanzanite. (Roskin 2002)

Tiffany began to sell tanzanite jewellery again in June 2002, saying that:

> Based on lack of credible evidence linking tanzanite and terrorism as well as progress to further controls over trade in tanzanite, Tiffany & Co. has resumed the sale of jewelry containing this gemstone. (Roskin 2002)

Furthermore, the American Gem Trade Association announced in October 2002 that tanzanite was to become a December birthstone, along with turquoise and zircon. Although other stones had been added to the list, news reports circulated that tanzanite became the first gemstone to be added to the birthstone list since 1912 (Weldon 2002). The demand side of the tanzanite story became healthy again.

A jeweller from St Maarten in the Caribbean said, 'In the 1980s no one had heard of tanzanite. Three years ago (1999) everyone started to carry it' (Roberts 2002). Tourists in the Caribbean are now well versed in tanzanite. As one couple's website said:

> Do we look like tourists? We're guessing Key West has seen a tourist or two in her day. We did a lot of shopping and returned to the ship loaded down with key lime souvenirs (and a little birthday tanzanite). (Bundy and Schilling 2002)

According to many reports, 80 per cent of all tanzanite has gone to buyers in the United States (see Roskin 2002; Hübschle 2004). Most of the rest of it goes to the Caribbean. There, it is bought by cruise ship passengers, many of whom are North Americans. Diamonds International began doing business in 1986 in St Thomas, and now has over 100 stores in the Caribbean, Florida, Mexico, Las Vegas and Alaska. Seizing on the interest in tanzanite, Diamonds International created a separate chain of shops, Tanzanite International. There are several competitors selling tanzanite on Duval, the main shopping street in Key West. Some jewellery stores post large signs outside the shop advertising the sale of tanzanite. One such is the Jewel Port, which, on its signs and on the internet, claimed it had the 'Best Collection of Tanzanite, (10%–70% off select items)'. One jeweller there told me in 2004 that tanzanite was the gemstone most cruise ship passengers were interested in buying. Some of those passengers become informal experts on tanzanite. One former cruise ship passenger wrote in the following about a 2003 cruise:

Day 5: Grand Cayman

Just behind Columbian Emeralds there are a number of shops situated around a courtyard. For those interested in Cigars, there is a very nice shop located in this area, as well as a shop called Island Pleasures. This is also a jewelry store but they sell a lot of loose Tanzanite. Normally Tanzanite can sell for up to $700 a carat but I was able to but [sic] a pair of 2 (identical in color and shape – 1 ½ carats total) for $299.00. Loose stones also do not have to be declared on your Customs forms. After some other souvenier [sic] shopping we went back to the ship. (McClure 2003)

In November 2005, one cruise ship passenger, 'Irish Boy', posted the following query to fellow readers of a cruise ship message board:

Hi Gang

Just got off the Victory & Imagination in October and if I ever hear the word Tanzanite again I will jump off the nearest bridge … what is this stuff I never heard of it before but never heard so much about it for 7 days on the cruise, do the cruise lines own a mine or what … observations please. :confused. (Irish Boy 2005)

A travel writer observed that tanzanite had become so popular with the cruise ship industry that it appeared as the only gemstone listed in a glossary of nautical terms, along with 'port' and 'starboard' (Clausing 2008). She went on to say that:

Tanzanite is a beautiful blue or purple stone seemingly found only in two places in the world: Tanzania, and cruise ships. You may find it in shops at home, but it seems to be a particular obsession in cruising circles— on some sailings you can't swing a shuffleboard cue without hitting a boutique specializing in the stone.

Even in Ketchikan and Skagway, Alaska, cruise ship passengers are briefed by 'port lecturers' on tanzanite acquisition at local gem stores. In turn, the stores are expected to pay the cruise lines a certain percentage of their profits (Glass 2012). One jeweller said he had paid US$25,000 and was expected to give 10 per cent of his earnings to the cruise shopping programs in addition. Complaints from customers and local businesses came to the attention of the consumer protection authorities for the state of Alaska; in 2013, the state of Alaska moved to require port lecturers to reveal the fact that they are advertising for these jewellery companies (Gutierrez 2013).

In October 2005, rings at one store in New Hampshire in the US were retailing at US$139 for tanzanite set with diamonds; a bracelet of tanzanite and diamonds was US$399. By January 2007, the same store was selling a tanzanite ring set with diamonds for US$799, and a bracelet of tanzanite and diamonds cost US$1,400, while anzanite pendants were selling for US$200–300. By January 2008, earrings and pendants were selling for US$400 each, while a set containing a ring, earrings, and a pendant was US$199. Demand had declined by 2008, a salesperson said, because tanzanite was not marketed as widely, and it was less available. Another jeweller in New Hampshire said he usually bought his tanzanite from a gem dealer in Boston, Massachusetts, but that dealer now spends more time buying sapphires in Sri Lanka than tanzanite in Arusha, Tanzania.

A carat is one-fifth of a gram. According to one jeweller in Massachusetts, cut tanzanite sells for between US$500 and US$3,000 a carat (2016 prices), depending on the cut, clarity and the quality of the stone. Cut diamonds can sell for US$5,500–6,000 a carat. Tanzanite occurs naturally in a variety of sizes and it often requires preparation by cutting and polishing. I have seen an unpolished and uncut 20-carat stone, valued at US$2,300, and a photo of a smaller cut tanzanite that was sold for US$8,000. It should be noted that the terms 'precious' and 'semi-precious' for gemstones are used less frequently now than in the past. For instance, a good-quality semiprecious tourmaline is usually more expensive than a lesser-quality diamond.

The global downturn in the economy after 2008 has had an effect on commodity production. One example is provided by Tiffany, the New York jeweller. When Wall Street provided end-of-year bonuses, Tiffany's Wall Street store was visibly advertising its wares. Holiday sales were an important source of income, but soft sales in 2009, as well as, for example, in 2015, led Tiffany to reduce its staff (Beilfuss 2016). Tanzanite International and its parent company, Diamonds International, continue to serve the cruise ship trade in the Caribbean and Alaska, but the stand-alone Tanzanite International store in Key West, Florida, eventually closed. Jewellery store owners, with whom I talked between 2009 and 2015, reported that they do not get as many requests for tanzanite as formerly, and order it only as needed. Three jewellery stores each had only five or six small stones on hand.

Mining in Mererani

> To allow the market mechanism to be sole director of the fate of human beings and their natural environment, indeed, even of the amount and use of purchasing power, would result in the demolition of society. (Polanyi 1957: 73)

The mines at Mererani were nationalised by the Tanzanian Government in 1971, and the State Mining Corporation, or STAMICO, took over production. This occurred during President Julius Nyerere's period of Ujamaa, Tanzania's version of socialism. Despite his and the program's best intentions, nationalisation failed to produce economic development. The movement's efforts to avoid the social injustices of free markets in the West were a test of an economic system that did not have the support of an infrastructure, both national and international, to keep it going. State mining efforts and production declined steadily at Mererani, while small-scale mining increased rapidly.

By 1990, the Tanzanian Government tried to increase control over the small-scale miners, and divided the area into four blocks. Block A was licensed by Kilimanjaro Mines Ltd, Blocks B and D were reserved for small-scale miners and Block C was taken by Graphtan Limited, a graphite mining company. Graphtan ceased production in 1996, and AFGEM acquired the licence. According to the TanzaniteOne website, AFGEM began tanzanite production in 2001. TanzaniteOne, based in Bermuda but with directors also associated with AFGEM, took over control of Block C in 2004.

Disputes arose when the small-scale miners expanded tunnels out and under other blocks, in particular into Block C, the AFGEM/TanzaniteOne block (Hernandez 2003: 10). In 2001, conflict erupted between small-scale miners and AFGEM. Gunfire was exchanged between AFGEM guards and miners in April that year, and an AFGEM building was bombed during Easter weekend (Ngowi 2001: 32). By March 2002, the Tanzanian mining minister had accused AFGEM of inviting the *Wall Street Journal* reporters to Mererani to investigate the al-Qaeda connection. AFGEM sought good public relations by providing water to

Maasai people and cattle during a December 2003 drought (*Arusha Times* 2003). In December 2006, the Manyara regional police commander gave TanzaniteOne, AFGEM's successor, the authority to use 'reasonable force' to protect its mining concession from artisanal miners, particularly those who tunnel underground into Block C. 'Reasonable force' included the use of air guns and live bullets. This action, the police commander said, was necessary because the small-scale miners used explosives underground, not only to frighten the TanzaniteOne security guards, but also to blow off grills that were placed to keep out the small-scale miners (Ihucha 2006).

Small-scale mining operations, while apparently chaotic, appear to be increasingly self-regulating. Bryceson et al. (2013) argue that small-scale miners have provided a democratising force in countries that are becoming increasingly 'mineralised'—that is, reliant on mining for economic improvement as traditional agriculture becomes less dependable as a source of income. The government has not acquired much control over operations. As miners arrived to try their luck, they began to register and licence themselves through informal and formal mining associations. According to Hernandez (2003), rights to plots of lands that were 25 m x 25 m have been given out to these miners. More recently, the small-scale miners have pushed the government to impose restrictions on export of unprocessed tanzanite in order to provide more jobs for Tanzanian processors. In a hearing in Mererani in January 2008, representatives of the miners asked for in-country annual auctions to ensure that international gemstone buyers would leave their money in Tanzania and, presumably, not deal with sources in India or elsewhere (Juma 2008).

Social and environmental issues in tanzanite production

> [T]he alleged commodity 'labor power' cannot be shoved about, used indiscriminately, or even left unused, without affecting also the human individual who happens to be the bearer of this peculiar commodity. (Polanyi 1957: 73)

Tanzanite mining, as with mining elsewhere, relies on the labour of the young and agile. Predominantly male, these young people have left subsistence agriculture behind, hoping to find gems. Small-scale mining occupies approximately half a million Tanzanians (Kinabo 2003) or more. Eftimie et al. (2012) reported that there were 550,000 small-scale miners in Tanzania, of whom 25 per cent were women. According to a miner, Mzee Nyumbani (a pseudonym), and others, there is substantial migration of young Arusha, Meru and Pare to the area of the tanzanite mines (Sheridan 2005). Some become farmers, growing maize and beans, while others work in the mines. The urban area of Mererani has grown from 37,109 people in 2002, according to the Tanzanian Housing and Population Census (United Republic of Tanzania 2002) to 50,800 in 2012 (Tageo.com 2012). A report on the Tanzanian mining sector said that 79 per cent of inhabitants of Mererani were people who moved in between 1985 and 1995 from other regions of Tanzania; those surveyed had earlier been farmers and civil servants (Mwalyosi 2004: 12).

For some Tanzanians, the system works to their benefit. Kelsall (2004: 47) describes Asfoku Mollel, a Maasai who entered the mining trade early and now owns several pits, reportedly worth hundreds of millions of Tanzanian shillings (as of early 2016, TZ2,100 = US$1). Mollel has invested in an hotel, a bus company, a football team and real estate.

Mzee Nyumbani described his own experiences as a Pare miner to Michael Sheridan, an anthropologist, as follows:

> In 1978, I started to dig for these gemstones [tourmaline?]. I was able to build a house in the town of Landanai, and I was able to buy first one vehicle then another. I continued to mine until I reached where I am today, now I have 15 vehicles and 32 houses. This kind of success from gemstones is what draws many youths from all over Tanzania, and this is indeed the reason why Pare agriculture is left to old people … Many youths who mine don't need much skill because it is their strength that is important. (Sheridan 2005)

Nyumbani went on to say that after the Europeans left, their equipment stopped working properly, increasing the need for physical labour. 'There in the bush near Arusha now there are about 20,000 people who are mining for gemstones every day' (ibid.).

According to Groves (2000), 11 per cent of the young miners in Mererani had not been to school and 75 per cent were under 16, the legal age for mining in Tanzania. A 2003 South African report placed the number of child labourers in Mererani at 22,500 (Kinabo 2003). Many of these children do not actually work in the mines but are hired to bring water and food to the miners, or, according to this report, to spy on the miners to ensure the miners will not steal the gemstones. Child miners, some as young as seven, are reported to number between 2,800 and 3,200 (ibid.: 48).

An International Labour Organization report (ILO 2003) on international child labour issues profiled a young former miner, Hamisi, who spoke of his time in the mines:

> I nearly suffocated inside the pits due to an inadequate supply of oxygen …
> At the mining sites and in the township we were called 'nyokas,' or 'snake boys,' since we literally crawled along the small tunnels underground just like snakes.

He said he often worked up to 18 hours a day with only one meal of buns and boiled or cooked cassava. According to the report:

> Children who work under the supervision of pit owners at the Mererani mines are paid around US$ 0.60–1.20 when they are assigned some tasks. Some children try their luck sorting the gravel left over by pit owners and can make considerably more on the extremely rare occasions that they find a gemstone. Their earnings in such cases may range from US$ 24.57–122.60.

Gem Slaves: Tanzanite's Child Labour, a short film made by the UN Office for the Coordination of Humanitarian Affairs, was released in 2006. The film stated that 4,000 child miners worked in the mines every day (IRIN News 2006b). The film was quickly criticised by some Tanzanians, and by TanzaniteOne, for the figures used for child labour (Newman 2007).

In 2011, the TanzaniteOne mining company donated TZ3 million to an orphanage that had brought child miners back to school (*Arusha Times* 2011). The director of the orphanage at the time was Dorah Mushi, a former member of parliament and one of the founders of the Tanzanite Women Miners Development Association in 2004.

Conditions in the mines are difficult for those of any age. In April 1998, rainwater flooded the mines and more than 50—some sources later reported up to 200—miners died (BBC 1998). On 22 June 2002, the BBC reported that as many as 42 miners had died in the tanzanite mines. This time, a fresh air pump had failed. Following the air pump tragedy, the Tanzanian Government temporarily halted all mining activities at Mererani (BBC 2002). Things were not much improved when miners returned. In April 2004, the *Arusha Times* reported that on 19 April two young men were found dead in an unused mine in Block B. They had never been employed by the owner, according to the paper, and the police thought that they had probably suffocated because of lack of air in the tunnel (*Arusha Times* 2004b).

There continue to be other environmental consequences as well. Those who have money in Mererani can afford to buy bush meat, reported by the *Arusha Times* to come from poached animals, some of whom are killed when they move across Arusha National Park boundaries. Zebra, bushbuck, giraffe and buffalo meat have all been confiscated or found at markets both in Mererani and in Sanya Juu, a town to the north (Nakora 2004).

Bråtveit et al. (2003) reported in the *Annals of Occupational Hygiene* that exposure to dust in Mererani's 300 privately owned mines was high. The 15,000 miners, working occasionally at a depth of 500 m, are usually without protection. Small-scale miners are at high risk for chronic silicosis, with 'total' dust at a median level of 28.4 mg/m^3 (Bråtveit et al. 2003: 235). Given these conditions, it is no wonder that the miners engage in risky behaviour. The *Arusha Times* in July 2003 (Nyaumame 2003) reported that the:

> Simanjiro District Commissioner, Mr. Fillemon Shelutete said the town's residents are noted for smoking *bhang*, chewing *khat* and drinking the illicit liquor popularly known as *'gongo'*.

To make conditions somewhat more difficult, Mererani, as with many mining towns, has a larger number of males compared to females. A Tanzanian census found that by 2002 there were 21,000 males and 16,000 females in the region (United Republic of Tanzania 2002). Given the spread of HIV/AIDS, the health workers were concerned—a 2012 UNICEF report indicated that 1.5 million Tanzanians (of a total population of 48 million), or 3 per cent, were HIV positive. For adults aged 15–49, the number was 5.1 per cent. In the Mererani area, the figure may be substantially higher. On 1 June 2004, Daniel Yona, the minister of energy and minerals, opened a four-day workshop for the small-scale miners of Mererani. He was quoted as saying that 60 per cent of all inhabitants of Mererani were HIV positive, and that there were at least 500 female sex workers in the town (*Arusha Times* 2004a). In 2005, the prime minister at the time, Edward Lowassa, said Mererani had a HIV prevalence rate of 16.4 per cent, in comparison to the national average of 7 per cent of all adults (IRIN News 2006a). The 60 per cent figure was most likely misheard in place of the 16 per cent cited by the prime minister, but there is no way to confirm it at present.

Missionaries have increasingly spent time in Mererani, establishing schools and bringing in health clinics to help with the problem (see Rothery and Rothery 2004). In 2004, women formed the Tanzanite Women Miners Development Association, or TWMDU; in 2005, women were finally allowed to enter the mines (Kondo 2005). They now have a strong voice in discussing conditions at the mines. There is also now a dispute committee in Mererani, which attempts to mediate conflicts between miners (ibid.). In March 2015, Jumanne Ngoma, the Meru cattle herder who had said he was the first to find tanzanite, was awarded the Order of the United Republic, second class, by President Jakaya Kikwete (Mwakyusa 2015). In what could be called a victory for local Tanzanians, on 26 November 2014, Richland Resources announced the conditional sale of its Tanzanian mining operations to Sky Associates Group Limited, an Arusha company. Richland said it planned to focus on developing the Capricorn Sapphire mine in Queensland, Australia. Then, on 2 February 2015, Richland Resources reported it was shutting down its Tanzanite Experience retail operations in Tanzania (Interactive Investor 2015). In April 2015, the Arusha store and museum was reopened, featuring a full-scale tanzanite mining shaft. According to their website, its goal is to educate tourists about tanzanite mining, and they sell certified tanzanite there and at another location in Arusha and in Ngorogoro (Philemon 2015). In the summer of 2015, Minister of Energy and Minerals George Simbachawene announced an initiative to conserve Mererani itself as a tourism destination (*The Guardian/ IPP Media* 2015). On the face of it, tanzanite looked as though it could have been profitable. Ihucha (2016) reported that export revenues were US$38 million, and worldwide revenues were US$500 million a year. But, in January 2016, TanzaniteOne, now controlled by Sky Associates of Arusha, laid off half of its workers. Citing reduced production of tanzanite (50 per cent during fall 2015), and expenses in buying mining equipment and pay, they let go 618 out of over 1,200 workers. One of the directors also said that increased illegal mining had led to this decision (Brummer 2016). In a way, the small-scale miners had won.

Conclusion

The complicated fictional web of land, labour and money that has been spun around tanzanite is similar to that woven around other commodities. Diamonds, rubies, lumber, water, air, manmade objects, natural objects and knowledge are all caught in the webs of human significance that humans have created, and from which humans find it difficult to extricate themselves. There remain unanswered questions concerning tanzanite: Why was the US State Department so dismissive of the connection to al-Qaeda made by Block and Pearl? What will become of the small-scale miners if the Tanzanian Government does indeed gain control over the mining operations and the cutting and export of stones? What will happen when the supply of tanzanite runs out? Tanzanian resilience and entrepreneurial ability keeps these people afloat. It will take a combination of local control, government monies and continuing investment to make the situation easier for the inhabitants of Mererani, and for those who wish to join them.

As Polanyi (1957: 73) has said:

> Robbed of the protective covering of cultural institutions, human beings would perish from the effects of social exposure; they would die as the victims of acute social dislocation through vice, perversion, crime, and starvation … But no society could stand the effects of such a system of crude fictions even for the shortest stretch of time unless its human and natural substance as well as its business organization was protected against the ravages of this satanic mill.

Acknowledgements

I thank Michael Sheridan, Paul Gross, Gideon Njenga and others in Tanzania for their various roles in thinking through the complexities of the commodification of tanzanite and of life in Tanzania. Krisan Evenson of Plymouth State University helped with the complexities of words.

References

Appadurai, A. (ed.), 1986. *The Social Life of Things: Commodities in Cultural Perspective.* Cambridge: Cambridge University Press. doi. org/10.1017/CBO9780511819582

Arusha Times, 2003. 'Water Almost as Rare as Tanzanite.' No. 299, 6–12 December. Viewed at www.arushatimes.co.tz/2003/48/local_news _5.htm (site discontinued)

Arusha Times, 2004a. 'Mererani Suffers Ravages of AIDS.' No. 323, 5–11 June. Viewed at www.arushatimes.co.tz/2004/22/front_page_1.htm (site discontinued)

Arusha Times, 2004b. 'Two Found Dead in Mererani Quarry.' No. 317, 24–30 April. Viewed at www.arushatimes.co.tz/2004/16/front_page _3.htm (site discontinued)

Arusha Times, 2011. 'Tanzanite-One Employees Assist Mererani's Good-Hope Orphanage.' 30 April – 6 May. Viewed at www.arushatimes. co.tz/2011/16/Local%20News_2.htm (site discontinued)

Babb, L.A., 2013. *Emerald City: The Birth and Evolution of an Indian Gemstone Industry.* Albany, NY: State University of New York Press.

BBC, 1998. 'Over 50 Dead in Tanzanian Mining Disaster.' BBC News, 14 April. Available at news.bbc.co.uk/2/hi/africa/77864.stm

BBC, 2002. 'Tanzania suspends gem mining.' BBC News, 22 June. Available at news.bbc.co.uk/1/hi/world/africa/2059985.stm

Beard, M. and H. Kondo, 2002. 'Industry Introduces Tanzanite Tracking.' *Colored Stone.* March/April. Viewed at www.tucsonshowguide.com/ stories/mar02/tanzanite.cfm (site discontinued)

Beilfuss, L., 2016. 'Tiffany Cuts Guidance and Staff after Weak Holiday.' Available at www.marketwatch.com/story/tiffany-cuts-guidance-and-staff-after-weak-holiday-2016-01-19

Block, R. and D. Pearl, 2001. 'Underground Trade: Much-Smuggled Gem Called Tanzanite Helps Bin Laden Supporters.' *The Wall Street Journal*, 16 November, p. A1.

Bråtveit, M., B.E. Moen, Y.J.S. Mashalla and H. Maalim, 2003. 'Dust Exposure during Small-Scale Mining in Tanzania: A Pilot Study.' *Annals of Occupational Hygiene* 47: 235–40.

Brummer, D. 2016. 'Tanzanite Miner Sheds 600 Jobs.' IDEX, 25 January. Available at www.idexonline.com/FullArticle?Id=41562

Bryceson, D., E. Fisher, J. Jønsson and R. Maipopo (eds), 2013. *Mining and Social Transformation in Africa: Mineralizing and Democratizing Trends in Artisanal Production.* London: Routledge.

Bundy, T. and D. Schilling, 2002. 'Day 7, Key West, Caribbean Cruise.' Available at www.bundlings.com/caribbean07.htm

Campbell, G., 2002. *Blood Diamonds: Tracing the Deadly Path of the World's Most Precious Stones*. Boulder: Westview Press.

Clausing, N., 2008. 'Don't Call it a Boat: Cruise Ship Vocab 101.' Travelocity. Viewed at leisure.travelocity.com/DestGuides/ (site discontinued)

Colored Stone, 2005. 'TanzaniteOne Launches New Supply Strategy.' January/February. Viewed at www.colored-stone.com/stories/jan05/tanzaniteone.cfm (site discontinued)

Donahue, K.C., 2003. 'Conceptualizing Social Ecology: The Logics of an Anthropological Practice.' Paper presented at the XV International Congress of the Anthropological and Ethnological Sciences, Florence, Italy, 5–12 July.

Donahue, K.C., 2007. *Slave of Allah: Zacarias Moussaoui vs. The USA*. London: Pluto Press.

Double-Tongued Dictionary, 2007. 'Push Present.' Available at www.doubletongued.org/index.php/dictionary/push_present/

Douglas, M. and B. Isherwood, 1979. *The World of Goods*. New York: Basic Books.

Eftimie, A., K. Heller, J. Strongman, J. Hinton, K. Lahiri-Dutt, N. Mutemeri, C. Insouvanh, M. Godet Sambo and S. Wagner, 2012. *Gender Dimensions of Artisanal and Small-Scale Mining: A Rapid Assessment Tool*. Washington, DC: World Bank Group's Oil, Gas and Mining Unit. Available at siteresources.worldbank.org/INTEXTINDWOM/Resources/Gender_and_ASM_Toolkit.pdf

Farah, D., 2001. 'Al Qaeda Cash Tied to Diamond Trade. Sale of Gems From Sierra Leone Rebels Raised Millions, Sources Say.' *The Washington Post*, 2 November, p. A1.

Farah, D., 2004a. *Blood from Stones: The Secret Financial Network of Terror*. New York: Broadway Books.

Farah, D., 2004b. 'The Role of Conflict Diamonds in Al Qaeda's Financial Structure.' Social Science Research Council, Global Security and Cooperation. Viewed at www.ssrc.org/programs/gsc/gsc_activities/farah.page (site discontinued)

Glass, N., 2012. 'Cruise Ships Financially Exploit Onshore Stores.' The Blog, *Huffington Post*, 21 May. Available at www.huffingtonpost.com/nicole-glass/cruise-ships-financially-_b_1531590.html

Groves, L., 2000. 'Child Miners in Mererani, Tanzania: An Anthropological Perspective.' Mimeo, University of Edinburgh.

Gutierrez, A. 2013. 'Alaska Tries to Curb Cruise Ship Kickbacks.' Alaska Public Media, 22 August. Available at www.alaskapublic.org/2013/08/22/alaska-tries-to-curb-cruise-ship-kickbacks/

Helliesen, M.S., 2012. 'Tangled Up in Blue: Tanzanite Mining and Conflict in Mererani, Tanzania.' *Critical African Studies* 4(7): 58–93. doi.org/10.1080/21681392.2012.10597799

Hernandez, A., 2003. 'Mining Cadastre in Tanzania.' FIG Working Week 2003, Paris, France, 13–17 April.

Hübschle, A., 2004. 'Unholy Alliance? Assessing the Links Between Organized Criminals and Terrorists in Southern Africa.' ISS Paper 93, Institute for Security Studies.

Human, R., 2006. 'Tanzanite Trouble.' *The New Internationalist*, Issue 388: 22–23.

Ihucha, A., 2006. 'Govt Allows TanzanineOne (sic) to use Live Bullets on Intruders.' IPP Media, 30 December. Viewed at www.ippmedia.com/ipp/guardian/2006/12/30/81343.html (site discontinued)

Ihucha, A., 2016. 'Gemstone Firms Sheds 600 Jobs to Cut Costs.' *The Citizen*, 11 January. Available at www.thecitizen.co.tz/News/Business/Gemstone-firm-sheds-600-jobs-to-cut-costs/-/1840414/3028766/-/5cbwjdz/-/index.html

Interactive Investor, 2015. 'Richland Resources.' Interactive Investor, 2 February. Available at www.iii.co.uk/research/LSE:RLD/news/item/1368139/closing-tanzanite-experience-retail-operations

International Labour Organization (ILO), 2003. 'Child Labour Stories, Hamisi.' Available at www.ilo.org/wcmsp5/groups/public/---ed_norm /---declaration/documents/publication/wcms_decl_fs_44_en.pdf

IRIN News, 2006a. 'Tanzania: AIDS Education as Rare as Tanzanite.' IRIN News, 4 July. Available at www.irinnews.org/report/39723/ tanzania-aids-education-as-rare-as-tanzanite

IRIN News, 2006b. 'Gem Slaves: Tanzanite's Child Labour.' IRIN News, 6 September. Available at www.irinnews.org/Report.aspx? ReportId=61004

Irish Boy, 2005. 'What the heck is Tanzanite!!!' Cruise Critic, 14 November. Available at boards.cruisecritic.com/showthread.php?t=255939

Juma, M., 2008. 'Commission Hears Small-Scale Miners Grievances.' *Arusha Times*, Issue 501, 19–25 January. Viewed at www.arushatimes. co.tz/front_page_2.htm (site discontinued)

Kelsall, T., 2004. 'Contentious Politics, Local Governance, and the Self: A Tanzanian Case Study.' Nordiska Afrikaninstitutet, Research Report 129, Uppsala, Sweden. Available at nai.diva-portal.org/smash/record. jsf?pid=diva2%3A240552&dswid=9561

Kinabo, C., 2003. 'Women's Engagement and Child Labour in Small-Scale Mining – Tanzanian Case Study.' *Urban Health and Development Bulletin* 6(4): 46–56.

Kondo, H., 2002. 'Minister Accuses AFGEM of Starting Tanzanite-al Qaeda Story.' *Colored Stone*, March/April 2002. Viewed at www.tucson showguide.com/stories/mar02/afgemnews.cfm (site discontinued)

Kondo, H., 2005. 'Tanzania Mining, Interrupted.' *The Ganoksin Project*, in association with *Colored Stone*, April 2005. Available at www. ganoksin.com/borisat/nenam/tanzania-mines.htm

Lange, S., 2006. 'Benefit Streams from Mining in Tanzania: Case Studies from Geira and Mererani.' CMI Report 2006: 11. Bergen: Chr. Michelsen Institute.

Maharaj, D., 2002. 'Gem Tied to Terror Loses Sparkle.' *The Los Angeles Times*, 20 March, p. A4.

Markon, J., 2002. 'Gemstone Dealers Named in Suit over Sept. 11.' *The Wall Street Journal*, 15 February, p. B1.

McClure, M., 2003. 'Rhapsody of the Seas Cruise Review.' Available at www.cruisereviews.com/RoyalCaribbean/RhapsodyoftheSeas58.htm

Mwakyusa, A., 2015. 'Tanzania: Kikwete Honours Brave Female TPDF Member.' *Tanzania Daily News*, 28 April. Available at allafrica.com/stories/201504280246.html

Mwalyosi, R.B.B., 2004. 'Impact Assessment and the Mining Industry: Perspectives from Tanzania.' International Association of Impact Assessment Conference, April. Available at www.tzonline.org/pdf/impactassessmentandtheminingindustry.pdf

Nakora, H., 2004. 'Game Meat Carnival Flourishes in Mererani.' *Arusha Times*, No. 333, 14–20 August. Viewed at www.arushatimes.co.tz/2004/32/local_news_5.htm (site discontinued)

Newman, J. 2007. 'Singing the Blues.' *Robb Report*, 1 May. Available at robbreport.com/Jewlery/Feature-Singing-the-Blues

Ngowi, H.P., 2001. 'Attracting New Foreign Direct Investments to Tanzania.' *Tanzanet Journal* 1(2): 23–39.

Nones, J., 2005. 'TanzaniteOne Unlocking the Riches of Mt. Kilimanjaro.' *Resource Investor*, 30 August. Viewed at www.resourceinvestor.com/pebble.asp?relid=12517 (site discontinued)

Nyaumame, S., 2003. 'Drugs, Alcohol, Threaten Lives in Mererani.' *Arusha Times*, 19–25 July. Viewed at www.arushatimes.co.tz/.../courts_&_crime_1.htm (site discontinued)

Philemon, L., 2013. 'Small Miners ask Government to Reopen Mines.' IPP Media. Viewed at www.ippmedia.com/frontend/?l=57536 (site discontinued)

Philemon, L., 2015. 'Tanzanite Tourist Attraction Centre Launched in Arusha.' IPP Media. Viewed at www.ippmedia.com/frontend/index.php/ol%3D23/javascript/page_home.js?l=79767 (site discontinued)

Polanyi, K., 1957 [1944]. *The Great Transformation: The Political and Economic Origins of our Time*. R. M. MacIver, ed. Boston: Beacon Press.

Roberts, C., 2002. 'Great Things to Do in St. Maarten: Shop.' Viewed at www.eaglelatitudes.com/current/article.html?id=7 (site discontinued)

Roskin, G., 2002. 'At Tiffany, Tanzanite is Back in the Case.' *Jewelers Circular Keystone*, 1 September. Viewed at static.highbeam.com/j/jewelerscircularkeystone/september012002/attiffanytanzaniteisback inthecasegemnotesbriefarti/ (site discontinued)

Rothery, M. and L. Rothery, 2004. 'Greetings from Tanzania.' Viewed at www.missionoz.com/NLet/June2004.htm (site discontinued)

Schroeder, R.A., 2010. 'Tanzanite as a Conflict Gem: Certifying a Secure Commodity Chain in Tanzania.' *Geoforum* 41: 56–65. doi. org/10.1016/j.geoforum.2009.02.005

Sheridan, M., 2005. Interview Notes. Personal communication to Katherine Donahue, April.

Tageo.com. 2012. 'Population, United Republic of Tanzania.' Available at www.tageo.com/index-e-tz-cities-TZ.htm

Tanzanite Foundation, n.d. 'Tanzanite and Tiffany.' Viewed at www.tan zanitefoundation.com/tanzanite_and_tiffany.html (site discontinued)

Tanzanite Gallery, 2007. 'Tanzanite: The Celebration of New Life.' *Tanzanite History*. Viewed at www.tanzanitegallery.com/acatalog/Taznanite_History.html (site discontinued)

Tanzanite Gems, 2002. 'Update on Tanzanite.' Available at tanzanitegems. com/tanzanite2.php

TanzaniteOne, 2007. 'The Story.' Viewed at www.tanzaniteone.com/tanzaniteone-tanzanite-story.asp (site discontinued)

TanzaniteOne, 2008. 'Maasai woman with baby.' Viewed at www.tanzaniteone.com/imagelibrary/19---Maasai-wearing-tanzani.jpg (site discontinued)

The Guardian/IPP Media, 2015. 'Govt Grapples to Turn Tanzanite Mining into New Tourism Destination.' 24 August. Available at en.africa time.com/tanzanie/articles/govt-grapples-turn-tanzanite-mining-new-tourism-destination

Time, 1969. 'New and Hard to Come By.' *Time Magazine*, 24 January. Available at www.time.com/time/magazine/article/0,9171,900582,00. html?promoid=googlep

United Nations Children's Fund (UNICEF), 2012. 'United Republic of Tanzania, Statistics.' Available at www.unicef.org/infobycountry/ tanzania_statistics.html

United Republic of Tanzania, 2002. 'Housing and Population Census.' Viewed at www.nbs.go.tz/Village_Statistics/PDF_files/Table _Manyara.pdf (site discontinued)

Weldon, R., 2002. 'Birth of a Birthstone.' *Professional Jeweler Magazine Archives*, December. Available at www.professionaljeweler.com/ archives/articles/2002/dec02/1202gn2.html

World Bank, 2014. 'Tanzania.' Available at data.worldbank.org/country/ tanzania

4

Agrarian distress and gemstone mining in India: The political economy of survival

Arnab Roy Chowdhury and Kuntala Lahiri-Dutt

Approaching Kalahandi

We reached the village Hinjlibahal in the Kalahandi district of Odisha[1] and noted the grazing fields were dried and desiccated; parched surroundings familiar to the district. This is where even a casual visitor cannot help but notice the small pockmarks on the face of the land. These are remnants of holes that have been dug by villagers searching for gemstones. As they are left unfilled, they formed muddy puddles after the mild rains.

When questioned about the condition of these fields, a villager[2] said, 'They were searching for *ratna* (gemstones)'. When asked about the people searching for gems, he said, 'Most of them are inhabitants from this village'. He also added that some time back, somebody from the village had a chance encounter with a very large piece of *manik* (ruby/ corundum ore) while tilling the paddy field. He said the person sold the gems for about INR20,000 (about US$300) and was hiding somewhere. Apparently, the amount earned from selling the gem was sizeable, and

---

1   Formerly known as Orissa.
2   A Kulita farmer interviewed in Hinjlibahal on 12 June 2015. Kulita and Agariya are farming castes in this region.

he had to go into hiding to avoid losing it or having to loan it to others. The news spread immediately, and many others began to dig their fields and the lands around the village.

We asked the villager what he did for a living. 'I am a "*chasi*" (peasant)', he said. 'Why don't you try your luck also?' we asked. When asked repeatedly, he said:

> I do it sometimes to earn extra incomes but not always, I am old now, mining is unpredictable and dangerous … Young people do it, sometimes for themselves or mostly working with the bigger *chasis* and *gauntias*[3] in their fields. Most villagers are poor and live on a daily basis on meagre incomes. Smarter ones have left the village after earning some money. The rest spend the money as soon as it is earned, sometimes for innocuous and necessary causes such as to repay debt, buy family groceries, buying food etc. and sometimes on vices [he winked his eye] such as … on *modo* (alcohol) and *randis* (prostitutes).

We asked another in the village whether he considered himself a peasant or a miner. This young Khond *adivasi*[4] man, who works part-time as a speculative gemstone digger, mostly for himself, also owns a small piece of land for paddy cultivation, which provides subsistence for his family. Slightly irritated by our question, he proffered an almost philosophical discourse on mining.

> We are also *chasis*, sometimes we dig for paddy and sometimes for stone. Mother Nature gives us paddy from the soil after we give our energy and sweat in it. We dig it and then sow the paddy, grow it with care, harvest it, husk it and process the paddy before we eat or sell it, we cannot eat paddy directly from the plant right? … In the same manner, we dig and put our energy and sweat into the soil and Mother Nature bestows us with these gemstones, we wash them, cut them and polish them, and sell them, gemstones are like flowers of the soil, beautiful, variously coloured … but we add value to them … otherwise what value these stones have anyway?

> We still live on our soil. Yes, mining is unpredictable. Sometimes, you find stones and sometimes you don't, but so is farming here, sometimes there is drought, no rain … sometimes we have food and sometimes we don't. So what is the difference?[5]

---

3   Traditional and erstwhile village headmen, the positional status is no longer valid; however, politically, socially, and economically they are still influential.

4   An umbrella term used to denote a heterogeneous group of tribes and indigenous group of people in India.

5   Between the peasant and a miner in this context, he implied.

We were dumbstruck and did not have any suitable response to match his explanations, which drew an interesting simile between agriculture and mining. Indeed, he is absolutely right in pondering what ascribes value to gemstones. In a study on sapphire mining in Madagascar, Walsh (2010) wrote that it is an example of ultimate 'commodification of fetishes'. Various stories, narratives and geographic origin of stones add 'value' in a sense to mined sapphires (which are varieties of corundum), which otherwise can be easily produced in laboratories. But the international market value attached to naturally produced sapphire, termed as a 'work of god', cannot be produced in laboratories (ibid.: 111).

A substantial reserve of ruby weighing about 5,348 tonnes is located in the state of Odisha, mainly in Kalahandi and Nuapada districts (Das and Mohanty 2014: 1). It is one of the poorest, driest agricultural regions of eastern India. Starvation deaths and farmers' suicides, not its gemstone reserves, brought the region to global attention.

Although traditionally the Khonds knew the art of gemstone mining, they practised it during the non-farming, drier season between the *kharif*[6] and the *rabi*[7] crops after the rains stopped. The produce was brought to the Nagavamsi King families who ruled Kalahandi, and sold to the local landlords or *zamindars*.[8] The seasonal activity was adopted to diversify their livelihood bases. Agriculture, particularly monsoon-fed paddy cultivation, remained the mainstay of economic life. Like other *adivasis* in this region who are subsistence farmers, peasants and landless labourers (Anonymous 1988), the terms *adivasi* and *chasi* conventionally represented their identity. As more villagers take up mining, their identity becomes complex, making it difficult to express through a single, generic term or profession. One question then arises: How do the villagers of Hinjlibahal like to represent themselves? We attempt to address this question in the conclusions.

We argue that agrarian distress and the frequent occurrences of drought for over a century initially led small peasants and labourers to practise artisanal mining, and allowed them to switch and shift between the two occupations seasonally. This occurred in the political–economic background of an extremely polarised and exploitative class structure,

---

6   Monsoon crop, that lasts from April to October.
7   Winter crop, sown in winter and harvested in spring.
8   In a similar vein, diamond washing has been traditionally done by Savara in Sambalpur and the Gonds and Kols in Madhya Pradesh (Biswas 1994: 401).

traditionally advantageous to those with larger agricultural landholdings. We also show that since the liberalisation of the Indian economy in the 1990s, opening up the region to foreign investments has brought in speculative and opportunistic traders from outside. Investments have connected the artisanal gemstone miners, as minute and insignificant labour, to a global market far beyond their comprehension. As national and international market chains generated interest and demands for gemstones, the *adivasis* and so-called 'lower caste' communities were assimilated into the 'value chain' primarily as labourers. The 'gem rush' of Kalahandi created more inequities than before.

Mining by peasants

Artisanal and small-scale mining (ASM) practices are informal modes of mineral extraction, mainly carried out by indigenous people, peasants and farmers globally (Hentschel et al. 2002). Artisanal mining covers relatively unorganised, clandestine, informal mining activities with a very low level of mechanisation mainly carried out by indigenous people and rural proletariats in common land or forest lands. Small-scale mining, on the other hand, is mechanised at a medium level, with organised and synchronised activities by a significant number of labourers, and generally controlled by a licensing authority. In our study region, gemstone mining includes scavenging, washing from alluvial deposits, digging in shallow trenches or inclines and surface mining such as quarries. It also includes small-scale mines such as unmechanised underground and surface mining.

In the absence of a clear definition of informal mining in India, production processes that involve manual methods such as panning and scavenging are considered 'artisanal'. In reality, a very large segment of all mining activities in India fit the definition of ASM (Deb et al. 2008). We define such mining as 'informal' to imply that it is part of the informal sector of the Indian economy. The characteristics of informal economy that Hart (1987) describes fit well with the informal mining economy. Small-sized production, low productivity, free entry and exit, labour-intensive production, activities determined by the casual labour market and the proliferation of self-employment are some characteristics of the informal economy and informal mining alike.

It becomes necessary to notionally distinguish between legal, illegal and non-legal aspects of informal mining in India. In issues of community subsoil rights in the case of *Panchayat Extension of Scheduled Areas Act 1996* (in the 5th and 6th schedules) of the state, it allows indigenous community ownership of mineral resources, such as artisanal coal mining in Meghalaya. Then there are 'illegal' artisanal mining cases on commons, state and private lands. However, scavenging gemstones from the banks of Mahanadi River near Hirakud Dam in Odisha or panning gold from the riverbank sand in Subarnarekha River in Jharkhand can be called non-legal or probably extra-legal, as no laws specifically criminalise these activities. Mining is mainly for livelihood purposes, but miners make very little money. Many miners live hand to mouth, and are extremely poor (Deb et al. 2008).

For some, artisanal mining is a continuation of their traditional livelihood. Others are thrown out of their community spaces by sudden economic and political changes of predatory globalisation, bolstered by large corporations in partnership with the state, causing large-scale displacement and forceful appropriation of land, water and forest resources of the poor through 'accumulation by dispossession' (Harvey 2004). Increasingly, large numbers of people in rural areas have lost their land, livelihood and cultural life-worlds due to a rise in land grab. They are pushed from traditional agricultural practices into the informal sector by global capitalist forces and designs. There is a surge in the number of development refugees (most of whom are erstwhile peasants) in informal mining in various parts of the developing world (Lahiri-Dutt 2007). Also, there is increasing tendency among the agriculturists to diversify through other temporary wage labour jobs, or even by leaving farming in favour of other jobs in rural areas, or migrating to the cities. This leads to what is known as deagrarianisation and de-peasantisation (Banchirigah and Hilson 2010).

Thus, poverty induced by factors over which the peasants and farmers have no control is one of the chief causes of the rural proletariat shifting to wage labour in mining. Banchirigah (2006: 165) explains how World Bank reforms in Africa led to rural impoverishment, and forced a large number of peasants into informal mining. The World Bank's International Roundtable on Artisanal Mining summary report, published in 1996, identified a link for the first time between rural impoverishment and growth of ASM. The roundtable delegates openly discussed that 'to a large extent informal mining is a poverty driven activity' (Barry 1996). The United

Nations (UN) and the Department for International Development (DFID) have since publicised these ideas. Labonne (2003: 131), speaking on behalf of the United Nations Economic Commission for Africa, said, 'Because artisanal mining is largely driven by poverty, it has grown as an economic activity, complementing more traditional forms of rural subsistence earnings'. A Mining, Minerals, and Sustainable Development (MMSD) report observed that ASM activities in rural areas in many cases represent the most promising, if not the only, income opportunity available, and is often a strategy of diversifying the livelihood basket (MMSD 2002: 314). Banchirigah (2006: 166) argues that organisations that condemn the existence of ASM and poverty are also those responsible for it. In sub-Saharan Africa, policies pursued by the World Bank, International Monetary Fund and DFID have given rise to a dual mining economy. The large-scale mining industry is flourishing and is often monopolised by multinational corporate players. Conversely, the expansion of large-scale mining, land acquisition, decreased land for peasant farming, loss of jobs, structural adjustment programs and privatisation of state enterprises has forced many into illegal ASM, operating in the interstitial spaces of capital—in conflict and in negotiation with it. In India, and in particular Kalahandi, initiating reform has a cascading malefic effect in forcing people to artisanal mining.

Bryceson (2002) argued that income diversification of most rural dwellers is for 'meeting daily needs amidst declining returns to commercial agriculture'. Many families have retained their base in subsistence farming and at the same time experimented with diverse non-farm livelihoods. These livelihood forms and choices are dependent on various agro-ecological, geographical and historical factors. Bryceson called this phenomenon a 'multiplex livelihood', and states that this diversification has, nevertheless, not produced a viable gainful employment, serious technical innovations, increase in purchasing power or improvements in forms of welfare. Citing Madulu (1998), she gave the example of Mwanza region of Tanzania where agrarian values run deep; however, the region has experienced a more recent boom in diamond and gold-mining. Eighty per cent of people involved in mining also farm in Mabuki village. When the farmers are asked who they think among their sons and daughters were 'successful' in life, in most cases they name those who are farmers. Here, pursuing farming and having large families are seen as markers of social status.

Amidst increasing conflict between agriculture and mining is some literature that discusses 'mining–agriculture complementarity', whereby their coexistence generates socio-economic development. That is, on the one hand, mining competes with agriculture for land, water and labour, but pollutes agricultural water. On the other hand, mining generates income supplements for farmers to buy fertilisers or hire labourers and send their children for higher education (Hilson and Garforth 2012, 2013).

For example, the Bolivian Gold cooperatives workers have diversified their economy by combining agricultural activities with mining. This complementarity gives rise to a low degree of conflict and polarisation between agriculture and mining. This also suggests a relatively low mobility of mine workers where they are also attached to agriculture (McMahon et al. 1999).

Yeboah (2014) asserted there must be better understanding of the dynamics of mining and agriculture interaction. Understanding the nexus between the two is essential for producing healthy interaction so that mining does not jeopardise agriculture. However, the catch is that ASM is significantly more income-generating than agriculture, but it is less sustainable. Besides, large-scale mining generates significant 'externalities' such as pollution and displacement. Displaced farmers often try to regenerate their income from 'illegal' artisanal mining (Slack 2013). It is generally young farmers who choose ASM; farmers older than 50 usually avoid ASM because of its unsustainable nature (Yeboah 2014).

Despite its income-generating potential, Yeboah (2014) argued that mining cannot really replace agriculture; it is not an alternative economic activity, but is complementary. Citing the example of Ghana, Yeboah stated that many cocoa farmers and non-farming miners often get into conflicts due to claims of ownership over land that was customarily owned by farmers, without any documented rights. The local government usually intervenes to give these lands to non-farmer miners at low prices. Yeboah quoted one of the farmers who was interviewed—'These houses that you see were all built with cocoa money, this is what feeds us'—and argued that farmers still have significant affinity for farming.

In the case of Kalahandi, we find both sentiments in support of agriculture and mining present in the people, but differentiated along class lines. The poor who work as mine labourers still imagine themselves as farmers, and the rich send their children to higher education with this money. So, the complementarity only helps the rich in this case.

Therefore, it is necessary to understand that unregulated and unlicensed mineral extraction by syndicates on a massive scale must be differentiated from ASM for managing livelihood. It would be unfair to label the poor person with 'illegality of action', who somehow manages their life because of a lack of other choices (Bhanumathi 2003; Vagholikar et al. 2003).

However, official state policies or civil society often talk about formalising these spheres, thus regularising business production and supply chains. Siegel and Veiga (2009) use De Soto's theory of 'extra-legality' to explain informal mining. In the extra-legality framework, formalisation is a process of absorbing existing customary practices, developed informally by different communities into the mainstream of a country's legal and economic affairs.

However, there is a reverse current of logic that flows within the capitalist system as well, particularly in its neoliberal varieties, which casualises and informalises labour, and accumulates capital on the back of this cheap reserve pool of informal labour. In explaining the persistence of informality, Verbrugge (2015) argued that there are certain barriers and impediments to formalisation in informal mining. Some of them are certainly political and legalistic entry barriers; however, there is a complementary economic logic of mining pursued by vested interests. Referring to his own work, he illustrates that ASM in the Philippines can be seen as the 'product of a transition away from capital-intensive large-scale mining to a flexible regime of accumulation built around the exploitation of informal ASM labour' (ibid.: 1024). This is a new capitalist logic of flexible accumulation (Verbrugge 2015). To conceptualise it further, Tsing (2009), in a slightly different context, talked about the emergence of 'supply chain capitalism'. Supply chain capitalism occurs when large corporate houses accumulate capital and create value by colonising the commons and natural resources by integrating global supply chains. That way, informal production of commodities are connected to the more institutionalised channels and centres of capital. By doing this, capitalists reduce the overhead costs of employing a permanent workforce, create a wide continental crossing scale of supply chains and convert exploitation of environmental commons and natural resources as 'externalities' of the capital. This is the new model of capitalism that very aptly fits in understanding the coloured gemstone mining and production industry. This mode of integrating the informal chains of artisanal mining into global capitalist supply chains occurs throughout the world, and the Kalahandi gemstones are no exception.

Lahiri-Dutt (2014: 6) suggested that five major factors can be identified as operating at the confluence of the State and the Capital (with capitalised 'S' and 'C') that are pushing peasants out of agriculture to mining: state reform and attracting foreign direct investment; poor productivity of agriculture; the state trying to maximise revenue from mineral extractive industries following a development model that equates 'mining' with development; environmental degradation at regional and local levels due to various interlocked climactic patterns giving rise to the displacement and rise of numbers of 'environmental refugees'; and incentive to earn cash income to meet the rising price of commodities. All these lead to the social formation of 'extractive peasants', who were formerly peasants but are now engaged in various ASM activities for subsistence. Throughout Africa, Latin America and South Asia, this pattern is becoming increasingly obvious.

In Kalahandi, historical agricultural distress and low productivity was compounded by exploitative social structures that hindered social mobility. These play a huge role in setting the stage for farmers to practise artisanal mining in the initial phase. Later, that process was accelerated by the liberalisation of the economy after the 1990s, which attracted a vastly larger number of peasants to artisanal mining. The presence of gemstones in Kalahandi was little known to the outsiders; the region is a trove of many varieties of gemstone such as cat's eye (*lahsunia*), sapphire (*pukhraaj/neelam*), aquamarine (*beruz*), emerald (*panna*) and garnets (*gomed*). Historically, the place is called Karunda Mandal, or the abode of corundum[9] (rubies). The important corundum- and ruby-bearing zones in Kalahandi are Jhillingdara, Hinjlibahal, Banjipadar and Kermunda. Kalahandi also produces some very good quality rubies. Most of these are eluvial and alluvial corundum that can be mined artisanally, and can be scavenged from the soil surface directly or by making small ditches (Das and Mohanty 2014: 1). The whole of western Odisha along the Mahanadi riverbank is historically known for its gemstone deposits and

---

9   Slightly low-quality rubies, which are often cut into cabochon shape and are sold as the 'Indian Star Ruby'.

diamonds.[10] The Sambalpur group of diamonds found in Mahanadi riverbanks, washed down from the lamprosite rocks of Madhya Pradesh, have been well recorded in history (Satapathy and Goswami 2006).

In ancient history, Kalahandi was ruled by the Nagavamsi rulers, whose family deity is aptly known as Manikeshwari (the goddess of rubies). Archaeological evidence from the nearby Asurgarh fort shows that these kings traded in gemstones. It is the Khond *adivasis* who used to mine, wash and polish the stones for the kings (Deo 1987).

Kalahandi district is surrounded by the districts Nuapada, Bolangir, Kandhamal, Rayagada, Koraput and Nawrangapur, and shares a part of the border with Raipur in Chhattisgarh (Mohanty 2010). The district was originally constituted by five *zamindaris* (landlords): Karlapat, Lanjigarh, Kashipur, Mahul-Patana and Madanpur-Rampur. Currently, the district comprises an area of 11,835 km² (Pati 1999: 345).

Kalahandi has a high percentage of *adivasis* and scheduled caste population, at 28.65 and 17.67 per cent respectively. About 45 per cent of *adivasis* are Khond and the rest belong to the Gond and Savara tribes. The Khonds are mainly divided along two clan lines: the Dongariya and the Kathuria. The former mostly inhabit the hills and practise shifting cultivation, whereas the latter is involved in cultivating plains (Padel and Das 2010).

Agrarian distress in Kalahandi

Dalkhaire;
desare kala akala
Ghara duara chhadi bidese ghara
Dalkhaire peta kaje harabara.

(O leaf eater, drought occurred in the country and sent us abroad, beyond homeland, unrest for belly, O leaf eater) (Local folk song quoted in Mishra 2011: 5)

The popular fable in Kalahandi tells the story of Indra, the god of rain, who was once angry with the people of Kalahandi, punishing them by withholding rain for 12 years. This prolonged drought resulted in an unprecedented famine in the region: people and livestock died of starvation.

10   Artisanal mining of alluvial diamonds occurs in some places of India. In Africa, artisanal mining of diamonds from riverbeds employing divers is common (Van Bockstael and Vlassenroot 2008).

The story then says that there was an old farmer, who summoned his sons, asking them to stop merrymaking now that he had become old like 'mature leaves of the dry tree' and might die soon. So he told them they must learn the technique of cultivation (Mishra 2011: 3–4). So he took his sons and bullocks to till the field, but the earth was dry and quite difficult to penetrate. So he took his sons to the riverbed, which was sandy and relatively softer. They started ploughing the riverbed; this would give them an idea about how to plough the field in rainy season. Seeing the idiosyncrasy of these farmers, Indra descended from heaven in the guise of a Brahmin. He asked the old man why he was ploughing the riverbed. In response, the old man said that he knew it was futile to plough the sand at the riverbank, but he should not forget his occupation. According to him, 'parents must teach their descendants their parental occupation', in the hope of securing their future, 'otherwise where would they go and what would they do after their parents are gone'. This was an enlightening realisation for Indra. He returned to his home in heaven and ordered his four sons (four clouds) to learn how to precipitate rain on earth. Now the barren earth of Kalahandi was pouring with rain. The clever farmer thus emotionally roused the rain god and made him sympathetic to the people and extracted rain (Misha 2011).

The conditions of drought and famine in Kalahandi are a stark reality, a recurring historical event and leitmotif in the life and livelihood of Kalahandi farmers. Therefore, the story's traces and utterances abound, and are embedded in the mythical, cultural and social life of the people. However, famines and droughts are not merely 'events'; they are processes of slow emaciation, starvation, loss of energy and death of an entire community, and they have totally devastated the rural areas of Kalahandi slowly over the years (Mishra 2005).

Nevertheless, since 1985—paradoxically when the drought was relatively stabilised—the discourse of drought, famine and hunger entered national and international consciousness. It was in the 1990s that journalists started using the term 'Kalahandi syndrome' to depict periodic food shortages, decreasing income levels and starvation amidst plenty that continues here, despite significant and prolonged development interventions by the state and civil society. Kalahandi is not poor in resources. Outsiders flock here in huge numbers to tap its mining and forest resources; however, this does not contribute to any local development.

Bouts of severe drought occurred in this region in 1866, 1868, 1884 and 1897. One of the most devastating droughts occurred in 1899 and in a way broke the backbone of the economy of this region (Mishra 2011). This famine, also known as Chhappan Salar Durbhikshya, affected people to such an extent that, even today, if a child cries for food, the mother says, 'why are you hankering like a drought-affected child of Chhappan sal?' (Mishra 2011: 2).

In the late colonial period from 1919 to 1920, the drought was followed by diseases like cholera and influenza due to lack of food and resources. Drought occurred intermittently from 1922 to 1967, making three-quarters of the crop fail, which created a reserve army of landless, unemployed agricultural workers. Finally, drought became relatively stabilised in the 1980s. However, it had devastating economic, political, psychological and corporeal consequences.

The landed class were the worst affected by the drought of later years. Due to social prestige of the class and status, they were not able to work as manual labour. Drought in a way created downward mobility of class rank and file; it pushed the classes such as rich peasants, middle peasants and other middle classes to lower positions. These categories of people turned into landless labourers in other's farmlands, colloquially called the Sukhbasis, meaning one who lives happily—the connotation is that one who has no possessions to worry about eventually lives happily (Mishra 2005, 2011). Each year, people have left the area en masse, permanently. Altogether, about 10,000 people, mostly from scheduled caste and tribe categories, have moved out of the Bolangir–Kalahandi area to Raipur in Chhattisgarh (Mishra 2011).

Despite droughts, famines and devastation, Kalahandi surprisingly is not rainfall poor. Data show that from 1977 to 1988, rainfall was about 1,255 mm; by 1991, this rose to 2,247 mm. In this period, there was above average (than other districts of Odisha) production of food grains (Pradhan 1993: 1085).

This creates a conundrum for this region, the answer to which lies in its political, economic and social history. The drought that occurs in Kalahandi is not meteorological or agricultural drought. It is a situation of human-induced drought and scarcity. The two main reasons behind this drought are poor management of the irrigation system and oppressive feudal social structures that created a polarised society—whereby landholding, and consequently economic and political power, lies with the upper class, who are also the upper caste. These issues are a deep chasm that polarise the society in such extremely unequal power poles, which disorientates and demoralises the oppressed.

Kalahandi is a paddy-cultivating region. Paddy is largely a water-dependent crop, so a large number of tanks, reservoirs, ponds and wells were constructed here for that purpose. The onus of maintaining these water bodies was with the local communities, and the *gauntia*s (village heads) and the kings in the pre-colonial period.

With the advent of British colonialism, which introduced permanent settlement, a series of changes in land settlement and a set of new intermediaries were introduced in 1883, 1885 and finally in 1904–05 (Anonymous 1985). That created a drastic transformation in the land relations, social structure and power dynamics within and between various communities. This also permanently alienated the *adivasis* from their land and forest rights (Mishra and Rao 1992). Managing common resources were previously the responsibility of the rural feudal classes and communities. With the usurpation of 'common' land, water and forest bodies, managing them became nobody's duty. So the reservoirs, no longer managed by the older feudal nobilities and communities, became neglected and dilapidated structures.

In post-colonial times, this trend continued with improper water planning. This is compounded by the fact that Kalahandi has a spell of 'resource curse': rich in resources and hidden from the public eye under the shadow of a poor state like Odisha, outsiders and businessmen flock here for opportunities such as gem mining and trade (Pradhan 1993).

Agrarian class structure in Kalahandi

Kalahandi historically has a very exploitative rural class structure: on the one hand, the trade and business is controlled by urban outsiders, mostly Marwaris; on the other hand, the power in the agrarian class structure is monopolised by erstwhile *gauntia* (both Brahmin and non-Brahmin) and *zamindar* families. In Kalahandi, the *gauntias* or the village headmen are at the top of the rural class (and also caste) hierarchy (Sahu et al. 2004).

Their power was initially bestowed by pre-colonial states for collecting revenue at the village level. This was later consolidated and legitimised by the colonial government through land revenue settlements. They usually controlled the best lands and also gained control over wastelands. They leased these lands on behalf of colonisers, and villagers who did not own any land were dependent on them. Slowly they started wielding immense power in the region (Sahu et al. 2014: 216).

Kalahandi was pre-colonially ruled by the Nagavamsi rulers, who brought in Brahmins[11] and Kulitas from Sambalpur and Raipur region to Kalahandi. The Brahmins were mainly brought in to create a hegemonic state effect over the *adivasi* collective consciousness and bring them within the fold of Hinduism. Their presence ensured that the large transformation towards a peasantised and Hinduised caste and class-based hierarchical society emerged with a degree of social legitimacy. Striking consensus in an intense conflict-ridden field was to ensure maximised surplus production and tapping of resources in a peaceful process. The Kulitas were known to be hardworking and industrious farmers who could produce surplus grain and food for the population and revenue for the state. These people displaced most of the *adivasis* and the lower castes in the region (aided by the pre-colonial rulers and by the British colonisers) from their lands into marginal spaces (Pati 1999: 346).

The pre-colonial period saw a Brahmin, Kshatriya and Kulita alliance. Later, the colonial state legitimised this oppressive class structure and became the fourth and most powerful entity in this alliance. Hence, in the colonial period, increasing discrimination against the *adivasis* continued. In the process of dispossession, intense conflict occurred, such as the Kandha (same as Khond) rebellions of 1882, mostly directed against the colonial state and the Kulitas (Pati 1999: 347).

The oppressive agrarian rural structure and class base created a polarisation of land holdings, mostly owned by the erstwhile *gauntias* families and the rich Kulita and Agariya farmers, causing the poor peasants and tribes to become extremely indebted historically. Distress sale is a common phenomenon in Kalahandi that leads to enforced commercialisation of land, labour and crops. In the villages, the credit is informal and non-institutionalised. The *gauntias* and the rich peasants, who were also the moneylenders, offered such credit. This kind of borrowing would lead

---

11 The ritually highest caste in the Indian caste hierarchy.

to distress sale, where '[t]he crop in the field was usually purchased by the moneylender himself during harvest time at a nominal rate and the price was adjusted against loan including the interest on the credit' (Government of Orissa 1980, cited in Sahu et al. 2004: 216).

Kalahandi was merged with Orissa on 1 January 1948. However, the *gauntia* system and its exploitative structure remained. When this was finally abolished in 1956, they had reconfigured, restructured and consolidated their power in the villages. Their power remained intact in post-colonial Orissa, in the scenario of limited possibility of land reform (Pati 1999: 355).

After independence, the Orissa congress government tried to improve the position of these small holders (*ryots*)—who did not hold any title to the land, and were vulnerable to moneylenders—through protective legislations. However, it failed to create much change in the rural society (Sahu et al. 2004: 215–16).[12]

Extreme cases of distress have led to a population of farm labourers steeped in loans and mortgages. Given a chance, people would migrate easily to other adjoining states, with non-tribals being the first to migrate. The tribes are too attached to their lands; they also migrate, but mostly in extreme situations (Banik 1998). Poverty, debt and scarcity lead to starvation, death and suicides of farmers in significant numbers. From 2009 to 2015, a sizeable number of farmers took their lives (*Dharitri* 2009, 2015).

Relief is a usual thing that comes to salvage the drought and famine-stricken. Through relief, the state reinforces its paternalistic responsibilities. The state purportedly provides relief of some kind, with the hope that it would trickle down to the poor, even when it is siphoned off in the middle through various channels of corruption. A number of non-government organisations have emerged in this process. In addition, there is a politics of relief that ensures by the time it is undertaken or the materials reach the community, its back is already broken and it is disempowered, which is visible in their destitution and starvation. The local people jokingly call the drought relief measures the *teesra fasal* (the 'third crop') (Sainath 1996).

---

12   Even in 1976, bonded labour existed in Kalahandi (Sahu et al. 2004: 216).

Kalahandi is in the central location of the geography of hunger in western Odisha. By the time drought relief is available, already a 'complex progression of coping mechanism' and strategies are enacted. The Dongariya Khond face acute food shortages for about three to five months a year, mainly from May to June in the post-sowing monsoon period, and again in March by the time the *kharif* crop is already exhausted (Mohapatra 2012: 57). During this period, they eat dried, powdered and detoxified mango seed kernels, make a gruel with millet called *mandia* and consume liquor derived from the *mowha* (*Madhuka longifolia*) flower collected from the forest (Pati 1999: 348). They also consume berries, mushrooms, tubers, leaves and tamarind, mainly from the forest. The forest base is also shrinking due to depletion from large-scale, aggressive mining activities in Niyamgiri. Many families go into debt because of loans incurred to buy food (Mohapatra 2012: 57).

The post-reform gem 'rush' in Kalahandi

After liberalisation in 1991 and the consequent economic boom, the gems and jewellery industry rapidly grew in India due to demand from outside. From 1993 to 1998, the Odisha branch of the Geological Survey of India carried out a high-resolution airborne survey in collaboration with World Geoscience Corporation of Perth, Australia. This survey generated data for more than 330,000 line kilometres covering an area of 75,000 km² in hard rock terrain of the state (Directorate of Geology n.d.). It identified gem resources and 'mother lodes'—the main area of concentration of gemstone ores in western Odisha. The survey was done to identify and tap the main gemstone-rich areas through capitalised and mechanised mining. This rumour circulated within the gem and production industry, and huge interest grew in western Odisha. Gems had always been produced locally at a limited scale until then, mostly controlled by the class and caste elites for regional markets. The stones were mined by indigenous tribes and other so-called scheduled caste labourers. Speculative investors, businessmen, brokers and agents arriving in this region after the 1990s sparked a gem rush. Many people, irrespective of caste and class, started coming into the business to make the most of the emerging opportunities, using their existing traditional knowledge of gem production (Roy Chowdhury and Lahiri-Dutt 2016).

They came with different levels of investment of capital and knowledge, in different capacities, to participate in the trade. Before the gem rush, tribal people were digging for the caste and class elites, mostly in off seasons at a relatively restricted and limited production scale. With the gem rush, the Khonds, other *adivasis* and other marginalised and so-called lower castes started digging for themselves individually and in groups, and also sometimes for illegal syndicates. Some also formed smaller groups of three to four people working as independent entrepreneurs; many of these groups have started working full-time in artisanal mining (Roy Chowdhury and Lahiri-Dutt 2016).

The increasing interest in mining was also in a way the last straw that poor, marginalised people could have grabbed at, against failing agriculture and economies of transition. That was the only option left for many of them before they would have migrated out of the region temporarily or permanently. In a way, agrarian distress, together with the reform, pushed and pulled a significant number of erstwhile marginal rural proletariat into artisanal mining. It became an activity in which they became doubly vulnerable to exploitation of various kinds—on one hand by the traditional elites and, on the other hand, by large corporations who pushed them out from their lands (Roy Chowdhury and Lahiri-Dutt 2016).

This area has recently started attracting many large corporate interests. A Special Economic Zone (SEZ) for gems and jewellery manufacturing is planned to be opened in the area by the Geetanjali group, one of the oldest companies in this industry. Chhattisgarh Futuristic Infrastructure Development Company has also proposed to open an SEZ for gems and jewellery manufacturing in Bhubaneswar, covering 150–200-acre lands with an initial investment of about US$45 million. Additionally, the Oriental Timex company has acquired mining rights for black granite in the Malkangiri district of western Odisha. Multilateral development organisations such as the United Nations Development Programme have assisted the Odisha Government in setting up a semi-mechanised panning and screening facility, with mobile washing plants on the banks of the Mahanadi River (Government of Orissa 2009). Around 1991, the state of Odisha established a gem-testing laboratory in the Directorate of Geology under the guidance of the United Nations, which tests various kinds of stones at a very cheap rate (Biswas 1994).

The Government of Odisha's Department of Steel and Mines is planning a survey of 80,707 km² in northwest and central Odisha to find traces of coal, gold and gemstones. It has sought the assistance of the Odisha branch of the Geological Survey of India (GSI). They are also planning to survey the Kalahandi district for gemstones. Somewhat related to these mineral explorations, the state is also planning to draft a state-specific policy on SEZs, corporate social responsibility and industrial policy resolution. A total amount of approximately US$1.7 million has been allocated for this survey (*Business Standard* 2014). The Odisha branch of GSI revealed that they have found significant gemstone deposits of topaz, tourmaline, agate and minerals like gold in the Boudh–Ramgarh area, along the Mahanadi and Tel riverbeds (Sahoo 2014). Through plans and programs, the state is aiming to extend and consolidate its extractive terrains in collaboration with capitalist firms. However, the global coloured gemstone production economy is slightly different from other mineral sources or even the diamond economy.

It is necessary to understand the process of mining rough stones and the social life that gems acquire after processing. After gems are mined, they pass through a chain of miners, brokers and wholesalers before being cut, polished and sold to the retail market. Many people in the coloured gemstone industry operate in the informal market, and are therefore difficult to track down. In contrast to diamonds, coloured gemstones are produced from relatively small, low-cost operations with a few dominant monopolisers in the market. Artisanal mining accounts for about 80–90 per cent of gemstone production (Cross et al. 2010).

The extremely fragmented nature of the gemstone market at every value chain makes it very difficult to identify significant actors. The global trade of coloured gemstones was roughly US$4 billion in 2008. Trade analysts say that this market has grown considerably since 2003. Most deposits of coloured gemstones are low yielding, and therefore mainly mined and produced through artisanal methods. The non-diamond coloured gemstone industries are cyclical and register a high point of production when a deposit is discovered, then go to a low point when it is exhausted. Different countries have emerged as prominent locations of coloured gemstone markets: Brazil in the 1950s and 1960s, Tanzania in the 1970s, and Afghanistan in the 1980s (Cross et al.: 17). This current phase will likely open up a high production point of coloured gemstones in the history of western Odisha.

Officially, 605 mining leases have been sanctioned in Odisha, of which only 16 to 21 are for precious and semi-precious gemstones (Murthy and Giri Rao 2006). Among these, 13 legally leased gemstone mines are situated in Kalahandi (Government of Orissa 2009). But official gem mines are scarce in Odisha, so the main sources of gems are mined artisanally.

The mining pits are mostly dug in agricultural lands. Mining pit owners are usually rich farmers—mostly Kulita, Agariya or Mali[13] by caste. Many pits are also owned by Brahmin *gauntias*. Most mining activities are carried out clandestinely in pits hidden in the agricultural fields.[14]

Most of the *adivasis* and the so-called scheduled caste miners, who do not hold any land, work in artisanal mining as wage labourers employed by pit owners to dig. Their chore is the same as agricultural labourers tilling fields, but they earn a little more in wages (Roy Chowdhury and Lahiri-Dutt 2016).

Some pits are also dug in jungles, inaccessible mountain areas and common lands by individuals, loosely organised groups or highly organised syndicates. The syndicates also employ the *adivasis* and the lower-caste population as wage labourers in the mine. They operate at night in relatively difficult areas of the forest, clandestinely hiding their activities from forest officials and the police.[15] Women's participation focuses on washing stones and sorting uncut pieces. Most labourers live hand to mouth, eating *pakhala* (fermented rice soaked in water) in good seasons and *mandia* in lean seasons. They are relatively better paid than the farm labourers. But their job is also risky in terms of health, life and legal issues involved.[16]

The process of mining is very rudimentary and primitive, using shovels and pickaxes. Because of the bluntness of these heavy instruments, sometimes stones get cracked internally and good-quality stones are not always produced (Roy Chowdhury and Lahiri-Dutt 2016). Pits are usually dug vertically like wells, about 1–3 m deep. Sometimes the pits are dug horizontally over a large area, as miners find good caches this way. Sometimes, small holes are carved in pit walls horizontally, just enough

---

13   Caste of gardeners.
14   Interview with a Khond local opinion leader in Thuamul Rampur, 14 July 2015.
15   Interview with a policeman in Bhawanipatna, Kalahandi, 1 July 2015.
16   Interview with a Khond miner in Kesinga, Kalahandi, 14 June 2015.

to crawl into, for a person to work inside. These pits are generally built without any reinforcements, and they often collapse after rains. Accidents and deaths of miners are common in these pits. But people take these risks for daily subsistence money and in hopes of getting lucky.

Sometimes people do get lucky. A Kulita man from Kalahandi, now staying in Sambalpur, narrated how in 1994 his father, while crossing the Tel River on a boat, found something shining underneath the water in the shallow, sandy riverbank. He descended into this knee-deep water and picked up the piece. It was a large chunk of diamond. Later, he sold the stone to a Marwari businessman in Sambalpur, who paid him INR1 million (about US$17,000). His father invested the money in building a house in Sambalpur. They currently live there and he has started his own gem business.[17] There are several rags-to-riches stories in Kalahandi. Many rich peasants and erstwhile *gauntias* send their sons and daughters to be educated in the most elite institutes in India or abroad under their own sponsorship.

Various kinds of stones are produced from these mines—ruby, corundum, aquamarine, cat's eye, sapphires (mostly white and yellow) and many other semi-precious stones. Alexandrite[18] mining is also being pursued in Tel River valley near the Kalahandi–Chhattisgarh border (Sahu 2013).

These stones are usually bought at a wholesale rate from the mines by middlemen employed by bigger businessmen, who mostly come from Rajasthan and operate their base from the adjoining Katabanji in Bolangir (Banerjee 1993). These stones gain a social and economic life of their own. They are sorted and go through a process of cleaving (giving an initial cut or shape), calibrating, cutting and bruiting (polishing) mostly on the local or regional lapidary. The best stones travel through Raipur and Jharkhand, and usually end up in Rajasthan's international gem market. Here, stones are further cut and sorted, and sold directly from the retailer, either loose or more usually embedded in jewellery produced there or exported outside the country.[19]

---

17 Interview with a gem shop owner in Sambalpur on 20 June 2015.

18 First discovered April 1834 in Russia, it is named after the Czar Alexander II. It demonstrates a property of dual colour under different wavelengths of light (see www.gemstone.org/index. php?option=com_content&view=article&id=127:sapphire&catid=1:gem-by-gem&Itemid=14).

19 Discussion with Jamal (name changed), the key informant in Sambalpur, on 20 May 2015.

As the businesses at the local level are mostly run by the caste and class elites, they have some degree of power, legitimacy and control over the local administration, which mostly act as docile and silent observers, for which they are paid. The local politicians also get a share of the trade, often given to them during elections. People stay docile; whoever tries to raise a voice is intimidated and 'taken care of' (Banerjee 1993).

Conclusion

Historically, Kalahandi's gemstone deposits have been artisanally mined by the Khonds, other *adivasi* and so-called low-caste labourers who mostly work seasonally for the feudal class of people such as *gauntias*, *zamindars* and the kings, and later in the fields of rich Kulita farmers. In conditions of severe drought and agrarian distress, this form of diversification of income was a survival strategy. Drought was so severe that it pushed many farmers out of the state to seek employment. But after the 1990s, there was a rediscovery of gem mining through interests initiated by economic reforms and resources in this region. In a pre-existing scenario of agrarian distress and knowledge of gemstone mining and production, there began and increase in demand and corporate interest, and opportunistic tradesmen and businessmen from outside the state. The anarchic opportunistic moment of gem rush absorbed both the poor and the elites to the extractive geography of gemstones in Kalahandi. In this process, the caste and class elites, and the merchant outsiders accumulated most of the capital by exploiting the informal labour of the *adivasis*. The marginalised and the vulnerable, in turn, were self-exploited for a slightly better life and relatively higher wages, as a survival strategy against extremely rapid economic, political and social transitions. Agriculture is still undertaken here along with mining, but this complementarity only helps the caste and class elites who are large landholders and who mine their land as well as use it for agriculture.

Revisiting the question briefly that we raised before: How do the villagers of Hinjlibahal like to represent themselves? Their identity is internally fractured now. They always practised cultivation, whether in the valley or in the hills. Mining was a lean season activity that supplemented cropping. They were always proud to be identified as a *chasi* or a peasant. Now, on the one hand agriculture is in distress, and on the other there is the anticipation of reform, and slow reform itself, in mining. These issues are pushing and pulling them out of their traditional occupation. These political and economic transformations are changing their life-world, and affecting the internal psychological world they are grappling with. They are part of the gem-mining labour process, but face an identity crisis and disenchantment collectively, due to the level of greed and fragmentation it produces socially. So they cling to their peasant identity that gave them a sense of honour, and are rationalising, renegotiating and trying to make sense of their transforming world.

As Banchirigah (2006) has shown in sub-Saharan Africa, economic reform fuelled an artisanal mining rush because it induced poverty. So, policies that are created for growth in the name of the poor itself tend to pauperise them. However, economic reform can trigger different societal responses, and the processes of pauperisation might take different routes in various contexts—as in the case of Kalahandi, reform of mining is at the same time pushing and pulling peasants into artisanal mining.

However, this is just the beginning. Corporatisation and land grabbing for gem mining has not yet been significant in Kalahandi. This is a transitional phase, as this space probably would soon be taken over by multinational gem mining and manufacturing houses for the creation of SEZs. As of now, companies are slowly moving in to colonise this space. The local elites and businessmen would probably adapt to these shifting political and economic situations in their own ways. The accumulation of capital would probably then be continued by companies by exploiting the *adivasis* as labourers. However, gemstones are finite resources, exhausted within a very short time; therefore, sustaining a life on this basis would remain as unpredictable and precarious as ever for these marginalised labourers.

References

Anonymous, 1985. 'Drought and Poverty: A Report from Kalahandi Orissa.' *Economic and Political Weekly* 20(44): 1857–60.

Anonymous, 1988. 'Kalahandi: A Stark Picture.' *Economic and Political Weekly* 23(18): 886–8.

Banchirigah, S.M., 2006. 'How Have Reforms Fuelled the Expansion of Artisanal Mining? Evidence from Sub-Saharan Africa.' *Resources Policy* 31(3): 165–71. doi.org/10.1016/j.resourpol.2006.12.001

Banchirigah, S.M. and G. Hilson, 2010. 'De-agrarianization, Re-agrarianization and Local Economic Development: Re-orientating Livelihoods in African Artisanal Mining Communities.' *Policy Science* 43: 157–80. doi.org/10.1007/s11077-009-9091-5

Banerjee, R., 1993. 'Orissa: Discovery of Rich Deposits of Precious Stones Lures Villagers Towards Illegal Mining.' *India Today*, 15 November. Available at indiatoday.intoday.in/story/orissa-discovery-of-rich-deposits-of-precious-stones-lures-villagers-towards-illegal-mining/1/303345.html

Banik, D., 1998. 'India's Freedom from Famine: The Case of Kalahandi.' *Contemporary South Asia* 7(3): 265–81. doi.org/10.1080/09584939808719844

Barry, M. (ed.), 1996. 'Regularizing Informal Mining: A Summary of the Proceedings of the International Roundtable on Artisanal Mining.' Industry and Energy Department Occasional Paper No. 6. Washington DC: World Bank, pp. 17–19.

Bhanumathi, K., 2003. 'Labour and Women in Mining.' Background paper presented at the Mines, Minerals and People for the Indian Women and Mining seminar, Delhi, April. Available at www.minesandcommunities.org/article.php?a=858

Biswas, A.K., 1994. 'Gem Minerals in Pre-modern India.' *Indian Journal of History of Science* 29(3): 389–420.

Bryceson, D.F., 2002. 'Multiplex Livelihoods in Rural Africa: Recasting the Terms and Conditions of Gainful Employment.' *The Journal of Modern African Studies* 40(1): 1–28. doi.org/10.1017/S0022278X01003792

Business Standard, 2014. 'Odisha to Survey in 80,707 sq km Area for Traces of Gold, Coal, Gemstone.' *Business Standard*, 27 July. Available at wap.business-standard.com/article/economy-policy/odisha-to-survey-in-80-707-sq-km-area-for-traces-of-gold-coal-gemstone-114072700737_1.html

Cross, J., S. van der Wal and E. de Haan, 2010. 'Rough Cut: Sustainability Issues in the Coloured Gemstone Industries.' Amsterdam: Centre for Research on Multinational Corporations (SOMO).

Das, S.K. and J.K. Mohanty, 2014. 'Characterisation of Eluvial Corundum (Ruby) from Kermunda, Kalahandi District, Odisha, India.' *Journal of Geology and Geosciences* 3: 180.

Deb, M., G. Tiwari and K. Lahiri-Dutt, 2008. 'Artisanal and Small-Scale Mining in India: Selected Studies and an Overview of the Issues.' *International Journal of Mining, Reclamation and Environment* 22(3): 194–209. doi.org/10.1080/17480930701679574

Deo, P.K., 1987. 'Why Kalahandi is called Karond or Kharonde?' *The Orissa Historical Research Journal* 31: 9–14.

Dharitri, 2009. 'Farmer Suicide in Kalahandi District of Western Odisha.' 18 November.

Dharitri, 2015. 'Farmer Attempts Suicide in Kalahandi.' 5 March.

Directorate of Geology, n.d. Available at orissaminerals.gov.in/Geology/Default.aspx

Government of Orissa, 2009. *Compendium of Mineral Resources of Orissa*. Department of Steel and Mines, Government of Orissa.

Hart, K.J., 1987. 'Informal economy.' In J. Eatwell, M. Milgate and P. Newman (eds), *The New Palgrave Dictionary of Economics*. London: Palgrave MacMillan. doi.org/10.1057/978-1-349-95121-5_804-1

Harvey, D., 2004. 'The "New" Imperialism: Accumulation by Dispossession.' *Socialist Register* 40: 63–87.

Hentschel, T., F. Hruschka and M. Priester, 2002. *Global Report on Artisanal & Small-Scale Mining*. Minerals Mining and Sustainable Development, no. 70. International Institute for Environment and Development, World Business Council for Sustainable Development.

Hilson, G. and C. Garforth, 2012. '"Agricultural Poverty" and the Expansion of Artisanal Mining in Sub-Saharan Africa: Experiences from Southwest Mali and Southeast Ghana.' *Population Resources Policy Review* 31(3): 435–64. doi.org/10.1007/s11113-012-9229-6

Hilson, G. and C.J. Garforth, 2013. '"Everyone Now is Concentrating on the Mining": Drivers and Implications of Changing Agrarian Patterns in the Eastern Region of Ghana.' *The Journal of Development Studies* 49(3): 348–62.

Labonne, B., 2003. 'Seminar on Artisanal and Small-Scale Mining in Africa: Identifying Best Practices and Building the Sustainable Livelihoods of Communities.' In G. Hilson (ed.), *The Socioeconomic Impacts of Artisanal and Small-Scale Mining in Developing Countries.* The Netherlands: A.A. Balkema. doi.org/10.1201/9780203971284. ch9

Lahiri-Dutt, K., 2007. 'Illegal Coal Mining in Eastern India: Rethinking Legitimacy and Limits of Justice.' *Economic and Political Weekly* 42(49): 57–67.

Lahiri-Dutt, K., 2014. 'Extracting Peasants from the Field: Rushing for a Livelihood?' Working Paper Series No. 216. Asia Research Institute, National University of Singapore.

Madulu, N.F., 1998. 'Changing Lifestyles in Farming Societies of Sukumaland: Kwimba District, Tanzania.' Working Paper No. 27. Dar es Salaam: Institute of Resource Assessment, and Leiden: African Studies Centre.

McMahon, G., J.L. Evia, A. Pascó-Font and J. Miguel Sánchez, 1999. 'An Environmental Study of Artisanal, Small, and Medium Mining in Bolivia, Chile, and Peru.' World Bank Technical Paper No. 429. Washington DC: The Word Bank.

Mining, Minerals and Sustainable Development (MMSD), 2002. *Breaking New Ground.* London: Earthscan Publications.

Mishra, A., 2005. 'Local Perceptions of Famine: Study of a Village in Orissa.' *Economic and Political Weekly* 40(6): 572–8.

Mishra, M.K., 2011. *Kalahandi Drought Lore.* Solo, Central Java, Indonesia: Centre for River Basin Organizations and Management.

Mishra, D. and R.S. Rao, 1992. 'Hunger in Kalahandi: Blinkered Understanding.' *Economic and Political Weekly* 27(24–25): 1245–6.

Mohanty, L. (ed.), 2010. 'Kalahandi District.' *Orissa Review (Census Special)* 67(5): 127–30.

Mohapatra, G., 2012. 'Hunger and Coping Strategies Among Kondh Tribe in Kalahandi District, Odisha (Eastern India).' *Transcience* 3(2): 51–60.

Murthy, A.A. and Y. Giri Rao, 2006. *Status Paper on Mining Leases in Orissa*. Bhubaneswar: Vasundhara, Conservation and Livelihood Team.

Padel, F. and S. Das, 2010. *Out of this Earth: East India Adivasis and the Aluminium Cartel*. New Delhi: Orient BlackSwan.

Pati, B., 1999. 'Environment and Social History: Kalahandi, 1800–1950.' *Environment and History* 5(3): 345–59.

Pradhan, J., 1993. 'Drought in Kalahandi: The Real Story.' *Economic and Political Weekly* 28(22): 1084–8.

Roy Chowdhury, A. and K. Lahiri-Dutt, 2016. 'The Geophagous Peasants of Kalahandi: De-peasantisation and Artisanal Mining of Coloured Gemstones in India.' *Extractive Industries and Society*: 1–13. doi.org/ 10.1016/j.exis.2016.03.007

Sahoo, A.K., 2014. 'Gem Stones Deposits Found in Odisha River.' *Deccan Chronicle,* 13 July. Available at www.deccanchronicle.com/140713/ nation-current-affairs/article/gem-stones-deposits-found-odisha-river

Sahu, J.S., 2013. 'Kotwar Guarding Alexandrite Gemstone in Chhattisgarh!' *Daily Pioneer,* 21 April. Available at www.dailypioneer. com/state-editions/raipur/kotwar-guarding-alexandrite-gemstones-in-chhattisgarh.html

Sahu, G.B., S. Madheswaran and D. Rajasekhar, 2004. 'Credit Constraints and Distress Sales in Rural India: Evidence from Kalahandi District, Orissa.' *The Journal of Peasant Studies* 31(2): 210–41. doi. org/10.1080/0306615042000224285

Sainath, P., 1996. *Everybody Loves a Good Drought*. New Delhi: Penguin.

Satapathy, R.K. and S. Goswami, 2006. 'Mineral Potential of Orissa State: A Kaleidoscopic Review.' *Orissa Review* (May): 1–14.

Siegel, S. and M.M. Veiga, 2009. 'Artisanal and Small-Scale Mining as an Extralegal Economy: De Soto and the Redefinition of "formalization".' *Resources Policy* 34: 51–6. doi.org/10.1016/j.resourpol.2008.02.001

Slack, K., 2013. 'The Growing Battle Between Mining and Agriculture.' 29 March. International Network of Economics and Conflict, United States Institute of Peace. Viewed at inec.usip.org/blog/2013/mar/29/growing-battle-between-mining-and-agriculture (site discontinued)

Tsing, A., 2009. 'Supply Chains and the Human Condition.' *Rethinking Marxism: A Journal of Economics, Culture & Society* 21(2): 148–76.

Vagholikar, N., K.A. Moghe, and R. Dutta, 2003. *Undermining India: Impacts of Mining on Ecologically Sensitive Areas.* New Delhi: Kalpavriksh.

Van Bockstael, S. and K. Vlassenroot, 2008. 'Perspective and challenges.' In K. Vlassenroot and S. Van Bockstael (eds), *Setting the Scene: Perspectives on Artisanal Diamond Mining*, Amsterdam: Academia Press.

Verbrugge, B., 2015. 'The Economic Logic of Persistent Informality: Artisanal and Small Scale Mining in the Southern Philippines.' *Development and Change* 46(5): 1023–46. doi.org/10.1111/dech.12189

Walsh, A., 2010. 'The Commodification of Fetishes: Telling the Difference Between Natural and Synthetic Sapphires.' *American Ethnologist* 37(1): 98–114. doi.org/10.1111/j.1548-1425.2010.01244.x

Yeboah, S., 2014. '"Crops" or "Carats"? Interaction Between Gold Mining and Cocoa Production and the Livelihood Dilemma in Amansie Central District of Ghana.' 30 October, United Nations Research Institute for Social Development. Available at www.unrisd.org/UNRISD/website/newsview.nsf/(httpNews)/8ADA17334D5C44E8C1257D81004558F0?OpenDocument

5

The social ecology of artisanal mining: Between romanticisation and anathema

Saleem H. Ali

Artisanal and small-scale mining (ASM) has existed for millennia, and is ingrained in many cultural traditions. However, as this book's first four chapters have demonstrated, the activity has often faced challenges of acceptance by mainstream institutions because it occupies an interstitial space 'between the pick and the plough'. It is at once an extractive sector but is also practised as a seasonal activity of agrarian peasants. It may not have all the hallmarks of a formal enterprise, but it is also seldom anarchic plundering of a resource. Thus, in this chapter, I attempt to negotiate through these seemingly conflicting elements of ASM by offering a synthetic conceptual anchor for the preceding chapters. I am guided in this task by an acute recognition that environmental concerns about ASM would need to be addressed in any effective framing of its social development imperative.

Development donors have considered ASM as suitable for technical interventions to improve yield of minerals or alternative techniques for safer extraction. The World Bank and the Communities and Small-Scale Mining (CASM) program[1] was operational from 2000 to 2010, and developed a

---

1   Details of the CASM program of work can be found at World Bank (2008).

broad repertoire of information exchange in this arena. The United Nations Industrial Development Organization (UNIDO) and Swiss Development Agency's efforts to focus on the use of mercury in ASM gold-mining are examples of such undertakings. Mercury reduction efforts have been spurred by the advent of the Minamata Convention on Mercury Reduction that has thus far been signed by over 128 countries, and ratified by 88 (as of January 2018).[2] The Convention recognises that mercury usage in artisanal and small-scale mining will likely be a challenge for many more years to come, given the remote locations of the mining sites and the relatively low cost of mercury worldwide. More recently, United Nations Development Programme (UNDP) has also started to engage with low-value minerals and the role of ASM in quarrying of industrial and construction materials, particularly in African, Caribbean and Pacific Island states (See UNDP 2015 for details of the program).

The goal of this chapter is to make the argument that all these laudable efforts need to consider the overall trajectory from a broader socio-ecological sustainability lens. Regulation—from tax collection, to environmental health and safety concerns, to conflict determinism—has vexed development practitioners for decades. Given the current context of sustainable development and the advent in 2015 of the 17 Sustainable Development Goals (SDGs), ASM should be considered as part of a hybrid livelihood strategy or a transitional opportunity for catalysing development. Thus, it is imperative that we find a more meaningful theoretical construct that could have broader development planning activity in order to ameliorate the plight of miners and assuage the concerns of governments alike.

The relevance of social ecology theory

The term 'social ecology' has developed in two parallel intellectual traditions that have seldom communicated with each other, despite having congruent objectives. First, the concept of social ecology was developed in psychological discourse, particularly by Arnold Binder (1974), and found grounding at the University of California through the establishment of the School of Social Ecology. The core aspects of this approach were subsequently summarised in six principles by Dave Taylor (1999) as follows:

---

2   A frequently updated status of Minamata signatories is available at www.mercuryconvention.org/

- Identify a phenomenon as a social problem.
- View the problem from multiple levels and methods of analysis.
- Utilise and apply diverse theoretical perspectives.
- Recognise human–environment interactions as dynamic and active processes.
- Consider the social, historical, cultural and institutional contexts of people–environment relations.
- Understand people's lives in an everyday sense.

The field in this inception builds upon traditions of human ecology, which were developed in geography, anthropology and environmental history. The context was that humans were an essential part of ecological systems and that environmental problem-solving must consider the transformative role of human societies on the environment.

Second, the term was embraced by the 'green anarchist' scholar Murray Bookchin as a response to what he perceived as a hierarchical approach to ecological problem-solving that was being posited in policy-making circles. Instead of regulating our way to solutions, Bookchin and his protégés at the Institute for Social Ecology (based in Plainfield, Vermont, USA) were focused on social transformation through decentralised inculcation of ecological ethics. It is important to note that through his intellectual development of social ecology, Bookchin dismissed some of the absolutist elements of anarchist thought around electoral participation. Instead, 'green anarchy' focused on how grassroots community organisations, which had direct connections to the land and livelihoods, could contribute effectively in local governance.

Despite his atheism, there was perhaps an inadvertent element of the Catholic social teaching of 'subsidiarity' within the approach posited by Bookchin as his writings matured, whereby the lowest functional level of governance was to be given ascendancy. Artisanal and small-scale mining cooperatives and localised governance forms are in congruence with such a paradigm as well. Interestingly enough, the most recent encyclical on ecology issued by Pope Francis, entitled *Laudato si'* (translated from medieval Italian as 'Praise be to You', and subtitled 'On Care For Our Common Home') echoes these themes of subsidiarity. At the 2015 conference of the Transnational Institute for Social Ecology, Ian Bekker astutely observed an 'overlap between the message of *Laudato si'* and the Social Ecology project of Murray Bookchin' (2015: 1).

In the context of mining, it is particularly important to also consider that larger mining firms have been a frequent target of activism by Catholic charities. Overtures by some of the larger mining company executives to the Vatican in an effort to improve perception of their development impact with the church have achieved mixed results. On the one hand, the church continues to express strong concerns about the negative impact of large-scale mining in Catholic countries such as the Philippines.[3] On the other hand, three major mining companies have been engaged in dialogue with the Vatican over a two-year project through the Kellogg Innovation Network.[4] As with other social advocacy organisations, mining continues to pose a dilemma for the Vatican in terms of its long-term contribution towards sustainability. Minerals are noted as essential, and often their role as catalysts of development is recognised. Yet, the scale and scope of the activity remains contested at multiple levels.

Since ASM is largely a decentralised activity, and many of the major concerns ensue from its environmental impact, a social ecology approach, in either of its intellectual lineages, has potential for better policy formulation. The ongoing debates about the governance and benefits of artisanal and small-scale mining, and its relationship with large-scale mining, can benefit from what Bookchin (1995a) called 'dialectical naturalism'. Dialectical methods can broadly be conceived to have four fundamental principles,[5] which Bookchin embraced, but with modifications:

- Human activities are transient and finite, existing in the medium of time.
- Human behaviour is composed of contradictions (opposing forces).
- Incremental changes lead to turning points when one force overcomes its opponent force (quantitative change leads to qualitative change).
- Change occurs in spiral not circular form—hence there is an evolution process that comes forth when contradictory perspectives interact with each other.

---

3   A summary of the Pope's concerns about mining can be found on the Vatican Radio (2015) website.

4   For details of the multi-faith outreach effort by the industry, refer to the Kellogg Innovation Network (2015) website, which has reports released in 2015.

5   These principles can be derived from a variety of texts on dialectics. An excellent framing of the use of this approach in organisations in policy and planning can be found in Mitroff and Emschoff (1979).

Bookchin (2005) was quite conscious about differentiating his approach from what he termed Hegel's 'empyrean idealism' as well as what he called Marx's wooden, often scientistic dialectical materialism. His approach embraces science and society, as providing feedback loops towards more effective ecologically sustainable governance at a non-hierarchical localised level. Through the lens of social ecology, in its tradition of dialectical naturalism (Brincat and Gerber 2015), ASM has the potential to be structured as a decentralised activity, but it remains to be seen whether the ecological impacts can be managed within such a devolved paradigm. The Irvine social ecology framework provides us some guidance on how to better formulate our approach towards effective problem-solving of ASM. Through this process, some mechanisms of governance can be identified and the seemingly contradictory elements of 'thinking locally' about its social context but 'acting national or globally' on its ecological footprint can find some reconciliation. The spiralling trajectory of dialectical naturalism can thus lead us towards more ecologically viable livelihoods generated from ASM in the short term, and broader community development outcomes in the long term. However, such a pathway will require us to navigate in far greater detail specific aspects of ASM activity, in order to make the application of these lofty philosophical principles more palpable for decision makers.

Applying social ecology concepts to ASM governance

Artisanal mining is a highly arduous activity that is extremely difficult to govern. Thus, its role as a livelihood strategy needs to be considered in those terms, with a focus on improving the living condition of miners and affording improved government or community oversight. The remoteness and desperation of many communities in ASM make them easy targets for exploitative practices, including child labour and bondage. For many miners, particularly in the gemstone arena, this profession is a transitory career to find a prized stone, and then to move on to less risky professional pursuits. In other cases, the profession is seasonal, and coupled with agricultural practices and trade. Mining rushes can create a frenzy of resource extraction that can degrade land and water bodies to the point where these alternative livelihoods, particularly agriculture or fishing, become unviable.

Much of the literature on ASM acknowledges that it is often a seasonal activity and that it can be considered a resilience strategy against downturns in agricultural productivity due to natural or anthropogenic activities. Even vociferous supporters of ASM as a development strategy acknowledge that those miners who engage in the activity for most of their lives do so because of poverty (Hilson 2010). The mineral 'rush' phenomenon associated with ASM, which can draw people from other professions, can occur with gemstone deposits (Cartier 2009) but is less prevalent with gold-mining. The difficulty of governing ASM within a conventional administrative framework has been amply documented in the literature on the challenges of 'formalization' efforts by governments (Lahiri-Dutt 2004). As a result of this challenge, social scientists have struggled to provide a meaningful theoretical frame whereby ASM governance could be operationalised, particularly with consideration of its ecological and health and safety impact on miners and communities. Siegel and Veiga (2009) attempted to use Hernado De Soto's theory of 'extra-legality' as a means of formulating a more effective governance mechanism for ASM. However, De Soto's approach as applied to ASM in this context aspires to coopt customary practices and use those as an 'extra-legal' formalisation process, which also transitions artisanal miners to careers in small- and medium-scale mining. The direct ecological impacts are not as clearly problematised by De Soto, nor have they been effectively addressed in terms of land use remediation and the long-term ecological impact of ASM. Perhaps the approach recently posited by Capra and Mattei (2015) in their book *The Ecology of Law* can provide better guidance in this regard. In their view, in order to have functional economies involving natural resource extraction, we need to transform 'legal institutions from being machines of extraction, rooted in the mechanistic functioning of private property and state authority, into institutions based on ecological communities' (ibid.: 5). Governance and regulatory enforcement, through such 'ecological communities' that have appropriate education on natural constraints of extraction, is likely to be the best way to manage ASM.

Often ASM governance gets framed in terms of risk management at the level of the individual, as well as collectively at the level of society. Human and environmental health exposure is perhaps the most direct encapsulation of this risk trade-off. However, this aspect of ASM has been addressed through numerous studies on technical solutions related to ASM usage of mercury (Sippl and Selin 2012) and cyanide exposure (Hilson and Monhemius 2006) in gold-mining. Mercury-use reduction targets in ASM gained momentum with a UNIDO/UNDP Global Mercury Project from 2002–07 that had a focal area on ASM,[6] and particularly since the promulgation of the Minamata Convention on Mercury Reduction in 2013. Safety concerns are also being addressed through distribution of personal protective equipment by non-profit organisations; although this area still deserves far greater attention (Smith et al. 2016).

However, the ecological externalities generated by ASM remain an imponderable aspect of the 'extra-legal' approach, especially since many miners are migrants and have little traditional connection to the activity. Retorts that might reduce mercury exposure to the miner can still be a contaminant in the environment if improperly disposed. Where ecological factors have been included through a reclamation plan, such as the work of the Asia Foundation and the Swiss Development Corporation in Mongolia, there has been some local government enforcement interface (Lindberg and Purevjav 2013).

The social ecology approach allows us to consider this deficiency in the current frame of analysis around ASM. Since much of the criticism comes from the environmental damage and the negative impact on other potential land uses that ASM can cause, addressing those challenges will be absolutely critical. The career trajectories of miners appear to be highly heterogeneous. The pioneering study of ASM miners in Tanzania conducted by Bryceon and Jønsson (2010) showed that only 5 per cent of miners who had previously been involved in agriculture consider a return to agriculture. Most had a diverse set of interests in business enterprise and using ASM as a means of social mobility, which was not 'rash', even in the wake of a mineral 'rush'. The incentive mechanism, for ensuring that the land being mined was ecologically rehabilitated for future agrarian use, therefore appears even more limited. The miners may see their career path in the dialectical spiral that is negotiated through contrarian forces of short-term versus long-term gains. Ecological factors could be brought forth through appropriate regulatory mechanisms that respect the devolved nature of ASM at a social and economic development level, but modulate its environmental and health and safety impact.

Social ecology thus provides a hybrid approach to addressing the negative externalities of ASM, but without losing its core entrepreneurial 'high-risk high-reward' function in peasant societies. Mining cooperatives may be another important mechanism for implementing a more decentralised

---

6   Details on their ASM activities can be found on the website of UNIDO (n.d.).

approach to ASM in the spirit of subsidiarity. Some of the past challenges of cooperatives, as documented by Levin and Turay (2008) could also be addressed by a social ecology approach. ASM, particularly in the gemstone industry, is a highly individualistic activity and has the potential to reduce itself to anarchistic arguments of autonomy from any regulatory mechanism. Fostering the value of cooperation through access to markets has often not been an adequate incentive for developing cooperative systems. Ecological and public health–oriented norms have greater potential for providing an incentive for cooperative development. Some form of standard-setting and regulatory enforcement can be better justified through such means. The dialectical principle of quantitative changes leading to qualitative changes can be embraced in the context of ASM by carefully monitored environmental health and safety education programs, coupled with enforcement against non-compliance. The proverbial 'carrot and stick' can find conceptual harmony through Bookchin's naturalism (Biehl 2015) and eventually lead to a dialectical 'turning point', whereby those engaging in ASM see ecological consciousness as a necessary part of their social system.

Conclusion: The development imperative

One cannot escape the fact that contemporary ASM is largely a phenomenon confined to the developing world, and is an adaptation strategy for poverty alleviation. It provides a conduit of livelihoods for vulnerable groups in society, but can also be a means of exploiting their vulnerabilities. The gender dynamics of ASM have been well studied in this regard and must also constitute an important part of the development goals that a social ecology paradigm may enable for this activity. During the 15-year period from the promulgation of the Millennium Development Goals until their date for completion in 2015, there have been several donor-driven programs to provide capacity around artisanal and small-scale mining. Many of these programs have arisen because of concern about mercury usage in ASM, which has been rising as a percentage of total mercury use, due to the obsolescence of other mercury uses in chloro-alkali plants and electronics. A few programs have targeted the gemstone industry because of linkages to conflict or governance failures. The CASM network provided a clearinghouse for almost a decade for these various activities. We thus have 15 years of program evaluation that needs to be better evaluated and harnessed for future development planning, and could indeed be used to connect ASM activity to the new SDGs and the 2030 development agenda.

No doubt the overall national development impact of large-scale mining in terms of fiscal flows and royalties to communities is likely to dominate the conversation in donor circles. Therefore, it is also essential that large-scale mining should have a more constructive relationship with ASM. Although there are cases where ASM activity has been allowed on large corporate mining concessions, a systematic approach with clear development indicators in mind has not been followed. The social ecology approach also has the potential to link these scales of activity to clear development metrics at different levels of governance. While large-scale mining can provide fiscal flows and development investment at a macro scale, ASM has the potential to deliver specific livelihood opportunities at a micro level.

Ecological consciousness can be the bridging element in this regard, and the social ecology approach to framing ASM allows us to consider this development imperative more directly. It is worth noting that political ecology (Peluso and Watts 2001) discourse in this regard has largely focused on a critique of resource scarcity determinism in conflict. Thus, political ecology has traditionally had a more diagnostic frame of analysis, whereas social ecology has been more prescriptive in its analytical trajectory. However, there can be a more constructive link between political ecology and social ecology in the context of ASM's contribution to development. The diagnostic insights of political ecology around land use and livelihood decisions of miners can be moved to the next phase through social ecology by developing a hybridised model of sustainable development and poverty alleviation. Social transformation through ecological learning and multi-tiered governance solutions will be essential to realise the development outcome of ASM. Perhaps Agrawal's (2005) notion of 'environmentality' can be used in this regard, building on Foucault's seminal work on knowledge and power in governance beyond conventional political institutions (governmentality).[7] Such an approach has direct relevance to ecological and health impacts of highly devolved social activities such as ASM, whereby conventional governments recognise their limits of influence. The role of governments thus becomes one of enabling knowledge acquisition in communities. Governments can also provide a forum to negotiate the contradictions in development objectives, which the theory of dialectical naturalism suggests they will inevitably encounter.

---

7   Much of Foucault's work in this regard can be traced to his later years, particularly *The Government of Self and Others: Lectures at the Collège de France 1982–1983* (Foucault 2010).

This chapter has attempted to show that ASM cannot be romantically considered a manifestation of pristine poverty, and an ode to self-reliance and resourcefulness of the peasant. It is part of a broader human consumptive quest, and linked to the dominant capitalist industrial system, which we must find more pragmatic means of managing. No doubt it has the potential to provide economic resilience, but the ecological cost of the activity must be better internalised through a multi-level approach to its governance. Social ecology theory provides the most cogent way to grapple with this complex activity, which involves some level of inherent obsolescence, anarchist reward structures and considerable environmental cost. As we ponder the future of this challenging, entrepreneurial and adventurous arena of human endeavour, let us ensure that our decisions are driven by good research and deliberations with those desperate diggers who imperil their lives to bring us treasures from the bowels of the landscape.

References

Agrawal, A., 2005. *Environmentality: Technologies of Government and the Making of Subjects.* Durham NC: Duke University Press Books.

Bekker, I., 2015. 'On Pope Francis' *Laudato si'* and the ideas of Murray Bookchin.' Paper presented at the Transational Institute for Social Ecology, Patras, Greece.

Biehl, J., 2015. *Ecology or Catastrophe: The Life of Murray Bookchin.* New York and Oxford: Oxford University Press.

Binder, A. 1974. 'Programme in Social Ecology.' *Environmental Education at Post Secondary Level.* Vol. 1, Paris: OECD.

Bookchin, M., 1995a. *Social Anarchism or Lifestyle Anarchism: An Unbridgeable Chasm.* Edinburgh and San Francisco: AK Press.

Bookchin, M., 1995b. *Philosophy of Social Ecology.* 2nd edition. Montréal and New York: Black Rose Books.

Bookchin, M., 2005. *The Ecology of Freedom: The Emergence and Dissolution of Hierarchy.* Oakland, CA: AK Press.

Brincat, S. and D. Gerber, 2015. 'The Necessity of Dialectical Naturalism: Marcuse, Bookchin, and Dialectics in the Midst of Ecological Crises.' *Antipode* 47(4): 871–93. doi.org/10.1111/anti.12140

Bryceson, D.F. and J.B. Jønsson, 2010. 'Gold Digging Careers in Rural East Africa: Small-Scale Miners' Livelihood Choices.' *World Development* 38(3): 379–92. doi.org/10.1016/j.worlddev.2009.09.003

Capra, F. and U. Mattei, 2015. *The Ecology of Law: Toward a Legal System in Tune with Nature and Community.* Oakland, CA: Berrett-Koehler Publishers.

Cartier, L.E., 2009. 'Livelihoods and Production Cycles in the Malagasy Artisanal Ruby–Sapphire Trade: A Critical Examination.' *Resources Policy* 34(1–2): 80–6. doi.org/10.1016/j.resourpol.2008.02.003

Foucault, M., 2010. *The Government of Self and Others: Lectures at the Collège de France 1982–1983* (ed. Arnold I. Davidson, trans. Graham Burchell). New York: Palgrave Macmillan.

Hilson, G., 2010. '"Once a Miner, Always a Miner": Poverty and Livelihood Diversification in Akwatia, Ghana.' *Journal of Rural Studies* 26(3): 296–307. doi.org/10.1016/j.jrurstud.2010.01.002

Hilson, G. and A.J. Monhemius, 2006. 'Alternatives to Cyanide in the Gold Mining Industry: What Prospects for the Future.' *Journal of Cleaner Production* 14: 1158–67.

Kellogg Innovation Network, 2015. 'Kin Catalyst: Mining Company of the Future. Development Partner Framework.' Kinglobal.org. Available from www.kinglobal.org/mining-catalyst.html

Lahiri-Dutt, K., 2004. 'Informality in Mineral Resource Management in Asia: Raising Questions Relating to Community Economies and Sustainable Development.' *Natural Resources Forum* 28(2): 123–32. doi.org/10.1111/j.1477-8947.2004.00079.x

Levin, E. and A.B. Turay, 2008. 'Artisanal Diamond Cooperatives in Sierra Leone: Success or Failure?' Ottawa: Partnership Africa Canada. Available at www.africaportal.org/dspace/articles/artisanal-diamond-cooperatives-sierra-leone-success-or-failure

Lindberg, M.C. and B. Purevjav, 2013. 'A Green Model for Mine Reclamation in Mongolia.' The Asia Foundation, 17 April. Available at asiafoundation.org/in-asia/2013/04/17/a-green-model-for-mine-reclamation-in-mongolia/

Mitroff, I. and J.R. Emschoff, 1979. 'On Strategic Assumption-Making: A Dialectical Approach to Policy and Planning.' *Academy of Management Review* 4(1): 1–12.

Peluso, N.L. and M. Watts (eds), 2001. *Violent Environments*. Ithaca, NY: Cornell University Press.

Siegel, S and M.M. Veiga, 2009. 'Artisanal and Small-Scale Mining as an Extralegal Economy: De Soto and the Redefinition of "Formalization".' *Resources Policy* 34(1–2): 51–6. doi.org/10.1016/j.resourpol.2008.02.001

Sippl, K. and H. Selin, 2012. 'Global Policy for Local Livelihoods: Phasing Out Mercury in Artisanal and Small-Scale Gold Mining.' *Environment: Science and Policy for Sustainable Development* 54(3): 18–29. doi.org/10.1080/00139157.2012.673452

Smith, N., S.H. Ali, C. Bofinger and N. Collins, 2016. 'Human Health and Safety in Artisanal and Small-Scale Mining: An Integrated Approach to Risk Mitigation.' *Journal of Cleaner Production* 129: 43–52. doi.org/10.1016/j.jclepro.2016.04.124

Taylor, D., 1999. 'Begging for Change: A Social Ecological Study of Aggressive Panhandling and Social Control in Los Angeles.' University of California (PhD thesis).

United Nations Development Programme (UNDP), 2015. 'Small-Scale Mining to Help Boost Social and Economic Development in the African, Caribbean and Pacific Group of States.' UNDP News Center, 15 July. Available at www.undp.org/content/brussels/en/home/presscenter/pressreleases/2015/07/15/small-scale-mining-to-help-boost-social-and-economic-development-in-the-african-caribbean-and-pacific-group-of-states.html

United Nations Industrial Development Organization (UNIDO), n.d. 'What We Do: Mercury Programme.' Viewed at www.unido.org/en/what-we-do/environment/resource-efficient-and-low-carbon-industrial-production/watermanagement/mercury.html (site discontinued)

Vatican Radio, 2015. 'Pope Francis Says Mining Sector Needs Radical Paradigm Shift.' 17 July. Available at en.radiovaticana.va/news/2015/07/17/pope_francis_says_mining_sector_needs_radical_paradigm_shift/1159107

World Bank, 2008. 'Communities, Artisanal and Small-Scale Mining (CASM): CASM's Holistic Approach to Small-Scale Mining Aims to Transform this Activity from a Source of Conflict and Poverty into a Catalyst for Economic Growth and Sustainable Development.' Issue Brief. Washington DC: World Bank. Available at siteresources.worldbank.org/INTOGMC/Resources/CASMFACTSHEET.pdf

Section Two:
Precarious and
gendered labour

6

Theorising transit labour in informal mineral extraction processes

Ranabir Samaddar

Between the plough and the pick there is a range of labour forms, all marked by what is known as 'informal' conditions—that is, informal terms of work, informal agreements, informal labour processes, informal nature of the job performed, informal nature of relations between those who work, as well as between those who work and those who own capital, and at times even the informal nature of the supply chain including the front and back ends. Yet, we have to note at the outset that there are several instances where informal and formal conditions were woven into one another in the structure of industrial processes.[1] This variety is not accidental. The division of the economy into formal and informal sectors—based on the assumption that in certain sectors labour would be highly formalised while in others it would be to the contrary—has an epistemological problem involved in it. There is also a historical confusion in considering as eternal this binary division of work conditions and work organisations as formal and informal. Throughout the history of capitalism, work conditions and work organisations have been a mix of the two. Yet, contemporary capitalism utilises the cheap labour, and makes

---

1   On the work condition in the tannery industry in Tangra, Kolkata, see Samaddar and Dutta (1997).

use of the global supply chains, as never before. I argue that three changes are signified by the extensive presence of informal labour conditions in the economy as a whole—at global, national and sectoral scales—and in mining in particular. Seen from this perspective, informal, artisanal and small-scale mining becomes more easily legible.

The three changes point to a reconsideration of surplus value accumulation as outlined by Marx. First, it seems as though capital is ordaining the informal conditions of labour in almost all spheres of economy, including the so-called formal sphere. Second, these informal conditions are dominant in productive sectors that can be described broadly as 'extraction'. The extraction could be primarily of nature and natural resources, known as 'nature's capital', but in this particular instance it is know as mineral resource extraction. Third, labour deployed in the domain of extraction will be mostly mobile, migrant labour, moving from one place to another. It may be thus termed as labour in transit, transit labour or migrant labour; in other words, labour transiting from one site to another, and one form to another, resulting in multiplication of the forms of labour.[2] The phenomenon of transit labour constitutes one of the prominent dimensions of the functioning of the economy under neoliberal policies.

These three changes are interlinked. From opening up new areas for mining to building new towns, extraction seems to be the hallmark of the expansion of the neoliberal economic policies. Essentially, they constitute the background to a return to primitive accumulation in the contemporary times. Yet, this is not the age of primitive capitalism, because on one hand, organised, large-scale, centralised production systems are devouring small- and medium-scale producers, while on the other, cheap labour is being utilised extensively in decentralised and informal production processes. Seen from this perspective, the unique growth of informal

---

2   Samita Sen (2012: 4) defines transit labour as follows: 'How do we conceptualise "transit labour"? I would suggest that we see this at the intersection of two major conceptual grids characterising the understanding of labour in the present: first, transitional forms of labour, which are inextricably related to transitions in mode of production, involving change in forms of labour arrangements, shifts in, creation or closures of labour markets, and in types and structures of labour deployment; and, second, transitory labour, which may be considered in chronological/empirical frame to denote changing and shifting patterns of employment or, in a more particularised sense, may address questions of labour mobility, both physical and structural.'

Mouleshri Vyas (2012: 10) pointed out three features of the concept: migration as a continuing phenomenon; sectoral profile of labour demand and supply; and labour flexibility and the inbuilt problem of city and labour.

extractive industries that contribute to primitive accumulation no longer appears to be an insoluble paradox. However, there is a need to investigate what causes this return to what we know as 'nature', as the site of renewal of capital. What does it do to the labour form and reproduction of labour power? Also, how does this return become a condition for neoliberal economic growth?

Unstable division between formal and informal

In this context, a set of investigations may be proposed. First, how does the neoliberal economy revolving around the development of infrastructure and cities become a site of extraction, an extractive zone impacting on all other productive processes? Second, to the extent it becomes so, what is the role transit labour plays in this metamorphosis of the economy from being a site of industrial production and liberal citizenship to one of extraction and disenfranchised migrant labour? That is to say, what does it mean in terms of organisation of production, its capillary forms, and the multiplication of labour processes? These two questions are important because, taken together, they have a strategic stake in our understanding of global capitalism in the neoliberal milieu today.

To repeat, transit labour stands at the crossroads of the trends referred to above—namely, extraction and new spaces of extraction, development of infrastructure, informal labour conditions, multiplication of the forms of labour and the politics all these give shape to—and in turn these trends are shaped by politics. This chapter will attempt to explore some of these trends that are concealed in the figure of transit labour.

Extraction I: The dust bowls of Bellary and other mining sites

Bellary mines are situated near the ruins of Vijayanagara, the former capital of the Vijayanagara Empire that flourished during the fourteenth to sixteenth centuries in Karnataka, India. The international boom in demand for iron ore made India the third largest exporter of iron ore in the world; a third of these exports came from Bellary. The rush for iron ore is evident from the estimate, by the mining and geology department, that the Bellary region alone exported 15 million tonnes of high-quality iron ore worth US$67 million overseas, mainly to China. At the same time, the international price of iron-ore rose from US$17 per tonne in 2000–01 to its peak of US$75 per tonne in 2005–06. The demand from the Indian steel producers also grew, pulling export figures, which earlier made up 75 per cent of the total production, down to 60 per cent. The iron ore began to be supplied in greater volume to the Indian market, where steel-producing giant industries—Arcelor Mittal, Posco, Tata Steel and Jindal—consumed the ore in their plants. In 2008, steel prices doubled, surpassing US$700 per tonne. In view of the abnormal rise in demand for iron the world over, several countries banned or regulated exports in order to keep domestic prices at reasonable levels.[3] In response, the mining of iron ore became widespread in Bellary, symbolising the surge in demand for iron ore and the accompanying shift to privatisation and open-market economy in India.

Meanwhile, agrarian stagnation forced the landless agricultural labourers and marginal peasantry to look for other means of wage earnings. Migrant labour 'floated from mining plot to mining plot searching for sustenance in an informal system of contract labour in the mining triangle' (Bulgarelli 2014a). By 2005, the hectic scramble for iron ore led to social and ecological chaos in the area (*Deccan Herald* 2011). Most mining operations were undertaken by small mining companies, often operating without a licence, which did not follow any environmental or social regulations.[4] Local landlords supplied labourers to the contractors, who extracted commissions from the wages paid to the labourers. There was no direct transaction between the mine owners and the labourers, and this system of sub-contracting freed the actual employers of all responsibilities towards the labourer (Dey et al. 2013). The number of daily wage labourers rose above 70,000, of which half were children under the age of 14 and around 20,000 women (Bulgarelli 2014b). The working conditions of the workers were highly exploitative and the living condition was insanitary. Mining dust was rampant (Menon 2015). All these factors affected mine workers, several of whom developed serious and chronic illnesses.

---

3   On the China-led global demand for iron and steel, the general commodity boom in the last 10 years and the subsequent decline, see the report by Miller and Samuel (2016: B1), which stated: 'Between 2012 and 2014, for example, Phoenix-based Freeport-McMoran Inc., the biggest US mining company paid out $4.7 billion in dividends, according to securities filings.'
4   Taken from a collection of reports from the website dedicated to the mining scam: bellary0. hpage.co.in/reddy-bros_1024057.html (accessed on 12 October 2015; the site is no longer accessible).

The recent history of the resurgence of this kind of mushrooming growth of decentralised and informal mining is important to understand the phenomenon of transit labour. This resurgence has been evident in various mineral-rich states of India during the last decade. They involved encroachment of forest areas, underpayment of government royalties and violation of the land and forest rights of the indigenous people. They also unleashed violent protests, and the ruthless suppression of the protests by the state was evidence of the co-mingling of political and mining interests.[5]

Extraction II: Migrant workers in the coal mines of Meghalaya

Migrant labour, including from Nepal, makes coal mining in the ecologically sensitive Jaintia Hills located in the northeastern Indian state of Meghalaya possible. The hills are pockmarked with holes. Money lies at the bottom of steep, sheer holes dug 100–180 feet deep into the ground. Sudden rain, a tipped cart, a falling rock—just about anything can mean death in the hostile pits of the Jaintia Hills. Some estimate that a staggering 70,000 children from Nepal, Bangladesh, Assam, Bihar and Jharkhand work in these private mines. Typically, a labour camp shelters 25 miners. Mining is rewarding, for while a driver may not earn more than INR5,000, mining can fetch around INR8,000–10,000 a month.[6]

Around 100,000 metric tonnes of coal, worth around INR500 million, is extracted from the Jaintia mines every day, and the government receives a royalty of INR290 per tonne; the mine owners sell it for INR4,200 per tonne. Life is temporary in every way in the rat mines of Meghalaya. The government does not think that any special action is necessary in this situation. As one report quoted the Deputy Chief Minister Bindo Lanong (Majumdar 2010):

---

5   The UN Special Rapporteur's report on the human rights of the migrants distributed by the General Assembly on 3 April 2014 noted the risks of particular groups of migrant workers, and reminded us of the situation described here (United Nations 2014).

6   This section draws from a report in *Tehelka Magazine*, by Kunal Majumder (3 July 2010). Figures and citations have been taken from this report. In *Let me Speak,* Domitila Barrios (1978) describes the work of women miners known as the 'rock pile women', who were mostly widows of miners who died in the Bolivian mines or were killed in massacres. These women miners worked on the artificial hills of rocks that had been thrown from the pits, which thus had some mixture of ore and stone. The women had to break the rocks and separate the two (Barrios 1978: 104–6).

Jaintia Hills mines are completely obsolete and environmentally hazardous … Since I took up the responsibility of mines a year back, we have been drafting a policy to take care of these important issues … Prohibiting may not be necessary to include in the mining policy, but we will consider. Why I say it is not necessary is because it is already there in the Central labour legislation, which does not allow children below 18, especially small boys, to work … It is for the Department of Labour to take action. They should impose the labour laws, knowing that these little kids are employed there. We will take up this matter.

Meghalaya as a Sixth Schedule State has three autonomous district councils, which, by law, have the sole authority to lease and licence mines (Lahiri-Dutt 2016). Traditional institutions openly flout mining norms, and land is let out at will to private operators. Most mine owners procure gelatine sticks and detonators on the black market from licensed contractors. Accidents often go unreported. The government collects royalties from mine operators and even issues them receipts.

Extraction III: Sand and other materials

Sand is formed as sediments are brought by a river and get deposited downstream. In tropical, monsoon-fed India, the process is repeated every rainy season, when new sand is deposited on the riverbed. Himalayan rivers carry enormous quantities of sand, the deposits of which force the rivers to change their courses frequently. The removal of sand from the river is a necessity to keep the riverbed intact, so that the carrying capacity of the river is maintained.

However, unplanned sand removal, or indeed sand mining, has flourished throughout India in response to the construction boom. Construction needs have relegated environmental concerns to the bottom of the priority list. Sand mining involves money, and large profits are to be made on little capital investment. Rampant sand mining has baffled the government, encouraging it to produce detailed guidelines on sustainable sand mining in response, to monitor the rivers and to increase the penalty against illegal miners.[7] However, it has been unable to control sand mining (*Ei Shomoi* 2015).

7   See the discussion on IHRO@yahoogroups.com, 15–17 September 2015. See also Aggarwal (2015), and Government of India (2015). See for details of the state of sand mining, Ministry of Environment and Forests (2013).

In this context, we may recall how Marx, while discussing primitive accumulation, repeatedly invoked the association of blood with intense exploitation of labour.[8] Observers, when discussing mutation in rural labour forms, do not bring into their analysis questions of accumulation and assume that savagery in the mutation of labour form is a matter of the past. Living labour in the post-colonial context is marked by increasingly informal conditions of work. Marx's formulation on primitive accumulation refers to conditions of life when it has been reduced to the minimum, so that capital can emerge. In the post-colonial situation today, labour migrates from work to work, and the peasant becomes a semi-worker, then a full worker, only to return to till his/her small parcel of land or work in others' fields when industrial, semi-industrial, semi-manufacturing or extractive jobs become scarce. In this context, it is important to note that the footloose post-colonial labour is also a consequence of international investment chains in countries of the Global South in overwhelmingly export-oriented production systems. Wages are often low, the work force is markedly female and the labour supervision rules are strict and marked with violence. Primitive accumulation involves the process of separation of labourers from the means of production so that they become free wage labourers for the purpose of capitalist exploitation. This is not a natural development, but the result of violent confrontations. This process not only speaks of a past (the process of initial transition from the precapitalist to the capitalist mode of production), it continues to this day on a great scale in the post-colonial world. The human factor is always present in production, and capitalist accumulation must depend on the continuous separation of the labourer from the means of production. In informal mining, accumulation is transition (transiting the borders of production and circulation), while primitive accumulation is the specific mark of this transition, reminding us that the transition from say feudalism to capitalism did not happen as a natural process. We cannot take transition for granted merely because history happened that way. The 'extra-economic' factors are always present within the economic, and only in this way can an adequate understanding of capitalism become possible. Capital accumulation begins in this contradictory mode—economic, but violent. A post-colonial critique of the capitalist accumulation process requires treating primitive accumulation not as a process of the

---

8   Marx (1990: 875); however, we have no way of knowing if Marx had thought of such possibilities also, though we know he wrote of the prices of native scalps. Also see page 920, where Marx invoked the association of blood with formation of capital 11 times in Part VIII of *Capital*, where he discussed the issue of primitive accumulation.

past, but as something happening today. It also means that capital has to circumvent the dividing line between its own prehistory and history of its present. Human sacrifice in the interest of accumulation should not astonish us if we recall Federici (2004), who recently argued that primitive accumulation must incorporate the disciplining of women through a campaign of terror in the witch-hunts of the sixteenth and seventeenth centuries. The medieval woman gradually became domesticated, her labour mystified, making it pivotal for her husband to be put to work by capital. In Federici's words:

> The most important historical question ... is how to account for the execution of hundreds of thousands of 'witches' at the beginning of the modern era, and how to explain why the rise of capitalism was coeval with a war against women. (ibid.: 14)

It will therefore be important to ask how global (or national) mining capital benefits from the reification of labour, through the division of the latter between formal and informal, organised and unorganised, rural and urban, industrial and artisanal, legal and illegal, male and female, and licit and illicit (thus mining, quarrying, extracting, digging, ploughing, threshing, loading, carrying and finally dying, in ways other than what we think to be normal), because the final goal is always accumulation of capital.

Labour in artisanal and small-scale mining in India

The report on artisanal and small-scale mining in India (Chakravorty 2001: 68–70) noted the transient element in the making of an informal miner:

> On an average, the piece-rated workers, with long experience in any particular type of job, have the highest weekly income although it fluctuates considerably from season to season and is dependent upon availability of ores and minerals ... In those mining areas, where the migrant labourers coming from different districts or even States live side by side in the '*huttings*' changes occur in socio-cultural practices, religious customs and behavior patterns. Both the tribal and non-tribal workers reciprocate with each other in this aspect.

However, the report has very little to say on how transit labour becomes the most critical element in the economy of informal (small-scale, artisanal) mining. Yet, migration is the crucial factor in artisanal and small-scale mining. Peasant households supply the kind of migrant labour (frequently in the form of family labour) that contributes to accumulation. Without taking migration into consideration, it would be difficult to understand how peasant labour enters the capitalist dynamics of accumulation. To the same extent, without migration the transient forms of labour cannot emerge. And once again to the same extent, only transient forms of labour make possible the kind of exploitation that exists in informal and artisanal mining. Yet, the reification of political economy leads us to a discussion of variegated peasant labour, leaving out the salience of transit labour, particularly in the age of neoliberal capitalism when, besides machines producing machine, money is also producing money—a process requiring an endless process of extraction of all kinds of conceivable resources. With transit labour, the capitalist can now imagine everything as a resource to be extracted, processed and sold in the market, each stage in this process producing profit for the capitalist. The reification of labour is thus not astonishing.

Transit labour comprises the backbone of the informal sector of Indian economy. The Expert Group on Informal Sector Statistics, working as part of the National Commission on Enterprises in the Unorganised/Informal Sector, estimated in 2007 that the contribution of the informal sector to the net domestic product was 48 per cent. The informal sector was estimated to constitute about 55 per cent of the total gross domestic product. This included agriculture, mining, construction, trade, transport and storage, real estate, renting, business services and hotels and restaurants (Raveendran 2006).

According to a comprehensive report on the unorganised workers, about 60 per cent of migrants join the unorganised workforce (National Commission for Enterprises in the Unorganized Sector (NCEUIS) 2007: 94). This has been an established trend since colonial times (Sen 2004), and has continued in the post-colonial times (Mazumdar et al. 2013). The NCEUIS report at the same time commented that the figures could be misleading and the extent of migration for unorganised work may be much more. According to the 2001 Census, the total migrant population in the country was 315 million (Srivastava and Sasikumar 2005). Both the Census and the National Sample Survey Organisation (NSSO) data indicate that the rate of migration has increased. NCEUIS observed:

> This [migration] could be reflective of the impact of structural changes on availability of employment opportunities … [which have] led to greater mobility of workers, a welcome fact, if it arises out of choice and not sheer economic compulsion. (NCEUIS 2007: 95)

However, as we know, the very idea of 'choice' has little meaning for the Indian poor. According to one study, 56.7 per cent of total male migration for work in 2001 was interstate, and male migration for work was 37.6 per cent of the total male migration (Mukherji 2013: 204, Table 10.11). The report also added, 'If all low grade occupational groups of migrants are added up, then it comprises as high as 60 per cent low grade workers among males, and 65 per cent among females (of all durations)' (ibid.: 212).[9] Further, NCEUIS noted:

> Temporary or short duration migrants need special attention because they face instability in employment and are extremely poor. They are engaged in agricultural sector, seasonal industries or in the urban sector as casual labourers or self-employed. By all accounts, the numbers of such migrants is much larger than that estimated in the official sources. The NCRL [1991] estimated the number of seasonal migrants at 10 million in rural areas alone. Such migrants work in agriculture, plantations, brick kilns, quarries, construction sites and fish processing. Some estimates suggest that the total number of seasonal migrants in India could be in the range of 30 million.[10] (NCEUIS 2007: 97)

Methodologically, a crucial way to understand the phenomenon of transit labour would be to weave the empirical realities of migrant labour into the empirical realities of unorganised sectors. The *Second National Commission on Labour Report* (Ministry of Labour 2002) noted that the rate of accidents in the Indian mining industry was very high when compared with similar situations in other countries (Dharmalingam 1995; Dewan 2005).

---

9   The *India Labour Report* prepared by TeamLease and IIJT (2009: 48) states: 'Unorganized sector male wage employment is primarily in manufacturing, construction, trading and transport. For women, trading and transport can be replaced by domestic services. Depending on how we count, the total is around 70 million. These figures are from 2004–05. They must have increased since then and it is a considerable number. Hence, one should ask the question: How do these workers find out jobs are available and decide on temporary or permanent migration? The answer is simple. Barring limited instances of job offers at factory gates, there are only two channels: informal (family, caste, community) networks and labour contractors. This kind of information dissemination cannot be efficient, apart from commissions, exploitative or otherwise, paid to agents. Other than such dis-intermediation and information dissemination being inefficient, there can be no question of skill formation if recruitment is through such informal channels.'

10   See also Srivastava (2005) and NCEUIS (2007: 95–6), paragraphs 6.5 and 6.8. Sharma and Das (2009: 38) suggest that there will be greater growth of plantation industry in small sectors (small growers), drawing more employment there as compared to the larger sectors.

The category of self-employed workers (as distinct and perhaps counterposed to migrant workers) needs to be analysed rigorously. However, that is beyond the scope of this chapter at present. In a way, the self-employed is the most readily available labour, part of the light infantry. The only difference is that while transit labour is most readily available on sites of extraction, for the self-employed the labour process in some cases may be different. In any case, we do witness an intense extraction of labour power of the body in the case of the self-employed. One can say that the complementary worlds of the self-employed and migrant labour mirror in the capitalist universe the complementary relation of the peasant households and the labour involved in informal mining and quarrying (Siddiqui and Lahiri-Dutt 2015: 27–32). For a more nuanced study on the same theme, where she argues for a combination of the insights from labour studies with peasant studies, see Lahiri-Dutt (2014). See also, Lahiri-Dutt et al. (2014: 119), where the authors write, scolding analysts:

> In studying rural livelihoods, much has been written on the mobility of rural labour from one region to another, from rural to urban settings. Comparatively less scholarly attention has been directed towards inter sectoral movement as a livelihood choice, and even less attention has been paid by scholars of South East [and South] Asia to the mineral dependency of rural sedentary or shifting farmers and nomadic herders.

The examples in this chapter bring to light the importance of adopting a cross-sectoral view of rural mobility, with the hope of continuing further research on mineral-based rural livelihoods. The misinformed synonymity of informality with illegality affirms the absolute and contested ownership of mineral resources by the colonial state and, more recently, by corporatised mineral enterprises.

Theorising transit labour

We should be in a better position now to summarise our arguments as to why the concept of transit labour is important: it is because of the linkages between extraction, infrastructure and labour. Indeed, the concept of transit labour will help us assess the problematic of who joins the working class today. To clear any possible misunderstanding, this is not a study on informal, small-scale artisanal mining; it is also not a study on the workers employed in the informal sectors.

Migration is crucial as a theme to the problematic of the neoliberal economy today, and sectoral mobility constitutes the core of the present dynamics of primitive accumulation, which is the other site of financialisation of the economy. Migration directs us not only to the supply chain of a commodity, but draws our attention specifically to labour power as a commodity. The working class evolves; it is not a solid, homogenous crust of material to be preserved in a museum (Schmidt 2014). Many of the formal features of capitalism, such as formal free wage agreements, may not be enough to understand neoliberal capitalism, which is marked by an enormously heterogeneous and complex composition. One of the effective routes to understand this heterogeneity is to see how gender, caste, race, age, territory, occupational holds and skill act as fault lines. These fault lines point out not only the borders and boundaries of capital/labour, but also how migration of labour acts as the *deux ex machina* of modern capitalism to cross those borders.[11]

We can neither argue that the peasant mode of production remains resilient, nor can we predict its demise in the wake of capitalism. Whether peasant society exists, whether peasant labour is actually multifarious labour is beside the point—and after a point, meaningless. Such debate unnecessarily valorises a sociological category and ignores the central question under capitalism—namely, what happens to *labour*? It is that enquiry in which we must be involved today.

---

11   Thomas Nail has argued in his recent book, *The Figure of the Migrant* (2015), for a more historical approach to migration that will tell us of its varying structures and connections, and its importance to capitalism. It is interesting to note that Saskia Sassen recognises the role of migration in the evolution of the global city, specifically mentioning the role of immigration as a 'major process through which a new transnational political economy and trans-local household strategies are being constituted. It is one largely embedded in major cities insofar as these concentrate most immigrants, certainly in the developed world, whether in the United States, Japan, or Western Europe. It is, in my reading, one of the constitutive processes of globalization today, even though not recognized or represented as such in mainstream accounts of the global economy' (Sassen 2005: 39). However, she does not link this to the much broader phenomenon of migration (including within a country and therefore perhaps excluded from the globalisation process) and accumulation.

On the other hand, Mouleshri Vyas (2012: 12) writes, 'The interlocking of migration, sectoral labour requirements and employment opportunities, and governance and policy framework, create a web-like situation for informal labour—one where it is in flux, where there is constant negotiation and claim-making through both formal and informal mechanisms, resulting in some segments of labour always being in transit on an everyday basis in cities in various parts of the world. In those where poverty, inequality and the informal economy are in evidence, the dynamic takes place in a specific physical direction as well.'

In India, the discussion of the peasant mode of production was subsumed in the academic debate around the general question of mode of production (see Patnaik 1991; also Patnaik and Moyo 2011), and took place in the wake of the peasant struggles in the decades of the 1960s and 1970s. And as happens with academic debates, while peasant struggles in the old form slowly gave way to other forms of struggles, the academic debates hovered over dead or dying issues. In part, our obsession with the transition question was responsible for this, and Marxist analysis was kept confined to issues of two transitions: from feudalism to capitalism, and transition from the division of labour under artisanal production to factory-based organisation of labour (Custers 2015). That they could coexist and that a new capitalist reality could incorporate artisanal arrangements into a global economy was not considered. We are now in a time when capitalism is marked by what can be called a *production complex* consisting of the artisanal, manufacturing, large-scale factory organisation and technologically automated production chains.

To conclude, the concept of transit labour is delinked here from the concept of transition, and this chapter, in focusing on the concept, asks the readers to conceptualise transit labour as a process distinct from labour in situ.[12] The concept forces our attention to the process of circulation (as opposed to production) because it compels us to study the transient forms of labour occasioned and caused by borders and internal structural boundaries of capital. Migration sits at the heart of the concept of transit labour. Yet this notion will not be fully understandable and analysable unless we take into account the organic relation between migration and the persistence of the unorganised form of work in capitalism. This is the reason why this chapter focuses on the extractive sites of production in modern capitalism, because the extractive sites bring the phenomena

---

12 Byasdeb Dasgupta, commenting on the concept of transit labour wrote, 'an attempt to understand the very process of labour in transit as opposed to the traditional process of labour in situ in production processes and to unfold in its term the very transition of economy and society as it is taking shape against the backdrop of a globalised reality construed by the dictate of global capital. The question of transition is perhaps a never-ending process of evolution and negation and a journey which goes on and on in any social plane ... Representing labour in transit in terms of class processes we can say the work performed by transit workers fall in two categories—Fundamental Class Process and Subsumed Class Process categories ... Labour in transit is much more disaggregated, decentred ... We would like to portray labour in transit as footloose labour in the true sense of the term. It is from nowhere to nowhere the journey, the mobility, the transition is shaping the live-forms and livelihood risks of these men and women. The real transition at the micro level—in our rendition which class as well as need-based transition—should be understood in the broader perspective of resistance to global capital and the current waves of globalisation' (2012: 24–5).

of migration, financialisation, accumulation and the commodity chain together. Labour is the ultimate commodity chain, and presents the most inscrutable supply chain of a commodity. Transit labour represents this truth. Marx (1962: 92) wrote, scoffing at those who ignored the particularities of circulation:

> The general forms of the movement P ... P is the form of reproduction and unlike M ... M, does not indicate the self-expansion of value as the object of the process. This form makes it therefore so much easier for classical Political Economy to ignore the definite capitalistic form of the process of production and depict production as such as the purpose of the process, namely that as much as possible must be produced and as cheaply as possible, and that the product must be exchanged for the greatest variety of other products, partly for the renewal of production (M-C), partly for consumption (m-c).

> It is then possible to overlook the peculiarities of money and money-capital, for M and m appear here as transient media of circulation. The entire process seems simple and natural, i.e., possesses the naturalness of a shallow rationalism.

References

Aggarwal, M., 2015. 'New Rules to Curb Illegal Sand Mining.' *Livemint*, 1 July. Available at www.livemint.com/Politics/lFs1vaw0 PHEMHizHgMk83N/New-rules-to-curb-illegal-sand-mining.html

Barrios, D., 1978. *Let Me Speak* (trans. Victoria Ortiz). New York: Monthly Review Press.

Bulgarelli, M., 2014a. 'Bellary's Mines (India).' Available at marcobulgarelli. com/bellarys-mines-india/

Bulgarelli, M., 2014b. 'Migrant Workers in India.' Available at marcobulgarelli.com/migrant-workers-in-india/

Chakravorty, S.L., 2001. 'Artisanal and Small-Scale Mining in India.' Policy Paper 78. Report for the Mining Minerals Sustainable Development Project of the International Institute for Environment and Development (IIED) and World Business Council for Sustainable Development.

Custers, P., 2015. 'Rethinking Marxism: The Manufacturing Phase in Europe.' *Frontier* 48(14–17), 11 October – 7 November. Available at frontierweekly.com/articles/vol-48/48-14-17/48-14-17-Rethinking%20Marxism.html

Dasgupta, B., 2012. 'Disinterring Labour in Transit in Terms of Class Processes.' In S. Sen, B. Dasgupta, Babu P. Remesh, and M. Vyas (eds), *Situating Transit Labour*. Calcutta Research Group research paper series. *Policies and Practices* 43.

Deccan Herald, 2011. 'Lokayukta Report on illegal mining in Karnataka.' Available at www.deccanherald.com/content/179715/download-lokayukta-report-illegal-mining.html|title=Download

Dewan, R., 2005. 'Gender Budget Perspectives on Macro and Micro Policies in Small Urban Manufacture in Greater Mumbai.' Discussion Paper 12. New Delhi: United Nations Development Programme.

Dey, I., S. Sen, and R. Samaddar, 2013. *Beyond Kolkata: Rajarhat and the Dystopia of Urban Imagination*. New Delhi and London: Routledge.

Dharmalingam, A., 1995. 'Conditions of Brick Workers in South Indian Village.' *Economic and Political Weekly* 30(47): 3014–18.

Ei Shomoi, 2015. 'Obaidha bali Khadan Niye Bitarka Trinamuler Ondore.' 24 August.

Federici, S., 2004. *Caliban and the Witch: Women, the Body and Primitive Accumulation*. Brooklyn, NY: Autonomedia.

Government of India, 2015. 'Statement by Environment Minister on Notification for Sustainable Sand and Minor Mineral Mining.' Press Information Bureau, Government of India, 24 September. Available at pib.nic.in/newsite/PrintRelease.aspx?relid=127174

Lahiri-Dutt, K., 2014. 'Extracting Peasants from the Fields: Rushing for a Livelihood?' Working Paper 216. Asia Research Institute, National University of Singapore.

Lahiri-Dutt, K., 2016. 'The Diverse Worlds of Coal in India: Energising the Nation, Energising Livelihoods.' In J. Goodman, J.P. Marshall and R. Pearse (eds), *Coal, Climate and Development: Comparative Perspectives*. Special Issue of *Energy Policy* 99: 203–13. Available at authors.elsevier.com/sd/article/S0301421516302762

Lahiri-Dutt, K., K. Alexander, and C. Insouvanh, 2014. 'Informal Mining in Livelihood Diversification: Mineral Dependence and Rural Communities in Lao PDR.' *South East Asia Research* 22(1): 103–22. doi.org/10.5367/sear.2014.0194

Majumder, K., 2010. 'Mining Sorrows: Half-life of the Coal Child.' *Tehelka Magazine* 7(26). Available at archive.tehelka.com/story_main 45.asp?filename=Ne030710coalchild.asp

Marx, Karl, 1962. *Capital: A Critique of Political Economy, Volume II. The Process of Circulation of Capital.* New York: International Publishers.

Marx, Karl, 1990. *Capital, Volume 1* (trans. Ben Fowkes). London: Penguin.

Mazumdar, I., N. Neetha and I. Agnihotri, 2013. 'Migration and Gender in India.' *Economic and Political Weekly* 48(10): 54–64.

Menon, J., 2015. 'A Bloody Scam that Shook Tamil Nadu.' *Times of India*, 27 September. Available at timesofindia.indiatimes.com/home/ sunday-times/deep-focus/A-bloody-scam-that-shook-Tamil-Nadu/ articleshow/49119913.cms

Miller, John W. and J. Samuel, 2016. 'Mining Companies Bury Dividends.' *The Wall Street Journal*, 10 December: pp. B1–B2.

Ministry of Environment and Forests, 2013. 'Report of the Ministry of Environment, Forests, and Climate Change.' Available at envfor.nic. in/public-information/report-moef-sand-mining

Ministry of Labour, 2002. *Second National Commission on Labour Report.* Government of India. Viewed at labour.nic.in/lcomm2/nlc_report.html (site discontinued)

Mukherji, S., 2013. *Migration in India: Links to Urbanisation, Regional Disparities, and Development Policies.* Jaipur and New Delhi: Rawat Publications.

Nail, T., 2015. *The Figure of the Migrant.* Stanford, CA: Stanford University Press.

National Commission for Enterprises in the Unorganized Sector (NCEUIS), 2007. *Report on Conditions of Work and Promotion of Livelihoods in the Unorganised Sector.* New Delhi: National Commission for Enterprises in the Unorganised Sector.

Patnaik, U. (ed.), 1991. *Agrarian Relations and Accumulation: The Mode of Production Debate in India.* New Delhi: Oxford University Press.

Patnaik, U. and S. Moyo, 2011. *The Agrarian Question in the Neoliberal Era: Primitive Accumulation and the Peasantry.* Cape Town: Pambazuka Press.

Raveendran, G., 2006. 'Paper 7, Estimation of Contribution of Informal Sector to GDP.' Agenda Item 3, Expert Group Meeting on Informal Sector Statistics, 11–12 May, New Delhi (Mimeo).

Samaddar, R. and D. Dutta, 1997. 'Knowing the Worker—The Tannery Majdur of Tangra.' In P. Banerjee and Y. Sato (eds), *Skill and Technological Change: Society and International Perspective.* New Delhi: Har Anand.

Sassen, S., 2005. 'The Global City: Introducing a Concept.' *Brown Journal of World Affairs* 11(2): 27–43.

Schmidt, I., 2014. 'The Downward march of Labour Halted? The Crisis of Neo-liberal Capitalism and the Remaking of Working Classes.' *Working USA: The Journal of Labour and Society* 17(1): 1–22.

Sen, S., 2004. '"Without His Consent?" Marriage and Women's Migration in Colonial India.' *International Labour and Working-Class History* 65: 77–104. doi.org/10.1017/S0147547904000067

Sen, S., 2012. 'Engaging with the Idea of Transit Labour.' In S. Sen, B. Dasgupta, Babu P. Remesh and M. Vyas (eds), *Situating Transit Labour.* Calcutta Research Group research paper series. *Policies and Practices* 43.

Sharma, K.R. and T.C. Das, 2009. *Globalization and Plantation Workers in Northeast India.* Delhi: Kalpaz Publications.

Siddiqui, Md. Z. and K. Lahiri-Dutt, 2015. 'Livelihoods of Marginal Mining and Quarrying Households in India.' *Economic and Political Weekly* (26–27): 27–32.

Srivastava, R., 2005. 'Bonded Labor in India: Its Incidence and Pattern.' In Focus Programme on Promoting the Declaration on Fundamental, International Labour Office, Cornell University.

Srivastava, R. and S.K. Sasikumar, 2005. 'An Overview of Migration in India, Its Impacts and Key Issues.' Paper presented at the Regional Conference on Migration, Development and Pro-Poor Policy Choices in Asia, Dhaka, Bangladesh.

TeamLease and IIJT, 2009. *India Labour Report.* Available at www.team lease.com/sites/default/files/resources/teamlease_labourreport2009.pdf

United Nations, 2014. 'Report of the Special Rapporteur on the human rights of migrants, François Crépeau with special emphasis on labour exploitation of migrants.' A/HRC/26/35. Available at www.ohchr.org/ Documents/Issues/SRMigrants/A.HRC.26.35.pdf

Vyas, M., 2012. 'Transit Labour in Mumbai City.' In S. Sen, B. Dasgupta, Babu P. Remesh and M. Vyas (eds), *Situating Transit Labour.* Calcutta Research Group research paper series. *Policies and Practices* 43.

7

A good business or a risky business: Health, safety and quality of life for women small-scale miners in PNG

Danellie Lynas

Alluvial and small-scale mining operations in Papua New Guinea (PNG) are collectively referred to as small-scale mining (SSM). Artisanal and small-scale mining (ASM), as this practice is most often referred to in the literature, describes an economically significant and growing area in more than 80 mineral-rich developing countries, and producing approximately 15–20 per cent of global minerals and metals (Jennings 2003; Buxton 2013). A 1999 International Labour Organization (ILO) estimate (Jennings 1999) places the number of people directly and indirectly dependent on ASM at 100 million. However, the worldwide mineral boom of the last decade, coupled with continued diversification of rural livelihoods suggests that this number is steadily increasing (World Health Organization (WHO) 2013). With no universal definition of ASM, it is usually distinguished as low-technology, labour-intensive mineral extraction and processing found across the developing world, and distinguished from industrial mining by low rates of production, lack of long-term planning, inadequate equipment and poor safety, health and environment conditions (Hilson 2002; Hilson and McQuilken 2014).

PNG alluvial gold-mining dates back to the late 1880s, when during the colonial era the industry was dominated by Europeans who employed PNG nationals (Hancock 1994; Susapu and Crispin 2001; Lole 2005). By the late 1960s, most Europeans had abandoned their leases and growing numbers of nationals commenced their own operations. Since independence in 1975, ASM has rapidly increased in scale, underpinning economic activity in much of rural and remote PNG (Moretti 2007; Papua New Guinea Mine Watch 2015b). While many miners are satisfied with non-mechanised mining to sustain their livelihood, others are looking for opportunities to form joint-venture operations that may provide necessary capital and skills to increase production levels and improve livelihoods (Crispin 2006; Moretti 2007). Unlike other parts of the world, PNG recognises SSM as a legal contributor to the national economy, with estimates of up to 90 per cent of alluvial gold production being extracted by rural-based miners using sluice boxes and panning dishes, with some mechanised operations using predominantly portable dredges, water pumps and excavators (Crispin 2003; Javia and Siop 2010). Mercury is widely used in PNG, with between 60 and 90 per cent of operations reportedly using it in capturing and amalgamating the gold (Crispin 2006; Leonhard 2015). Alluvial and small-scale mining is administered by the Mineral Resources Authority (MRA), and governed by the *Mining Act 1992* and the *Mining (Safety) Act & Regulations 2007*, which apply equally to alluvial and hard-rock mining. Current legislation recognises customary landowners, allowing non-mechanised alluvial mining on their land without the need to obtain a registered lease, while large-scale mining leases are required to have at least 51 per cent national ownership. While alluvial and placer gold is found almost everywhere in PNG, the remoteness of many of these sites makes it difficult both financially and technically to locate mechanised equipment on site. As a result, simple, non-mechanised techniques remain the most widely used.

Research indicates that most of the SSM population lacks the financial and educational capacity required to undertake fully mechanised or larger-scale mining operations. SSM is most often undertaken in family units of two to five people, with estimates indicating that this group comprises almost 99 per cent of total SSM operations (Lole 2005; Crispin 2006; Moramo 2015). Similar to other mineral-rich developing countries, alluvial mining in PNG is extensively practised as an alternative economic seasonal activity where, in times of economic stress, many miners live a subsistence lifestyle driven by commodity prices and basic needs, and switch between fishing and market gardening, and mining. Their level of involvement is often governed by festivities and family expenses, such as school and medical fees. Reports indicate that SSM is usually undertaken at will, without systematic planning of operations, employs rudimentary equipment and involves high labour intensity, with an average weekly gold recovery of approximately 2.5 grams (Moramo 2015). Additionally, PNG miners often undertake multiple income-generating activities at the same time (Lole 2005; Javia and Siop 2010; Moramo 2015).

While representing an important livelihood strategy for rural households, SSM generally encompasses a significant amount of risk and vulnerability, including problems with government licensing schemes, disputes over land ownership, environmental degradation and health and safety issues (Javia and Siop 2010; Lole 2005; Leonhard 2015; Moramo 2015). Recent press releases quote Morobe Province Governor Kasiga Kelly Naru as confirming 'alluvial mining is the source that measures the income and livelihoods of ordinary landowners involved in the activity', and for most miners it is a survival strategy for remote and rural communities. Morobe is one of the oldest mining areas of PNG. For miners seeking greater financial outcomes by entering into joint venture partnerships, acquiring a mining lease is a long process. The applicant must meet regulatory requirements by submitting detailed survey reports, development plans, compensation plans, environmental plans and financial capacity reports to the MRA for approval. In another press release, the MRA's managing director, Philip Samar, has been quoted as saying, 'Alluvial mining is a sleeping giant that has captured Mineral Resource Authority's (MRA) attention and the goal is to encourage and promote alluvial mining to meet Governments' medium and long term development plans because the bigger mining companies could not deliver alone'; further commenting, 'by 2020–30, alluvial mining would double the revenue for the Government coffers', and that as the government mining regulator, the MRA would design best practices to protect it (Papua New Guinea Mine Watch 2015a).

Health and safety

Definitions of occupational health and safety vary across authors. However, as defined by the WHO, 'occupational health deals with all aspects of health and safety in the workplace and has a strong focus on primary prevention of hazards' (WHO n.d.-b), with health defined as 'a state of complete physical, mental and social well-being and not merely the absence of disease or infirmity' (WHO n.d.-a). Across the developed world, the most effective health and safety management programs are based on strong risk management principles, underpinned by knowledge that provides the ability to assess and manage work-related hazards. While aspects of ASM health and safety have captured international attention and concern, primarily due to the use of mercury in gold processing, a lack of information exists on the other health and safety risks impacting these miners. An extensive literature review reveals that most health and safety issues faced by ASM workers are attributed to the illegal nature of the operations, in combination with competing socio-economic demands, lack of expertise and training in safety measures and inadequate equipment (Jennings 1999). Basic record-keeping and data collection remains minimal in ASM administration, largely due to miner illiteracy, workforce itinerancy and lack of human resource capacity within the regulatory bodies overseeing ASM operations (Minerals Commission Ghana 2014; Minerals Resources Authority 2015). Although widespread illegality within ASM generally leads to under-reporting of accidents and deaths, Jennings (1999) reports indicate that non-fatal deaths in ASM are up to six to seven times that of formal large-scale operations (Hinton et al. 2003). A study by Hentschel et al. (2002) found the most common occupational health and safety deficiencies in small-scale mining to be lack of awareness of the risks in mining, coupled with lack of education and training.

It is reported that many underground ASM accidents result from poor understanding of mine shaft ventilation practices, with asphyxiation frequently occurring as miners either work in shafts while compressors are operational or miscalculate and re-enter shafts before adequate ventilation and air quality have been re-established, post blasting (Gratz 2003). Reports from South America and sub-Sahara Africa indicate that most accidents among ASM operators are caused by rock falls and subsidence, use of poorly maintained equipment and poor safety practices, including non-compliance with wearing proper protective equipment. Inadequate understanding of mining and ventilation practices have been identified as the major causes of fatality with suffocation, uncontrolled explosions and being trapped or buried listed as the top three incidents leading to miner deaths. Other identified health issues include dust, food quality, extreme heat in the pits and shafts, mercury vapour inhalation, water-borne diseases, water-related problems, such as falls when the site is slippery, red eyes caused by muddy water, and body pains from lifting and carrying heavy loads. Concerns also include pit instability and collapse, load falls resulting in injury or death, rising water from the underground water table and flooding of pits, shafts and ghettos (Tschakert 2009; Lopez-Trueba 2014). A 2009 presentation by CEO Toni Aubynn to the Ghana Chamber of Mines listed health and safety as one of the four key challenges faced in regulatory management of ASM in Ghana.

Communities that have ASM operations nearby are in turn exposed to environmental safety and health hazards due to contamination of the larger environment through water run-off, air contamination, ground contamination, landslides and subsidence, as SSM operators rarely attempt land rehabilitation post mining operations (Hilson 2002; Hinton et al. 2003; Veiga and Baker 2004; Moretti 2007). Water contamination due to improper waste disposal, erosion in mining sites and mercury and cyanide poisoning have been linked to increased rates of cancer and skin lesions, while contamination in water systems also disrupts the aquatic ecosystem, eventually affecting humans (Veiga and Baker 2004; Moretti 2007).

Mercury and cyanide are the two most significant chemicals widely used in gold processing across most countries where ASM is practised. Mercury is poisonous to humans and other living organisms, and the effect is intensified through bioaccumulation. While the most significant side effect on the human body is neurotoxicity, mercury also crosses the placenta, causing foetal abnormalities, and it also affects the respiratory tract, digestive system, skin and eyes. The effect of mercury on the human body and the environment has been extensively researched and reported, and elimination of mercury from gold processing remains the focus of many funded programs, such as the World Bank, International Council for Mining and Metals (ICMM), Fairtrade Foundation and Alliance for Responsible Mining, Blacksmith Institute and, most recently, the Minamata Convention on Mercury Reduction. Many authors argue that the health effects of environmental pollution are disproportionately felt by women miners because of their role as primary carers responsible for the health of their families, and as agriculturalists and miners. Women's reproductive role increases their vulnerability to mercury and other heavy metals in water and food supplies, with mercury known to severely affect foetal development (Hinton et al. 2003; Lahiri-Dutt 2006; Van Hoecke 2006; Hayes 2008; Simatauw 2009; Jenkins 2014).

Women in ASM

Men and women are differently involved in, and differently affected by, mining practices, cultural practices and legislative practices, and often their role is dependent on the mineral commodity mined. Estimates indicate that the proportion of women engaged worldwide in ASM is approximately 30 per cent (Hilson 2002; Hinton et al. 2003). In Asian countries it is reported that women make up close to 50 per cent of miners, and in parts of Africa, anywhere between 40 and 100 per cent of the workforce are women (Jennings 1999; Hilson 2002; Lahiri-Dutt 2008). Current estimates indicate that over 100,000 persons are involved in alluvial mining across PNG, of which approximately 30 per cent are women (personal communication, A.B. Comparativo, Manager Small-Scale Mining Training Centre, Wau, PNG (SSMTC), 23 October 2015).

Traditionally, women are associated with the labour-intensive tasks of transporting and processing, involving high levels of manual activity to extract mineral remnants by hand from tailings, and panning and sluicing mud and sand to recover particles of gold, while yielding the lowest economic return (Hinton et al. 2003). Benefits from mined and processed resources frequently reside with the men, while the women often do not receive an independent wage, and earn less than their male counterparts for performing similar work. In some countries, women do own mines and mining equipment, but this is not common practice (Hentschel et al. 2002; Hilson 2002; Heemskerk 2003; Hinton et al. 2003; Van Hoecke 2006). Typically, women encounter difficulties obtaining development assistance, including access to bank credit, technical knowledge and skills development (Hilson 2002; Hinton et al. 2003; Eftimie et al. 2012). This poses significant challenges for women who are household heads, as mining is often their only economic option. A number of studies indicate that money generated by women miners contributes more directly to household well-being than that of the men, with women's income providing for food, education, medical expenses and replenishing agricultural cropping needs, while men's income is often spent on gambling, prostitution and alcohol (Jennings 1999; Krimbu 2005; Javia and Siop 2010). While women's mining contributions are often in ancillary roles such as food vendors, sex workers and other service providers, these roles are performed in addition to their domestic responsibilities, often making their contribution 'less visible', and therefore less recognised and valued (Lahiri-Dutt 2006, 2012). A comprehensive literature critique by Jenkins (2014) provides valuable insight into the ways women are affected by ASM mining activities, with the author highlighting a lack of in-depth analysis of women's diverse experiences as mine workers across a number of parameters, including gender impacts of mining, changing gender dynamics and gender inequalities in mining communities.

Access to education is critical to empowering women to derive greater benefits from ASM. However, in many countries, girls are still discriminated against in terms of education, mainly due to social and cultural barriers, and family expectations that they will assume domestic responsibilities. Additionally, cultural taboos and superstitions often exclude women from working in mines due to beliefs that they will bring bad luck by attracting bad spirits or offending the gold god (Hinton et al. 2003; Moretti 2006). While women are critical to the stability and cohesiveness of their communities and often act as change facilitators, they are often excluded from the decision-making process based on gender (Hinton et al. 2003). Lahiri-Dutt (2012) emphasised the importance of 'asking what mining really means to [women in] poorer communities'. In communities where women feel they do have the ability to influence decisions, they often lack access to knowledge that will better equip them to act in an informed capacity.

Current occupational health and safety issues in PNG

Small-scale mines in PNG experience many of the global health and safety issues. Trainers from the MRA-managed SSMTC located in Wau in Morobe Province, the oldest mining area of PNG, have identified the low education level of the miners, particularly in relation to technical knowledge and mining expertise, lack of knowledge of the dangers associated with mining operations, the rudimentary nature of the equipment used, low production outputs, the propensity of the miners to spend all that they make and not save and the seasonality of their operations as challenges to developing effective health and safety practices. Additional significant management challenges include the location and accessibility of operations, cultural practices and social beliefs, the technical and economic environments the mines operate in, including lack of access to available credit facilities, fair access to market and fair price for the gold they mine. Trainers indicate that records show at least five to six reported deaths per year in SSM largely due to unsafe work practices.

In an account of women alluvial and small-scale miners in PNG, Crispin highlighted the vital source of income that mining provides for many rural-based communities, arguing that 'access for women to the economic benefits, skills involved, and awareness of possible health and safety issues is important for community based mining to remain sustainable and to continue to contribute to economic survival and development' (2006: 256–7). Following fieldwork undertaken in the Mount Kaindi area, Moretti observed that local women 'face serious obstacles to full and equitable participation in mining' (2005: 133–4). These obstacles included cultural beliefs, land tenure practices, unequal control of household resources and gendered division of labour. In many regions of PNG, women are considered a dangerous presence on a mine site, with the belief that they can pollute the gold and anger *hikoapa*, the ancestral and nature spirits that guard the land and its riches (Moretti 2005).

As in many other developing countries, PNG women often take their children on-site, as there is nowhere else to leave them. Reports indicate that around 30 per cent of school-age children are working on SSM sites rather than attending school; however, extensive awareness campaigns have seen this number reduced to around 15 per cent (personal communication, A.B. Comparativo, Manager Small-Scale Mining Training Centre Wau, PNG, 23 October 2015). Girls make up the largest proportion of this number, often being kept back from school simply for gender reasons (Krimbu 2005). Child labour remains a significant issue in PNG, with many children leaving school in order to mine to support themselves and their families. Children on the site are exposed to dust, falling objects, machinery noise, heavy manual work and mercury vapour. Women and children are particularly exposed to drugs, communicable diseases such as HIV/AIDS and gender-based violence (GBV).

Typical of many SSM communities, PNG women mine on a seasonal basis, either at times of increased financial commitment such as payment of school and medical fees, or in conjunction with other seasonal income-earning activities such as market gardening, cropping and animal rearing. Lack of community access to clean running water and sanitation facilities substantially adds to the impact and challenge of managing health-related issues within the communities. While many communities accept the situation as a normal way of living, exposure to water-borne diseases such as malaria and cholera, and continued use of unsafe and often mercury-contaminated water for cooking, drinking and washing remains detrimental to the health of the whole community.

In a thesis completed on an Andean mining community in Bolivia, Lopez-Trueba (2014) makes a distinction between specific 'health and safety risks' and the broader range of dangers, uncertainties, risks and ambiguities associated with miners' livelihoods, referring to them as 'occupational uncertainties', and indirect causes of occupational health and safety risks. Taking this broader approach allows for further exploration and understanding of the health and safety issues from the perspective that these issues are rooted in the immediate and wider socio-political and economic contexts in which the miners live and their livelihoods take place. This discussion takes up this concept and explores health and safety issues among women small-scale miners and their communities in an alluvial gold-mining province in PNG.

Methodology

Ethnographic research provided the opportunity to better understand how gender-specific health and safety concerns impact on the lives of two groups of women involved in alluvial gold-mining in PNG. Two field trips were undertaken to the Wau area where local women actively engaged in small-scale mining in their local area were interviewed. One group comprised of 16 women and the other had 12. Group and individual semi-structured interviews, direct field observation, photos and video footage were used to gain an overall picture of the health and safety issues experienced by the group. An interpreter was used for the interviews. The questions asked in the semi-structured interviews covered basic parameters such as age, education level, years spent mining, why they had started and continued mining, if they would prefer to not be mining, and if so what they would prefer to do, how much time they spent mining, and how much time on agricultural work and family responsibilities, if the lease was held in their name, and how much longer they thought they would continue with mining. Specific questions were then asked regarding health and safety issues they encountered both in relation to their mining and to their family and community responsibilities. They were also asked if they would prefer if their children did not become miners and, if so, what kind of employment they would like them to have.

Women in SSM in PNG: Reflections from the field

> We have been neglected for a long time, even our parents were neglected. It is only now we are being recognised and recently there has been some change. We need to be recognised, I have had enough of working in rivers ... if I could do something else ... or if there was machinery we could use to help us so we don't need to work so hard, I could keep going.

Cultural diversity within PNG means the status and role of women varies, with some areas being matrilineal and some patrilineal, with economic and social power differing accordingly. Women in PNG are predominantly restricted by their lack of ownership over mining land, with almost all registered mining leases, tributary agreements and customary land being held by men and transmitted patrilineally. While in practice some women do hold certain secondary rights to land and resources of their kin, these rights are mostly claimed and exercised by their spouses and male relations (Moretti 2005).

This is evident from an extract from the group interviews:

> I have had no control over the lease since my husband died—others ['settlers'] have come in and we are trying to move them on. My son-in-law and grandsons make the decisions.

With this division of ownership rights come problems of domestic violence, rape, HIV/AIDS and lack of economic equality between women and men, as well as lack of adequate opportunities to obtain equality (Crispin 2006). Some of the women in the group reported being victims of domestic violence. Recent media reports from the Porgera mine area in PNG highlight the gang rape of women while scavenging mine site tailings (Papua New Guinea Mine Watch 2015a). Women in mining regions are particularly vulnerable to socio-economic hardship, and are largely forced towards mining for the economic survival of their families. An account by Macintyre (2006) of women in Lihir Province indicated that women take on multiple roles simultaneously, leading to them being overworked and overburdened. This was supported by the evidence gathered through the field narrative:

> My body is tired. I want to stop working, I will only work when we need extra money for school and medical help. (Older woman from SSM alluvial mining area in PNG)

Traditionally, women undertake heavy manual work, and a number of the interviewed group reported back and knee pain from shovelling, lifting and carrying heavy rocks. In most SSM in PNG alluvial gold-mining, women reported working long hours bending and twisting to pan and sieve, while standing or squatting in water often contaminated by mercury. Some of the group reported back pain and internal organ pain/uterine pain, and many of the women indicated they had to continue lifting heavy and awkward equipment and loads while pregnant. Most reported that they worked while pregnant and breastfeeding, and all reported concerns about miscarriages when working while pregnant. The women also described work practices leading to musculo-skeletal, joint and abdominal pain, infected cuts that were slow to heal and abrasions and bruising from stones falling on them. On observation, many of the women had abrasions on their hands and feet and broken toenails. A typical narrative is as follows:

> I work long hours in the river water, I have sores on my fingers that don't heal. Rubbing the gravel on the sluice boxes gives me a rash and blisters that take a long time to heal. I have sores on my feet from standing for a long time in the river.

Many of the women reported extreme tiredness, as their mining work is additional to their domestic responsibilities, and indicated that tiredness led them to undertake what they considered unsafe practices in order to complete their mine work to get back to their families. Typical narratives included:

> My body aches. I have back pain, my joints hurt. I have back pain, and pain in my shoulders and knees from using shovels and crowbars. My hands and fingers are sore from long hours using sieves. I have sores that don't heal from handling big stones. I get headaches from working long hours in the sun. I am exhausted.

> My back hurts all the time now and my insides hurt by the end of the day, I am standing in water all day, I have been doing this since I was seven years old—my body is just so tired, I don't think I can do this anymore. If there was another way to do this maybe I could keep working.

Mercury is widely used across alluvial mining activities in PNG, and all women interviewed reported concerns about mercury use and its effect on them and their families. Almost all reported symptoms attributable to exposure to mercury, and many admitted they burned mercury at home on the kitchen stove for amalgamation. It is often reported in literature that amalgamation is usually undertaken by the men; however, the women interviewed indicated that they were responsible for burning the mercury to amalgamate the gold they had panned during the day. Practised out of necessity rather than choice, recovered gold is amalgamated daily and sold to provide immediate money for household essentials. Many of the women reported eye problems from the mercury fumes, and many displayed physical signs of eye irritation such as redness and tears. However, most of the women indicated that they would continue the practice as it meant that at the end of the day they had money for food and other family necessities. Typical narratives included:

> I know I shouldn't burn the mercury on the stove at home, but the days are long and I am bending and panning all day, I get home tired. I get up early to get my housework done and my children to school, then I travel by truck to river, it is a long way. I spend long hours lifting, shovelling and throwing stones, it's a long day and by the time I get home I am exhausted. If the day is too long I bring home the gold and do it [the amalgamation] at home so I can sell it to pay for food.

> I get up at 6 am to do my housework and get my children ready for school. I then work until 3 or 4 pm collecting concentrate. I then need to amalgamate it. Sometimes I am still on site at 6 pm. I am too tired at the end of the day so I take it home and burn it at home.

Most of the women indicated that they lived in traditional houses, often without running water or sanitation facilities, meaning that in addition to their mining work they need to carry fresh water to their houses for washing and cooking. In a vicious cycle, the mining contamination of local water bodies means additional distances need to be travelled to collect water suitable for drinking and cooking. Transport limitations add to the day, with long hours on foot or travelling in the back of a truck to their mining lease or to shopping facilities. Some families reported living on the riverbank in makeshift housing to be close to their lease, and to protect it from either being mined by others or being robbed. A number of the women interviewed indicated that the time they had to spend mining impacted negatively on the time they could spend with their children, and on other home-related issues. Most reported that despite their concerns, financially they had no choice but to continue mining as they needed to have money to support their households. Some of the women interviewed indicated that the men would go away to sell the gold and often waste the money on alcohol and prostitutes. Of the women who had been able to give up mining or reduce their time spent mining, health and safety concerns were cited as the main reason for their decision.

A possible way forward

Women involved in ASM operations face huge socio-economic issues. The women indicated that accidents, injuries or lack of family support often forced them into mining operations. Without adequate knowledge of good safety and health practices, these women face work-related risks, forcing them into actions that by western standards would be viewed as unsafe and unacceptable. While in some areas few women are involved in mining, in other areas several generations may be seen mining together, including children, mothers, aunts and grandmothers. Safety records remain negligible, and as a result awareness of occupational health issues and prevention of accidents remains minimal. Women in these communities face competing responsibilities, including child rearing, domestic duties and subsistence farming, and often need to take their children with them to the mine site where they too are then exposed to health and safety issues associated with mining activity. However, recent figures have shown a positive trend indicating the number of school-age children on mine sites has reduced from approximately 30 per cent in 2005 to around 15 per cent, largely due to extensive awareness campaigns across the country highlighting the need for education opportunities to be maximised. Women do not have as many opportunities to acquire mining skills as their male counterparts, and so do not have equal access to the cash made through mining, despite playing an active role in the mining process. Education opportunities for women and girls in the mining areas are essential to their futures—many of the women interviewed lacked basic literacy stills. Essentially, the women lack access to information, microcredit and training.

Time is the most significant issue impeding these women from accessing opportunities to improve their livelihoods. The second-most significant issue for these women is their lack of literacy. The women who formed the discussion groups indicated that they could not attend training programs because the programs currently offered did not take into consideration their need to fulfil family and domestic obligations first and foremost, meaning they were not able to attend full-day or longer live-in programs. Thus, some issues to consider when developing training programs for women in remote mining communities are:

• Time constraints: the women are time poor, so delivery methods need to be flexible, sensitive and accommodating of their domestic and family obligations.

- High rates of illiteracy and lack of basic education (this needs to be incorporated into all training programs).
- Women stereotyped as not being technically savvy.
- Traditional cultural beliefs and values precluding them from certain mining activities, meaning they often miss out on the benefit streams that naturally flow to the men.
- Lack of business/financial management skills.
- Lack of understanding of health and safety issues: mercury dangers, ground stability, mining methods, GBV, child labour, HIV/AIDS, water-borne diseases.
- Lack of access to microcredit.

Most importantly, training programs must be seen by the women to:

- Add value to their daily life and livelihoods.
- Must lead to activities that are practical, tangible and sustainable, and designed specifically for the users.
- Build capacity and empower the women.
- Not expose them to higher levels of GBV.

The women in the interview groups expressed a desire for information on safer working practices to enable them to understand what was safe and what was not. However, their health and safety concerns extend beyond work-related illnesses, to other issues that impact their families and communities. In particular, the women wanted training related to:

- Access to simple machinery and the technical knowledge of how to operate and maintain it.
- Safer and less physical mining practices.
- Understanding the hazards associated with mining practices, and how to undertake safer mining practices.
- Understanding how to prevent injury to themselves and other family members engaged in mining.

All of the women interviewed expressed health and safety concerns, with many expressing the opinion that it was their responsibility to ensure the health and safety of their families, including male members of their family having a good understanding of safe working practices. While some of the women expressed a desire for an alternative source of income, for many of those interviewed mining was the only occupation they had generational experience in, and they wanted to remain miners, not wanting to re-skill as farmers or service providers. All indicated they would value a means of obtaining knowledge and information to help them understand and manage the health and safety issues they saw as major concerns influencing their family's welfare.

In 2005, the Assistant Director of the Small-Scale Mining Branch of the Department of Mines, PNG, contended that the fundamental safety issue in SSM in PNG was that most miners lacked knowledge and skills on health and safety issues. This lack of knowledge translated to the inability to identify hazards and underestimation of the risk of being injured or killed, as the primary goal was to access high-grade gold for economic survival. It would seem that 10 years on this still remains a significant challenge for the Small-Scale Mining Branch of the MRA. At the Alluvial Mining Congress in Lae, PNG, held on 29–30 September 2015, representatives from the MRA-managed SSMTC highlighted some of the health and safety issues associated with alluvial and small-scale mining, and the difficulties in reaching small-scale miners with training. Operating since 2009, the centre conducts live-in two-week courses covering a number of mining-related subjects, including occupational health and safety, and outreach training programs across many remote alluvial mining areas of PNG, including adjacent islands. A significant challenge for the centre is the low literacy level of alluvial and small-scale miners attending the courses. Courses are developed and conducted at three levels: illiterate, semi-literate and educated. Despite having a high practical component, financial constraints make it difficult to attract attendees. By 2015, the centre had trained over 4,500 alluvial and small-scale miners; however, trainers estimate over 90,000 miners remain untrained. While courses are available to both male and female participants, the programs are delivered either in a two-week training block requiring participants to live at the facility, or as a one-week outreach program. While all women interviewed expressed genuine concern at their lack of health and safety knowledge, all indicated that family responsibilities generally preclude them from attending formal training programs that require extensive time commitments. Women in general tend to stay away from outreach programs for miners.

In order to provide effective training programs, it is necessary to accurately document target audience concerns, and provide training for local people in their local language. When asked what kind of program would work best for them, the women indicated a preference for flexible and informal training programs lasting two to three hours per day over three days of the week. The program would need to incorporate a level of basic literacy training for the women. Development of a program such as this would have widespread application, not only to other communities in PNG, but also to other countries where women are engaged in ASM activities.

Conclusion

Health and safety issues are inextricably linked to community, economic and environmental aspects of SSM. Despite the significant health and safety issues related to small-scale mining, alluvial gold-mining remains attractive in PNG, as miners can receive up to 70–90 per cent of market value, compared to rural commodities, which are paid at a much lower rate and generally take longer to realise any financial return. While women's participation is estimated at around 30 per cent, men and women are differently impacted upon by mining practices, cultural practices and legislative practices. The interviews provided a rich and valuable source of gender-specific information on SSM-related health and safety issues, which have not been well documented to date.

These interviews provided valuable insight and information that will assist the development of targeted training programs to help women better understand and manage health and safety issues that affect them and their families, and are consequently a step forward to improving their overall quality of life.

Acknowledgements

The writer acknowledges the assistance of the Minerals Resources Authority (MRA) and the Small-Scale Mining Training Centre (SSMTC) located at Wau Province, and in particular the assistance of the Centre Manager Al B. Comparativo, Business Manager Otto Morabo, trainer Immaculate Javia and the mining women of Wau who participated in the interviews.

References

Aubynn, T., 2009. 'On Mainstreaming ASM Research Findings.' Paper presented to Ghana Chamber of Mines, Accra Ghana, 16 September.

Buxton, A., 2013. 'Responding to the Challenge of Artisanal and Small-Scale Mining: How Can Knowledge Networks Help?' London: International Institute for Environment and Development.

Crispin, G., 2003. 'Environmental Management in Small-Scale Mining in PNG.' *Journal of Cleaner Production* 11(2): 175–83. doi.org/10.1016/S0959-6526(02)00037-9

Crispin, G., 2006. 'Women in Small-Scale Gold Mining in Papua New Guinea.' In K. Lahiri-Dutt and M. Macintyre (eds), *Women Miners in Developing Countries: Pit Women and Others.* Aldershot: Ashgate.

Eftimie, A., K. Heller, J. Strongman, J. Hinton, K. Lahiri-Dutt, N. Mutemeri, C. Insouvanh, M. Godet Sambo and S. Wagner, 2012. *Gender Dimensions of Artisanal and Small-Scale Mining: A Rapid Assessment Tool.* Washington, DC: World Bank Group's Oil, Gas and Mining Unit. Available at siteresources.worldbank.org/INTEXTINDWOM/Resources/Gender_and_ASM_Toolkit.pdf

Gratz, T., 2003. 'Gold-Mining and Risk Management: A Case Study from Northern Benin.' *Ethnos: Journal of Anthropology* 68(2): 192–208. doi.org/10.1080/0014184032000097740

Hancock, G., 1994. 'The Economics of Small-Scale Mining in Papua New Guinea.' In A.K. Ghose (ed.), *Small-Scale Mining: A Global Overview.* Rotterdam: A.A. Balkema.

Hayes, K., 2008. 'Artisanal & Small-Scale Mining and Livelihoods in Africa.' Amsterdam: Common Fund for Commodities.

Heemskerk, M., 2003. 'Self-Employment and Poverty Alleviation: Women's Work in Artisanal Gold Mines.' *Human Organization* 62(1): 62–73. doi.org/10.17730/humo.62.1.5pv74nj41xldexd8

Hentschel, T., F. Hruschka and M. Priester, 2002. *Global Report on Artisanal & Small-Scale Mining.* Minerals Mining and Sustainable Development, no. 70. International Institute for Environment and Development, World Business Council for Sustainable Development.

Hilson, G., 2002. 'Small-Scale Mining and its Socio-economic Impact in Developing Countries.' *Natural Resources Forum* 26: 3–13. doi.org/10.1111/1477-8947.00002

Hilson, G. and J. McQuilken, 2014. 'Four Decades of Support for Artisanal and Small-Scale Mining in Sub-Sahara Africa: A Critical Review.' *Extractive Industries and Society* 1(1): 104–18. doi.org/10.1016/j.exis.2014.01.002

Hinton, J.J., M.M. Veiga and C. Beinhoff, 2003. 'Women and Artisanal Mining: Gender Roles and the Road Ahead.' In G. Hilson (ed.), *The Socio-Economic Impacts of Artisanal and Small-Scale Mining in Developing Countries*. The Netherlands: Swets & Zeitlinger B.V. Publishers. doi.org/10.1201/9780203971284.ch11

Javia, I. and P. Siop, 2010. 'Paper on Challenges and Achievements on Small Scale Mining and Gender: Papua New Guinea.' Paper presented at Women in Mining Conference SIDS-18. New York, 10 May.

Jenkins, K., 2014. 'Women, Mining and Development: An Emerging Research Agenda.' *Extractive Industry and Society* 1(2): 329–39. doi.org/10.1016/j.exis.2014.08.004

Jennings, N., 1999. 'Social and Labour Issues in Small-Scale Mines.' Report for discussion at the Tripartite Meeting on Social and Labour Issues in Small-Scale Mines. Geneva, Switzerland: International Labour Organization.

Jennings, N., 2003. 'Addressing Labour and Social Issues in Small-Scale Mining.' In G. Hilson (ed.), *The Socio-economic Impacts of Artisanal and Small-Scale Mining in Developing Countries*. The Netherlands: Swets & Zeitlinger B.V. Publishers. doi.org/10.1201/9780203971284.ch10

Krimbu, J., 2005. 'Gender Issues in Small-Scale Mining.' Paper presented at a workshop on Community and Small-Scale Mining: Sharing Experiences from the Asia-Pacific Region held in Manila, Philippines, 7–12 June. Viewed at www.casmite.org (site discontinued)

Lahiri-Dutt, K., 2006. 'Globalization and Women's Work in the Mine Pits of East Kalimantan, Indonesia.' In K. Lahiri-Dutt and M. MacIntyre (eds), *Women Miners in Developing Countries: Pit Women and Others*. Aldershot: Ashgate.

Lahiri-Dutt, K., 2008. 'Digging to Survive: Women's Livelihoods in South Asia's Small Mines and Quarries.' *South Asian Survey* 15(2): 217–44. doi.org/10.1177/097152310801500204

Lahiri-Dutt, K., 2012. 'Digging Women: Towards a New Agenda for Feminist Critiques of Mining.' *Gender Place and Culture* 19(2): 193–212. doi.org/10.1080/0966369X.2011.572433

Leonhard, S., 2015. 'MRA Small-Scale Mining Training Centre.' Paper presented at the 2nd PNG Alluvial Mining Convention, Lae, PNG, 29–30 September.

Lole, H., 2005. 'The Trend in Artisanal and Small-Scale Mining Development in Papua New Guinea.' Paper presented at a workshop on Community and State Interests in Small-Scale Mining: Sharing Experiences from the Asia-Pacific Region, Manila, Philippines, 7–12 June.

Lopez-Trueba, M., 2014. '"Looking at Risk with Both Eyes": Health and Safety in the Cerro Rico of Potosí (Bolivia).' University of Sussex (PhD thesis). Available at sro.sussex.ac.uk/51385

Macintyre, M., 2006. 'Women Working in the Mining Industry in Papua New Guinea: A Case Study from Lihir.' In K. Lahiri-Dutt and M. Macintyre (eds), *Women Miners in Developing Countries: Pit Women and Others.* Aldershot: Ashgate.

Minerals Commission Ghana, 2014. Report of the Performance of the Mining Industry (2014). Viewed at www.ghanachamberofmines.org/media/publications/Performance_of_the_Mining_Industry_Ghana_2014 (site discontinued)

Minerals Resources Authority, 2015. 'Corporate Plan, 2008–2013.' Viewed at www.mra.gov.pg/Portals/2/Publications/MRA%20Plan%20-%202008%202013.pdf (site discontinued)

Moramo, O., 2015. 'Small-Scale Alluvial Mining.' Paper presented at the 2nd PNG Alluvial Mining Convention, Lae, PNG, 29–30 September.

Moretti, D., 2005. 'Community and State Interests in Small-Scale Mining.' Paper presented at a workshop on Community and State Interests in Small-Scale Mining: Sharing Experiences from the Asia-Pacific Region, Manila, Philippines, 7–12 June.

Moretti, D., 2006. 'The Gender of the Gold: An Ethnographic and Historical Account of Women's Involvement in Artisanal and Small-Scale Mining in Mount Kaindi, Papua New Guinea.' *Oceania* 76(2): 133–49. doi.org/10.1002/j.1834-4461.2006.tb03041.x

Moretti, D., 2007. 'Report on the Japan Social Development Fund Project on Artisanal and Small-Scale Mining in Papua New Guinea.' Research report, Artisanal and Small-Scale Mining in Asia-Pacific Case Studies Series. Viewed at www.asmasiapacific.org (site discontinued)

Papua New Guinea Mine Watch, 2015a. 'ABG Not Doing Enough Awareness on Potential Dangers of Alluvial Mining.' Available at ramumine.wordpress.com/tag/alluvial-mining

Papua New Guinea Mine Watch, 2015b. 'Alluvial Mining a Sleeping Giant.' Available at ramumine.wordpress.com/tag/alluvial-mining/

Simatauw, M., 2009. 'The Polarisation of the People and the State on the Interests of the Political Economy and Women's Struggle to Defend their Existence: A Critique of Mining Policy in Indonesia.' In I. Macdonald and C. Rowland (eds), *Tunnel Vision: Women and Communities*. Fitzroy, Victoria: Oxfam Community Aid Abroad.

Susapu, B. and G. Crispin, 2001. 'Report on Small-Scale Mining in Papua New Guinea.' London: International Institute for Environment and Development.

Tschakert, P., 2009. 'Recognising and Nurturing Artisanal Mining as a Viable Livelihood.' *Resource Policy* 34(1/2): 24–31. doi. org/10.1016/j.resourpol.2008.05.007

Van Hoecke, E., 2006. 'The Invisible Work of Women in the Small Mines of Bolivia.' In K. Lahiri-Dutt and M. Macintyre (eds), *Women Miners in Developing Countries: Pit Women and Others*. Aldershot: Ashgate.

Veiga, M.M. and R. Baker, 2004. 'Protocols for Environmental and Health Assessment of Mercury Released by Artisanal and Small-Scale Gold Miners.' Report to the Global Mercury Project. Vienna: United Nations Industrial Development Organization.

World Health Organization (WHO), n.d-a. 'Constitution of WHO: Principles.' Available at www.who.int/about/mission/en/

World Health Organization (WHO), n.d-b. 'Occupational health.' Available at www.who.int/topics/occupational_health/en/

World Health Organization (WHO), 2013. 'Mercury and Health Factsheet.' Available at www.who.int/mediacentre/factsheets/fs361/en/

8

Rice, sapphires and cattle: Work lives of women artisanal and small-scale miners in Madagascar

Lynda Lawson

Strong global demand for coloured gemstones,[1] particularly in India and China, has led to a phenomenal expansion of the market in recent years (KPMG 2014). Levin (2012) suggests that globally, 80 per cent of gemstones are mined artisanally. In Africa, with the notable exception of Gemfields, a London-listed large-scale miner of gemstones in Zambia and Mozambique, most of the mining of coloured gemstones is poorly regulated and conducted by artisans using hand tools. Artisanal mining of gold has been widely investigated, but that of coloured gemstones has received comparatively less attention. The role of women in gemstone mining has received even less.

In the past 17 years, Madagascar has risen to become one of the world's largest producers of fine sapphires, ranging from the best blues to yellows, pinks, oranges and purples (Shigley et al. 2010). Large investments by the World Bank to professionalise mining and to create sustainable livelihoods through, for example, the creation of lapidary training centres have had limited success. For the most part, rough stones are shipped out of the

---

1   The technical term used for coloured stones—not diamonds—such as sapphires, rubies and emeralds, but also a wide range of other minerals such as aquamarine, garnet, amethyst, opal and citrine.

country to be cut in the large specialist gemstone-cutting centres in South Asia, and then onto showrooms in Hong Kong (Cartier 2009). My review of project documents showed that the role and unique contribution of Malagasy women in artisanal and small-scale mining (ASM) is all but ignored.

Madagascar remains one of the poorest countries on earth, with 88 per cent of the population living on US$1.25 a day (World Bank 2014), with most of the poor living in rural areas with high and chronic levels of food insecurity (World Bank 2011). There have been marked increases in poverty following the political crisis of 2009 and environmental shocks such as extended drought and cyclones. Poor economic development hampers Madagascar's capacity to respond to changing climate and unexpected shocks (ibid.). Madagascar is expected to be severely affected by climate change and attendant increases in cyclones, drought and floods, with significant decreases in its farming communities' capacity to produce staple crops (Harvey et al. 2014).

In 2013, Madagascar was officially declared a fragile state—that is, 'a state with weak capacity to carry out basic governance functions, and lacks the ability to develop mutually constructive relations with society. Fragile states are also more vulnerable to internal or external shocks such as economic crises or natural disasters' (Organisation for Economic Co-operation and Development (OECD) 2013). Madagascar combines extreme poverty and aid dependence with all the above-mentioned features of a fragile state.

In response to these challenges, ASM has provided a lifeline to an estimated 800,000 people (personal communication, Rupert Cook, August 2014), with a significant number of these people working in the gemstone sector. Roughly half of those working in ASM are thought to be women. Very little in revenue has returned to the state, or worse, most has been misappropriated (Duffy 2007; Cartier 2009). There are just a handful of scholarly publications that have studied ASM in Madagascar (Walsh 2003, 2012; Cartier 2009; Canavesio 2014), and there are a number of World Bank reports that are either dedicated to, or reference, ASM (Cook 2012). All make only passing reference to the role of women. The one exception is the work of Remy Canavesio (2013), a French geographer who spent five years researching gemstone mining and supply chains, and who returned to his thesis data to re-interrogate it from a gendered perspective.

To contextualise the discussion, we begin by reviewing the literature related to women and ASM, mostly in Africa, and then focus on women miners of gemstones in Madagascar. We outline the methodology used and then present cases.

Women in ASM: The African perspective

The literature on women in ASM over the past 20 years has not been extensive, and in many ways it reflects broader societal trends that have only recently begun to document and analyse women's important role in, and the impacts on, the extractive industries. Noestaller's (1987) work on ASM commissioned by the World Bank makes no reference to women or gender issues. However, in 1995, Béatrice Labonne of the World Bank presented on women in ASM at a special event on 'Women and Natural Resources' at the Fourth World Conference on Women in Beijing. Drawing on work by Esther Ofei-Aboagy on Ghanaian women miners, Labonne published one of the first papers in this area: 'Artisanal Mining: An Economic Stepping Stone for Women' (Labonne 1996). It paints a positive view of opportunities for women in ASM, but makes no reference to the specific health and safety issues faced by women on ASM sites. By contrast, in 2003, Heemskerk's detailed ethnographic study of the Maroon women artisanal miners working in rural Suriname found that if long-term social and health conditions were considered, work on informal mining was not likely to improve the quality of women's lives (Heemskerk 2003). Yakovleva (2007) presents a detailed, essentially descriptive, case study of the work, income, health and family of women miners in camps and villages in the Eastern Region of Ghana. She argues for gender mainstreaming of assistance for ASM. However, it is the work of Jennifer Hinton that sets the benchmark for the literature on ASM and women in Africa, both as a co-author when considering the impacts of mercury for women (Hinton et al. 2003), and in her broad and analytic review of mining and gender roles in her doctoral work (Hinton 2011). She brought to the fore key issues, such as the impact of the commodity and mine lifecycle on the work of women in ASM.

Hinton has warned against over simplifications and generalisations about women's roles (Hinton 2011); however, typically the work women do is less visible than that of men. They are not usually found digging the main ore bodies and going underground, but they may be found digging or panning around the edges, transporting, washing and processing. Malpeli and Chirico (2013) investigated 137 gold and diamond sites in West Africa over five years, collecting data directly from women at mine sites. They found that women's participation in mineral extraction was dependent on the thickness of the overburden and accessibility of the deposit, and that women were mostly involved in processing. Along with geomorphology, economic factors such as the value of the commodity being mined, the stage of the life of the mine, access to finance, the hierarchical organisation of the mine, land, permits and equipment also determined women's participation.

Women's work in ASM is often not clearly differentiated from other duties; for example, women may grind stones with kitchen equipment, and process gold with mercury as they cook, while their children play beside them. In Madagascar, it is common to see women sieving for sapphires next to others doing the family laundry. Thus, the women's role is often not valued, well renumerated or recognised and, as a result, women may not be included in any formal or informal census of those involved in ASM, and may not be considered by policymakers (Eftimie et al. 2012). Policy may impact men and women miners differently, and their specific needs may be neglected in gender-neutral policies (Hinton 2011). Hinton, in her doctoral study of gender and ASM in Uganda, argued that such gender-neutral policy actually negatively impacts on women in ASM, thus damaging the whole community (ibid.).

One thing that is well established across the globe, and particularly in Africa, is that women involved in ASM earn considerably less than men, even when doing the same task. In addition, they work longer hours since they still carry the burden of domestic work (Eftimie et al. 2012). For example, in a formal gold mine camp in Ghana, women who transport gold ore and water, and pound rocks, have salaries 60 per cent lower than men involved in digging (Akabzaa and Darimani 2001).

Women's role in ASM remains constrained by socio-economic and cultural barriers, which impact on resource rights and decision-making. Women are typically proffered land of less value, and women's capacity to benefit from ASM may be constrained by de jure and de facto inequity in access to land and property rights (Meinzen-Dick et al. 1997). In Kenya, for example, the traditional social system allows women access to, but no control over, the land; thus, their overall production is low (Amutabi and

Lutta Mukhebi, in Lahiri-Dutt 2008). Women have difficulty obtaining finance from banks and may require their husband's consent before obtaining a permit.

Women are often excluded from direct contact with more valuable deposits, and can be found digging on the less valuable sections of the lease while men pursue more lucrative seams underground. Where women are involved in the sale of minerals, they tend to deal with less lucrative sales deals. For example, Malagasy women traders buy and sell the smaller gemstones, while the larger, more precious loads are reserved for males acting in concert with other powerful males, such as the mayor and the local police chief. This limits their access to real financial power (Canavesio 2010). Where ASM has been formalised, women working in cooperatives, for example in Ethiopia, are paid less than men (personal communication, Solomon Negussie, December 2014). Even women mine owners experience gender bias, with men being reluctant to follow their orders, and women being forced to use male agents (Tallichet et al. 2003).

ASM has the potential to provide a livelihood, but there is no capacity for resilience and little savings in the case of an accident, illness or a natural disaster. Canavesio (2010) describes the tragic situation of women who have gone to the sapphire fields in Madagascar in the hope of finding an income for their families, have been unsuccessful and, not having the funds to return home, have died of starvation.

Women miners' identity

Critical feminist insights into the oppressed position of women miners and into their identity have been provided in the work of feminist geographer Lahiri-Dutt, who published on ASM in Asia in 2004. More recently, she has reframed the field by referring to peasant miners rather than artisanal miners (Lahiri-Dutt 2014)—this is particularly pertinent in rural Madagascar. Bryceson's work in East Africa on rural women's livelihoods began to reference ASM in the mid-1990s (Bryceson 1996). Scholars and international aid agencies began noting how deteriorating farm conditions were forcing women into ASM, and the unequal and unfair remuneration women were receiving (Yakovleva 2007). Bryceson's most recent work on ASM in Tanzania (Bryceson et al. 2014), in particular, has provided unique insights into the broader impacts of ASM—its democratising impacts, and its impacts on gender relations, marriage and casualised sex.

Questions about the identity of women miners are explored in a recent critical paper based in the Democratic Republic of Congo (DRC) (Bashwira et al. 2014). Gender-based violence towards women miners in conflict zones in the Eastern DRC led to non-government organisations attempting to move women from ASM into alternative livelihoods; in turn, the DRC Government banned pregnant women from ASM, despite the fact that there was little other work (ibid.). As Lahiri-Dutt notes, 'popularist and universalist conceptions of femininity and womanhood tend to normalise contested gender roles through protective legislation that operates against women's interests' (Lahiri-Dutt 2013: 224). It is vital that policymakers listen carefully. It is time to move beyond a stereotypical homogenous view of women in ASM to 'a far more real picture of diversity, opportunism and agency' (Mahy 2011: 61).

Jenkins (2014), in the most thorough review to date of women in mining (both large- and small-scale), argued that the role of women in ASM has generally been under-theorised and under-recognised, and that this is a key issue in terms of understanding the role of the mining sector in relation to development of poor communities in the Global South. A challenge for researchers is to go beyond factual accounts and single case studies of women's activities to 'develop strong critiques of the gendered dynamics and power relations at work' (ibid.: 32). Also, few studies make any connection between the broader question of feminism and women at work, or the impact of climate on rural livelihoods such as ASM.

Some of the most promising methodological approaches can be found in research on male small-scale miners, which considers small-scale miners' decision-making and 'career trajectories' (Bryceson and Jønsson 2010: 382). Bryceson and Jønsson (2010) addressed many of these concerns in a study of the lives of Tanzanian gold miners. They describe a 'coalescing career formation arising almost entirely from the small-scale miners' own organizational constructs and individual decision-making' (ibid.: 387). The research contains valuable and vivid detail and, although the sample did not include women miners, its comprehensive research design provides the kind of detail needed in future studies of women miners' work–life courses.

Women 'help the men'

'Women? They help the men.'[2]

In Madagascar, there are just a handful of scholarly publications that have investigated ASM (Walsh 2003; Canavesio 2011, 2013; Cartier 2009; Cook 2012), and only Walsh (2003) and Canavesio (2011) consider women's role in any detail, even though it is known that women are involved throughout the supply chain. Anthropologist Andrew Walsh's (2003) ethnographic studies of ASM miners in Northern Madagascar found that while young male miners often considered money earned from sapphires to be 'hot' money that had to be spent quickly on hedonistic pursuits, women miners tended to use 'cool' money, which was destined for buying houses and cattle. The women interviewed stated that the only way to get ahead was 'to put their money to work' (Walsh 2003: 294), spending their money wisely on long-term investments. A typical comment was that they were motivated to do this because men 'have made them suffer' (ibid.: 294), and that careful management of sapphire earnings would enable them to live independently of men. This resonates with Canavesio's (2013) comments about the women of the southwest that 'migrate in order to become richer, but they also look for a new life in a society where gender inequalities are smaller than in the other parts of the country' (Canavesio 2013: 1). This desire for emancipation also led some women to strategically marry foreign traders; such 'mine marriages' have also been noted in gold ASM communities (Bryceson et al. 2014) and in diamond trade communities in Angola, where De Boeck comments that 'mine marriages' tend to 'serve an economic, purely utilitarian purpose in the short term, with the woman involved oftentimes for advantageous financial outcome' (cited in Walsh 2003: 302).

Methodology

The methodology used in this study aims to give women, who have rarely had their voice heard, an opportunity to tell their story in their own words. This fits with the feminist theoretical framing of the project, as 'oral interviews are particularly valuable for uncovering women's experiences' (Anderson and Jack 1991: 11). It also fits with the vibrant oral tradition

---

2   This comment came from a prominent gemmologist working in the region.

of storytelling in Africa, and Madagascar in particular. The methodology draws on two research approaches: the Panos Oral testimony project (Panos Oral Testimony 2014), with its detailed method for collecting stories outlined in *Listening for a Change* (Slim and Thomson 1993) and *Methods of Life Course Research* (Giele and Elder 1998).

Life course research

Life course research is widely used across a range of disciplines, such as medicine, social sciences and development studies, and was elaborated by Janet Giele and Glen Elder (Giele and Elder 1998). The research approach grew out of an outstanding series of twentieth-century longitudinal studies of American life, such as Elder's painstaking and ground breaking study, *Children of the Great Depression* (1974), which followed the lives of 167 people born in Oakland, California, in the 1930s into the late 1960s (Elder 1974). From such research, Giele and Elder (1998) identified central themes that determine the shape of the life course—for example, location in time and place (cultural background)—and heuristics such as life transitions. This multifaceted data is collected primarily in the form of life histories through interviews; instruments such as life event calendars (Drasch and Matthes 2013) or diaries may be used. The life course research paradigm has been used successfully in the development context to investigate gender, the work–life course and livelihood strategies in a South Indian fish market (Hapke and Ayyankeril 2004).

Analysis of data from the interviews

Data from the interviews was translated, transcribed and analysed using content analysis. The categories used to analyse the data were both inductively derived—that is, allowed to emerge from the data in relation to the research question and scoping visits[3]—and deductively and iteratively determined based on insights from theory and literature, principally life course research (Giele and Elder 1998). The thematic categories were location in time and place, and climate (cultural background); linked lives—for example, family (social integration); human agency (individual goal orientation); and the timing of lives and climate change (strategic

---

3   For example, a category related to climate change emerged. ASM and climate change are not often linked in the literature; however, in interviews a number of women spoke of changes in rain patterns, which meant that they were no longer able to grow enough rice to feed their families, and this was a factor in their decision to take up sapphire mining.

adaptation). The heuristics to explain work–life course were life transitions or turning points, trajectories, sequences and life events (based on Giele and Elder 1998).

Research questions

Based on the literature and the theoretical and methodological orientation, the following research questions were constructed: What does the work–life course of a woman in gemstone mining look like? What motivates women to take up ASM? How have they learned their skills? What have been the turning points in their work–life course? How do they manage family? How has this work activity impacted on their health? Does it provide a reliable source of income?

Research process

Two preliminary scoping visits were made to the Ilakaka–Sakaraha region in 2014. Twelve women found working by the streams as miners of sapphires in the Ilakaka–Sakaraha region were interviewed in the field, with the assistance of a Malagasy research assistant. Photographs and participant observation were also used. For this study, we chose to concentrate on the interviews of women whose only livelihood was sapphire mining. Of the 12 women approached, seven were working full-time as miners and their interviews were selected for detailed analysis.

Extended semi-structured interviews were used to elicit narratives of work–life course history, and participants were asked to comment on specific questions in relation to the research question for each case.

Work–life stories of women sapphire miners

Alvine

'The four of us work together.'

Alvine is 17 years old, she is married and has a nine-month-old baby, and works in a group with her husband and cousins in a team. She has been mining for three years across different mine sites in a 50 km radius from Ilakaka, and is now working at Bekily.

Before I came to mine I used to grow rice, but because there was not enough rain and no harvest I started sapphire mining. Sometimes I find something, sometimes I don't. Some days I earn 10–20,000 Ar[4] a day but often I earn nothing. I use everything I earn on my daily expenses.

The four of us work together (my husband and cousins), the boys dig, the girls sieve. I like doing this because we might find something. I can earn much more from sapphires than from gold.

I have a problem with my back. If I get sick I just keep working, I can't afford to see a doctor.

Ravao

'But unfortunately God hasn't given me yet.'

Ravao is 20 years old and she came from the south. She has been mining in this area at Bevilany for six months. She has one child who is with her parents in Fort Dauphin, as she goes to school. Her husband is away working on the new site.

My mother grows rice in Fort Dauphin but it did not provide enough to sustain us. My mother paid for my husband and I to come here to mine. I will keep mining until I have enough, but I haven't found anything big yet. I chose sapphires because I don't know how to do gold. My brother taught me how to sieve for sapphires. I get a bit of money from sieving— sometimes 500 Ar[5] a day, but not every day. If I can't find anything, I go to bed with an empty stomach. If I find a big stone, I will buy a sewing machine and some gold jewellery. But unfortunately God hasn't given me yet.

Harena

'I mine because I want my children to have the same as others who have value, that have a better life.'

Harena is 24 years old and her family are from Androy, west of Fort Dauphin. She is married with three children aged between five and two-and-a-half. She kept sieving until she was seven months' pregnant. She went and had the baby with her parents, and after two months she came back to mining, breastfeeding the baby at the same time. She used to grow rice and manioc, but with no rain there was no harvest, so she

---

4   US$3–6.
5   US$0.16.

took up mining. She had been at Bekily just one week, but she had been mining for six years at places like Antsoha and Amboalano. She works for a 'boss', and sometimes they earn 10,000 Ar [6] for three days for three people.

The children live with her husband, who also mines sapphires about 10 km away. They only see each other when they find a sapphire. When asked if it was safe for her to live alone on the rush site she said, 'I do it because of poverty. We are both looking for money'.

She has a large sieve and she gives it to the men to use; she then sorts the second wash. Her brother died when the earth caved in on the mine when he was underground.

> I can live on sapphires. I prefer to mine sapphires rather than gold as they are more valuable and also I do not know how to pan for gold. I can get more for them to buy gold jewellery and cattle. The cattle I have bought are with the family. I have found a few small stones and sold them to the businessmen. I will continue to mine until I find a large sapphire. I mine because I want my children to have the same as others who have value, that have a better life.

Titae

> 'They belong to me, I bought them with my own money.'

She comes originally from Amboasary (60 km from Fort Dauphin). She doesn't have any children, and she has been mining for eight years in many locations around Ilakaka, but she has decided to stay in Bevilany on the Maninday River as she no longer wishes to move. Her husband works with her; he digs and brings stuff down for her to sieve. They have found one good stone. They bought four cattle with it. The money they earn is for both of them. If her husband finds, he shares what he finds.

> It's hard. We only do it because we are poor. If we can't find anything we sleep hungry. I will keep mining until God gives us enough. Then I will do many things—buy gold jewellery, dresses, cattle. The sieve cost 8,000 Ar[7] and the spade 3,000 Ar.[8] They belong to me, I bought them with my own money. It is me who will work using them and make money from them.

---

6   US$3.
7   US$2.60.
8   US$1.

Vola Julienne

'I want a sapphire for myself.'

She is 46 years old. She comes from Ilakaka where her parents had a hotel. She lived there from 1980, and she was there when the sapphire boom started. She started mining then. She has moved around at least five mine sites around Ilakaka and Sakaraha. She is now mining at the Ambarinakoko mine at Bevilany, on the Maninday River. She is married with two children. One is finishing school, and one is married.

> I mine because I am looking for money. I found a sapphire in 2004, I built a house and bought cattle and paid for my children's schooling.
>
> I want a sapphire for myself. The father of the children is dead, this husband is their stepfather and I need my own money to look after them. I will continue mining until I find something. I want to build another house to rent for my old age.

Golden Smile

'I have been mining for a long time.'

Golden Smile is 46 years old, and she works with her two sons, who mine underground. She lost one of her sons, and used the sapphire money to bury him.

She sieves by the river. At 11 o'clock she goes up to where her sons are working and collects soil to take and sieve by the river. It is a 200 m steep walk on slippery ground, and it is hard to keep her balance. She is barefoot. She has been mining for eight or nine years. She has worked at least five mine sites around Ilakaka and Sakaraha. She lives on the money from sapphire mining and is bringing up her granddaughter.

> Sapphire mining is an activity for men. I do it because I don't have a husband. No one taught me but I watched my sons. I watched, I watched, I watched and then I did it. I have found some stones and bought food and medical treatment. If God wants to tell me to stop, I will do it. As you see I cannot dig but I have no money. No one else gives us money. If I had money, I would start a second-hand clothes business.
>
> I don't feel good, but if I find some sapphires I'll buy some jewellery.

Oly

Oly is 25 and has four children under four. One has died. Oly is a Bara woman, one of the main ethnic groups living in the southwest. It is a group that has worked traditionally with cattle. She works with her husband; he digs for sapphire-bearing soil and stones, and she sieves by the river, either the soil he has mined or directly from river stones. They mine all year, but in the wet season they also grow rice.

> We have two rice crops a year. When the rice is ready we harvest and stock and then go back to sapphires. We get 20 big bags each year. We store it and resell when the prices go up we buy zebu. I don't want my kids to go to school, I want them to work with me. We want to buy a car and a gun.

Discussion

> 'Women are like a thread passed through the eye of a needle.'
> – Proverb from Masikoro (an ethnic group found in this mining region).

I have framed this discussion with the Masikoro proverb, which reflects the position of these women, who against enormous odds are seeking 'to make change happen for better' (Cornwall and Edwards 2014: 2). Each woman speaks of using sapphire money to do something for herself, to help her children, to buy a business or personal items. Their spades and seives are their own. There is a sense of agency and some pride in what they have achieved. However, the opportunities are very constrained by the eye of the needle of these women's circumstances in a society where, structurally and culturally, they often have little power. ASM mine sites are typically portrayed as masculine frontiers where women 'help the men'. By talking to these women and hearing their life stories, it is noticeable that these women do not identify themselves as just 'helping men'; some are working alongside men in work that may be identified as 'men's work', but they have very clear and diverse work–life trajectories. As Canavesio (2013) has argued, the rush sites of this region offer opportunities for women to find economic and personal emancipation.

Andrea Cornwall's (2003) work has been significant. She has investigated the failure of development projects, even those that claim to be using participatory approaches to listen to and incorporate women's voices. In her comprehensive review and critique of gender and participatory development, she argued, 'what is needed is strategies and tactics that take account of the power effects of difference, combining advocacy to lever open spaces for voice with processes that enable people to recognize and use their agency' (Cornwall 2003: 139). Despite extremely difficult circumstances, there is no way these women can be constrained into some kind of 'average Third World Woman' leading an 'essentially truncated life' (Mohanty 1988: 56). Mohanty argued that many well-meaning feminist development researchers emphasise the needs of third-world women, and fail to analyse the work they do, both in the formal and informal sectors. The analysis in the following section attempts to address this gap.

Location in time and place, and climate

The women in this study find themselves on the edge of a sapphire mining boom that began some 18 years ago, and is strongly controlled by foreign buyers of valuable stones. The buyers and those with control are predominantly, but not all, male. The easy-to-find stones—'grass' as they are called in Malagasy—have long gone, and younger, more mobile miners have moved to other areas. These women have not progressed to becoming stone traders, nor are there any opportunities for beneficiation or formal employment. They aspire to buy cattle, businesses and property, and to have some control of their circumstances. The ethnicity of the woman miner also emerges as a factor in relation to her motivation for mining. Madagascar has 18 different ethnic groups and, as is evoked in the proverb, many are oppressive to women. In particular, Bara women like Oly have few rights in relation to cattle and land ownership. Her comments seem unlike the others. She alone does not want to send her children to school, and instead of wanting to buy jewellery or a sewing machine, she wants to buy cattle and guns.

Turning points and strategic adaptation

A turning point in the work–life trajectory of all the miners, leading to the decision to start sapphire mining, was the persistence of poverty and drought. Changing climate patterns seem to be a significant catalyst to take up mining. Three women had moved from the south and southeast, because the rains there had failed, and they were no longer able to support the family by growing rice and manioc. This region of the south is dry and particularly prone to changing climate; food insecurity is extreme and sending family members to find work elsewhere is one strategic approach to survival (Harvey et al. 2014), and is an example of Elder's strategic adaptation.

Strategic adaptation is also seen in the way they have all moved across many different sites when new sapphires were discovered. The older women have moved across at least five different sites. They live in makeshift tents, with their only tools a shovel and sieve. They all hoped to move on from mining and wanted to buy cattle, a potent symbol of wealth in Malagasy rural life. Some hoped to establish other businesses such as second-hand clothes stalls, dressmaking or farming. They hoped for better lives for their children. Harena's comment sums it up: 'I want my children to have the same as others who have value, that have a better life.'

Linked lives

A strong factor influencing change in work–life trajectories are social and personal factors—the linking of other lives with our work decisions. For these women miners, there are a complex array of personal relationships at play in their work–life decisions. A number have very young children, and have given birth at a young age. In our sample, only Titae, Alvine and Oly are living and working alongside their partners, and Golden Smile does not have a partner but works with her sons. In the other cases, the husband and wife are hedging their bets by each working on a different site, and only seeing each other when they find a stone. Vola Julienne is using sapphire mining in a very strategic way to manage complex family issues—she needs to provide for her children from her first marriage as her second husband will not, and also to provide for her retirement.

Oly's case is different. She was working an area close to the original sapphire town of Ilakaka, not one of the more chaotic rush sites where the other women were found. Despite the great hardship of losing a child, she and her young family are moving ahead using a combination of mining, rice growing and cattle. She is a Bara woman, the cattle-based ethnic group of the southwest grasslands, and the original owners of the sapphire country around Ilakaka. They have been particularly impacted by the breakdown in law and order, and ruthless cattle rustlers. This is reflected in her wish to buy guns and transport. She is also very protective of her children and does not wish them to go to school, but wants them to work with the family.

Human agency (individual goal orientation)

All women display great courage and human agency to provide for themselves and their families. Women in this part of Madagascar are not in a strong position, 'especially the Bara women—they don't have a place' (personal communication, Mayor of Ambrinany and local doctor Alain Randrianirina, February 2014).

It has been remarked that artisanal mining provides an opportunity for women to have their own money, to use money more wisely than some of the male miners and to gain some independence in a system that is quite oppressive for women (Walsh 2003; Canavesio 2013). This is evident in some of the comments of the more mature women, Vola Julienne and Harena, who asserted quite forcibly in these relatively short interviews that mining could support them: 'I can live on sapphires'; 'I want a sapphire for myself. The father of the children is dead … I need my own money to look after them.'

Likewise, Titae is proud that she has her own tools and of the personal power that they give her: 'The sieve cost 8,000 Ar and the spade 3,000 Ar. They belong to me, I bought them with my own money. It is me who will work using them and make money from them.'

Age and time spent mining are also significant. The younger women had begun mining in their mid-teens and were working as a family team, with men collecting sapphire-bearing gravel from under the ground.

Older women like Vola Julienne are a little different, as she had found a large stone earlier in her life and this had permitted her to buy a home and cattle. However, she had married again and needed to make money to look after children from her first marriage, and also to provide for herself in retirement.

Mining of sapphire was both a means of survival in extremely precarious circumstances—a number of women spoke of going to bed hungry if they didn't find a stone—but also a means to finance other livelihoods in the future: to buy a sewing machine or to sell second-hand clothes. Almost all women made reference to God, a typical comment being, 'I will keep mining until God gives us enough.'

Conclusion

Andrea Cornwall (2003) investigated the failure of development projects to listen to and incorporate women's voices. In her comprehensive review and critique of gender and participatory development, she argues, 'what is needed is strategies and tactics that take account of the power effects of difference, combining advocacy to lever open spaces for voice with processes that enable people to recognize and use their agency' (Cornwall 2003: 139). Development agencies are returning to Madagascar after the four years of political instability, and there is interest in development activities for women in ASM. They would do well to heed this advice.

Cornwall and Edwards's (2014) recent work is a response to such critiques. It aims to explore in a more holistic way how women in different cultures experience change and empowerment in their lives and, in the spirit of true feminist research, it seeks to discover 'hidden pathways, the otherwise invisible routes that women travel on to empowerment' (Cornwall and Edwards 2014: ix).

Using the work–life course framework (Giele and Elder 1998) provides an in-depth and respectful basis from which to better understand the lives of women sapphire miners, and their 'hidden pathways' to empowerment. The diverse motivations of women and their individual agency it reveals are striking.

A predominant theme that emerges is of rural women from different ethnic groups, often from the south of Madagascar, taking up artisanal mining in response to deepening rural poverty and food insufficiency, caused by the failure of crops and changing climate. This is particularly significant in Madagascar, where the impacts of changing climate on small farmers are expected to be severe, and where preservation of its unique biodiversity is crucial. The issue of climate change and ASM has not been widely explored in the literature. The intersection on the pathway between women miners, food insecurity and changing climate in this research warrants further investigation.

References

Akabzaa, T. and A. Darimani, 2001. 'Impact of Mining Sector Investment in Ghana: A Study of the Tarkwa Mining Region.' *Third World Network*. Draft report prepared for Structural Adjustment Participation Review Initiative.

Anderson, K. and D. Jack, 1991. *Women's Words: The Feminist Practice of Oral History*. New York: Routledge.

Bashwira, M.-R., J. Cuvelier, D. Hilhorst and G. van der Haar, 2014. 'Not Only a Man's World: Women's Involvement in Artisanal Mining in Eastern DRC.' *Resources Policy* 40: 109–16. doi.org/10.1016/j.resourpol.2013.11.002

Bryceson, D., 1996. 'Deagrarianization and Rural Employment in Sub-Saharan Africa: A Sectoral Perspective.' *World Development* 24(1): 97–111. doi.org/10.1016/0305-750X(95)00119-W

Bryceson, D. and J. Jønsson, 2010. 'Gold Digging Careers in Rural East Africa: Small-Scale Miners' Livelihood Choices.' *World Development* 38(3): 379–92. doi.org/10.1016/j.worlddev.2009.09.003

Bryceson, D.F., J. Jønsson and H. Verbrugge, 2014. 'For Richer, for Poorer: Marriage and Casualized Sex in East African Artisanal Mining Settlements.' *Development and Change* 45(1): 79–104. doi.org/10.1111/dech.12067

Canavesio, R. 2010. 'Exploitation Informelle des Pierres Précieuses et Développement dans les Nouveaux Pays Producteurs – Le cas des Fronts Pionniers d'Ilakaka à Madagascar.' Université de Bordeaux (PhD thesis).

Canavesio, R., 2011. 'Croissance Economique des Pays Emergents et Géographie Mondiale des Pierres Pécieuses.' *Echo Geo* 17(11): 1–14.

Canavesio, R., 2013. 'Les Fronts Pionniers des Pierres Précieuses de Madagascar: Des Espaces d'émancipation pour les Femmes?' *Géocarrefour* 88(2): 119–29. doi.org/10.4000/geocarrefour.9046

Canavesio, R., 2014. 'Formal Mining Investments and Artisanal Mining in Southern Madagascar: Effects of Spontaneous Reactions and Adjustment Policies on Poverty Alleviation.' *Land Use Policy* 36: 145–54. doi.org/10.1016/j.landusepol.2013.08.001

Cartier, L., 2009. 'Livelihoods and Production Cycles in the Malagasy Artisanal Ruby–Sapphire Trade: a Critical Examination.' *Resources Policy* 34(1): 80–6. doi.org/10.1016/j.resourpol.2008.02.003

Cook, R., 2012. *Madagascar Case Study: Artisanal Mining in Protected Areas and a Response Toolkit.* Washington, DC: World Bank.

Cornwall, A., 2003. 'Whose Voices? Whose Choices? Reflections on Gender and Participatory Development.' *World Development* 31(8): 1325–42. doi.org/10.1016/S0305-750X(03)00086-X

Cornwall, A. and J. Edwards, 2014. *Feminisms, Empowerment and Development: Changing Women's Lives.* London: Zed Books.

Drasch, K. and B. Matthes, 2013. 'Improving Retrospective Life Course Data by Combining Modularized Self-Reports and Event History Calendars: Experiences from a Large Scale Survey.' *Quality and Quantity* 47 (2): 817–38. doi.org/10.1007/s11135-011-9568-0

Duffy, R., 2007. 'Gemstone Mining in Madagascar: Transnational Networks, Criminalisation and Global Integration.' *The Journal of Modern African Studies* 45(2): 185–206. doi.org/10.1017/S0022278X07002509

Eftimie, A., K. Heller, J. Strongman, J. Hinton, K. Lahiri-Dutt, N. Mutemeri, C. Insouvanh, M. Godet Sambo and S. Wagner, 2012. *Gender Dimensions of Artisanal and Small-Scale Mining: A Rapid Assessment Tool.* Washington, DC: World Bank Group's Oil, Gas and Mining Unit. Available at siteresources.worldbank.org/INTEXTINDWOM/Resources/Gender_and_ASM_Toolkit.pdf

Elder, G.H., 1974. *Children of the Great Depression: Social Change in Life Experience.* Chicago: University of Chicago Press.

Giele, J. and G. Elder (eds), 1998. *Methods of Life Course Research.* Thousand Oaks, California: Sage Publications.

Hapke, H.M. and D. Ayyankeril, 2004. 'Gender, the Work–Life Course, and Livelihood Strategies in a South Indian Fish Market.' *Gender, Place and Culture* 11(2): 229–56. doi.org/10.1080/0966369042000218473

Harvey, C.A., Z.L. Rakotobe, N.S. Rao, R. Dave, H. Razafimahatratra, R.H. Rabarijohn and J. MacKinnon, 2014. 'Extreme Vulnerability of Smallholder Farmers to Agricultural Risks and Climate Change in Madagascar.' *Philosophical Transactions of the Royal Society B: Biological Sciences* 369: 1639. doi.org/10.1098/rstb.2013.0089

Heemskerk, M., 2003. 'Self-Employment and Poverty Alleviation: Women's Work in Artisanal Gold Mines.' *Human Organization* 62(1): 62–73. doi.org/10.17730/humo.62.1.5pv74nj41xldexd8

Hinton, J.J., 2011. 'Gender Differentiated Impacts and Benefits of Artisanal Mining: Engendering Pathways out of Poverty: A Case Study of Katwe-Kabatooro Town Council, Uganda.' University of British Columbia (PhD thesis).

Hinton, J.J., M. Veiga and C. Beinhoff, 2003. 'Women, Mercury and Artisanal Gold Mining: Risk Communication and Mitigation.' *Journal des Physiques Archives IV France* 107: 617–20. doi.org/10.1051/jp4:20030379

Jenkins, K., 2014. 'Women, Mining and Development: An Emerging Research Agenda'. *The Extractive Industries and Society* 1(2): 329–39. doi.org/10.1016/j.exis.2014.08.004

KPMG, 2014. 'The Gobal Gems and Jewellery Industry—Vision 2015: Transforming for Growth.' A GJEPC–KPMG Report. The Gem & Jewellery Export Promotion Council.

Labonne, B., 1996. 'Artisanal Mining: An Economic Stepping Stone for Women.' *Natural Resources Forum* 20(2): 117–22 doi.org/10.1111/j.1477-8947.1996.tb00644.x

Lahiri-Dutt, K., 2004. 'Informality in Mineral Resource Management in Asia: Raising Questions Relating to Community Economies and Sustainable Development.' *Natural Resources Forum* 28(2): 123–32. doi.org/10.1111/j.1477-8947.2004.00079.x

Lahiri-Dutt, K., 2008. 'Digging to Survive: Women's Livelihoods in South Asia's Small Mines and Quarries.' *South Asian Survey* 15(2): 217–44. doi.org/10.1177/097152310801500204

Lahiri-Dutt, K., 2013. 'Bodies In/Out of Place: Hegemonic Masculinity and Kamins' Motherhood in Indian Coal Mines.' *South Asian History and Culture* 4(2): 213–29. doi.org/10.1080/19472498.2013.768846

Lahiri-Dutt, K., 2014. 'Extracting Peasants from the Fields: Rushing for a Livelihood.' Asia Institute Working Paper 216, National University of Singapore.

Levin, E., 2012. 'Mineral Sector Development.' Available at www.estell elevin.com/ourwork/mineral-sector-strategy/

Mahy, P., 2011. 'Sex Work and Livelihoods: Beyond the "Negative Impacts on Women" in Indonesian Mining.' In Kuntala Lahiri-Dutt (ed.), *Gendering the field: towards sustainable livelihoods for mining communities.* Canberra, ACT: ANU E Press.

Malpeli, K. C. and Chirico, P. G., 2013. 'The Influence of Geomorphology on the Role of Women at Artisanal and Small-Scale Mine Sites.' *Natural Resources Forum* 37(1): 43–54. doi.org/10.1111/1477-8947.12009

Meinzen-Dick, R.S., L.R. Brown, H.S. Feldstein and A.R. Quisumbing, 1997. 'Gender, Property Rights, and Natural Resources.' *World Development* 25(8): 1303–15. doi.org/10.1016/S0305-750X(97)00027-2

Mohanty, C.T., 1988. 'Under Western Eyes:Feminist Scholarship and Colonial Discourses.'*Feminist Review* 30: 61–8. doi.org/10.1057/fr.1988.42

Noetstaller, R. 1987. 'Small Scale Mining; A Review of the Issues.' World Bank Technical Paper 75. Washington DC: World Bank.

Organisation for Economic Co-operation and Development (OECD), 2013. *Fragile States 2013: Resource Flows and Trends in a Shifting World.* Washington, DC: OECD.

Panos Oral Testimony, 2014. *Pushed to the Edge.* Available at panos.org.uk/oral-testimonies/pushed-to-the-edge/

Shigley, J., A. Laurs, S. Elen, and D. Dirlam, 2010. 'Gem Localities of the 2000s.' *Gems and Gemology* 46(33): 188–216. doi.org/10.5741/GEMS.46.3.188

Slim, H. and P. Thomson, 1993. *Listening for a Change: Oral Testimony and Development*. London: Panos.

Tallichet, S., M. Redlin and R. Harris, 2003. 'What's a Woman to Do? Globalized Gender Inequality in Small-Scale Mining.' In G.M. Hilson (ed.), *The Socio-Economic Impacts of Artisanal and Small-Scale Mining in Developing Countries*. Netherlands: Swets & Zeitlinger B.V. Publishers.

Walsh, A., 2003. '"Hot Money" and Daring Consumption in a Northern Malagasy Sapphire-Mining Town.' *American Ethnologist* 30(2): 290–305. doi.org/10.1525/ae.2003.30.2.290

Walsh, A., 2012. 'After the Rush: Living with Uncertainty in a Malagasy Mining Town'. *Africa,* 82(2): 235–51. doi.org/10.1017/S0001972012000034

World Bank, 2011. *Vulnerability Risk and Adaptation to Climate Change Madagascar*. Washington, DC: World Bank.

World Bank, 2014. *Poverty and Equity Country Indicators World Bank*. Washington, DC: World Bank.

Yakovleva, N., 2007. 'Perspectives on Female Participation in Artisanal and Small-Scale Mining: A Case Study of Birim North District of Ghana.' *Resources Policy* 32(1–2): 29–41. doi.org/10.1016/j.resourpol.2007.03.002

9

Is it possible to integrate health and safety risk management into mechanised gold processing? A methodology for artisanal and small-scale mining communities in the Philippines

Gernelyn Logrosa, Maureen Hassall, David Cliff
and Carmel Bofinger

The large number of people involved in artisanal and small-scale mining (ASM) worldwide face significant health and safety issues that have an impact beyond the miners and their communities. In the Philippines alone, ASM sustains more than 500,000 miners and processors (Galvez 2012, cited in Verbrugge 2014). Its immense contribution—supplying 80 per cent of the country's gold supply, which earned PHP48.9 billion in 2010 (Mines and Geosciences Bureau 2016)—proves that ASM is thriving and is worthy of support.

The diversity within artisanal gold-mining is recognised by the Swiss Agency for Development and Cooperation (Hruschka 2011), and can range from formal and responsible ASM communities as pillars of the local economy to chaotic and uncontrollable mining sites, where negative impacts prevail. Four categories of ASM were recognised for purposeful macroeconomic and sectoral approaches by the World Bank (Weber-Fahr 2002): permanent ASM, seasonal ASM, rush-type ASM and shock-push ASM. These categories are in some way reflected in the primary gold-mining provinces in the Philippines—Benguet and Compostela Valley—which are discussed in the following sections.

In Benguet, ASM developed from the previous workings of the indigenous Igorot small-scale miners who originally worked the area. It is a classic example of how ASM in the Philippines had thrived long before the arrival of large-scale mines (Bugnosen 2002). Artisanal mining operations in Benguet are perceived as old and traditional in terms of mode of tunnelling and gold extraction, as compared to other ASM areas of the Philippines. The cyanidation process of gold extraction is used more widely by the small and corporate small-scale mining associations than amalgamation using mercury. In Benguet, the Ibaloi and Kan-kanaey miners are known as *abanteros*. Having inherited their skills from their ancestors, their knowledge of gold-mining has been shaped by years of experience. Some migrant miners from the lowlands have also learned these skills. Altogether, this hierarchy of non-industrialised miners form a community of ASM, without distinction between miners who use the traditional method and miners who employ mechanised and more advanced technologies. The provisions of the Philippine Presidential Decree No. 1899 allow corporations to also engage in small-scale mining (Alternative Forum for Research in Mindanao (AFRIM) 2012).

Traditional mining in Benguet is typically seasonal. The Kan-kanaey tribes in Benguet treat mining as a family-based activity, performed during agricultural off-seasons. The ore-processing stage is performed usually by the women, while the rest of the stages, such as ore extraction, milling, gravitation and panning to separate gold from the ore are performed by the male members of the family. Until 1996, the tools used were simple, such as iron chisels, double-sided iron hammers with a wooden handle, iron crowbars, iron shovels and battery-operated lamps. Initial breakage of the ores is achieved using improvised crushers. After this, grinding proceeds, as the ores are loaded into rod mills or ball mills. This is followed by smelting and processing wherein women do most of the work (Caballero 1996 as cited in Lu 2012). Households are reformed into workplaces as processing of the ore is made to fit into the household responsibilities. Inevitably, of course, the inherent risks from these typical working conditions emerge, as other members of the family are exposed to the hazards of mineral processing (Bugnosen 1998).

A different outlook is provided by the ASM community in Compostela Valley Province, also known as 'ComVal'. ComVal typifies a gold rush–type ASM practice that has sustained itself for more than 30 years. The brief settlement history of the ComVal uplands may offer a useful perspective in understanding its ASM context. Throughout the 1950s and 1960s, expansion of corporate logging and mining brought with it to ComVal the massive influx of migrant workers from Visayas and Luzon, which are located in the central and northern part of the Philippines. The gold rush in the 1970s and 1980s expedited the utilisation of both the migrant workforce and the unskilled 'indigenous labour' in large-scale companies, such as APEX mining in the municipality of Maco, and SABENA mines in New Bataan. However, the gold rush was short-lived. The downturn of large-scale mining in the 1980s forced the impetus towards the expansion and transformation of ASM through the creation of a skilled workforce reserve.

In terms of working practices and degrees of mechanisation and capitalisation, Verbrugge (2014) classified ASM in ComVal into three pertinent categories. The first category is river panning, which is an alluvial mining technique that captures coarse gold, in some cases gold nuggets, using a simple pan or sluice box. The miners rely on natural conditions, such as the passage of heavy rain, which tends to loosen gold-bearing dirt. The second category is self-financed tunnelling operations, in which a 'corpo' is formed by the miners and resources are pooled among themselves. The third category, which is supposedly the defining feature of ASM in ComVal, is the bigger tunnelling operations. Here, the operations are characterised by a much higher degree of mechanisation and the employment of heavy machinery, such as pneumatic drills, excavators, explosives, 2-MW diesel generators, water pumps and mine carts. Consequently, the degree of labour specialisation also increases, providing employment no longer limited to the usual portal guards, carpenters and electricians. Some cases reveal the hiring of chemists and geologists. A fourth category, which is not highlighted in Verbrugge's account but is still worth noting, is the 'banlas' or hydraulic mining, which uses high-pressure jets of water to dislodge rock material or loosen packed gold-bearing sediments.

The role of the state

It has been institutionally recognised that there are drivers that push ASM to continue performing their gold-processing techniques, despite the risks that have been largely discussed in literature, in terms of social and regulatory dimensions (Hentschel et al. 2002; Lahiri-Dutt 2004; Buxton 2013; Hilson et al. 2014). The support from government, local and foreign organisations and other institutions has, in part, been directed at finding technological solutions to ASM-related problems (Priester et al. 1993; Mutagwaba et al. 1997; Hentschel et al. 2002; Hruschka 2011). The most common example is the end-of-pipe technology, in forms of retort or improvised filters to address toxic mercury emissions. This initiative of solving ASM problems related to mercury through the use of technology is a model of technical solutions to reduce environmental impacts. There are also other modifications of the mineral-processing circuits to mitigate unwanted social and environmental impacts (Priester et al. 1993).

The mechanisation of gold processing holds promise in its potential to achieve cleaner and toxin-free ASM practices. However, Hruschka (2011) argues that mechanisation does not replace labour in ASM gold processing. The mechanisation today in ASM usually starts at a basic level, and its primary purpose is to ease physical work conditions and to increase productivity. Labour remains the second-most important production factor next to mineral deposits in ASM. As a result, ASM produces 'only' about 15 per cent of the worldwide gold production, but employs more than 80 per cent of the workforce (Hruschka 2011).

Scholars have so far focused on how technology use has an impact on labour, its regulation, on women's rights and informality (for instance, Nuwayhid 2004; Hermanus 2007; Marriott 2008; Lu 2012; Verbrugge and Besmanos 2016). However, there seems to be little discussion about health and safety and, in particular, there is no examination of how health and safety risk management approach could be used to improve ASM methods for processing gold.

Health and safety project

This study aims to investigate the health and safety conditions in ASM, and to integrate health and safety risk management into mechanised gold processing, through a gold-processing plant recently designed for ASM in the Philippines. It was a collaborative effort, initially made by a state university and a government department to improve ASM gold processing in the country. The technology offers the customary set of principles in gold processing. Innovative mechanisation is highlighted by putting in equipment and methods that stand as a contemporary and creative approach to the recent technical demands in ASM gold processing. The project seeks to develop a plant that overcomes the challenges associated with the two primary ASM gold-processing methods—cyanidation and amalgamation—which cause adverse effects on health and the environment. Another consideration is that ASM gold-processing sites may lack adequate facilities for detoxification treatment, thus compounding the health and environmental risks they cause. ASM gold-processing sites often lack adequate facilities for detoxification treatment, thus compounding the health and environmental risks they cause. Another consideration is the inefficiency of separation and recovery of gold using these processes that leads to major inefficiencies in material utilisation. Once this technical challenge is addressed, the economic opportunities can be enhanced.

Technical demands of the small mines industry have been becoming increasingly dynamic and complex. This is mainly due to the varying mineralogy and metallurgical characteristics of gold ore. Thus, mechanisation has been developed over time. The introduction of mechanisation to ASM was not limited to external influence, such as government projects (Mutagwaba et al. 1997) and foreign aid. In order to meet local conditions, artisanal and traditional techniques have been modified, even through unconventional channels (Priester et al. 1993; Hentschel et al. 2002). This modification was a definitive task among ingenious ASM miners or, as Rawls et al. (1999) described it, enterprising amateurs who characterised the gold rush technology.

This contemporary state of ASM working conditions may include more modern occupational health and safety hazards that may arise from the new machinery. Hentschel et al. (2002) discussed how frequently ASM miners try to modify the conventional equipment they have in order to fit their needs. Unfortunately, in many cases, safety features are suppressed; for example, in the water supply for drill hammers. Nevertheless, a surprisingly large variety of technical solutions to gold processing for ASM have been developed for their local activity.

The risks to health and safety from the gold-processing operations can be potentially increased if mechanisation is applied in the absence of complementary safety measures. In order to solve this, projects, machines, and equipment should be designed according to explicit health and safety considerations. Government and donor organisations should serve as a model of safe mechanisation by consciously integrating risk management into their projects. Good designs have provisions to easily identify hazards and risks early on. This can result in reliable predictions and proactive controls of the gold-processing technology, which adds value to the ASM's inherent ingenuity.

The 'risk' framework

Standards Australia/Standards New Zealand Committee (2009) suggests that the risk management model should ensure the integration and sustainability of health and safety. Such integration is prescribed by the international standard for risk, ISO 31000, as shown in Figure 9.1. The complex interactions between different risks, unique workplace conditions and recent evolution of mechanised technology together establish the context: the first step of the risk management process. This complexity requires that a range of different approaches must be employed to identify the risks. The next step is risk identification. In the case study research, the risk identification element involved collecting data in the form of fieldnotes, audio recordings, transcripts and surveys. Another data source included a set of on-site video recordings. From the data, which covered broad to specific issues of concern, risks were identified and categorised. After this, the risk analysis stage proceeded in agreement with the sequence of elements in the process, as shown in Figure 9.1.

In the risk analysis step, the data collected is analysed to rank the risks based on the severity and frequency of risks. These are prioritised in terms of seriousness and relevance to stakeholders. The risk analysis and evaluation process utilises the risk identification data to holistically inform the introduction and implementation of the proposed gold-processing plant. Risk evaluation is then performed to identify the more serious risks that need risk treatment. Therefore, the output of the risk assessment steps is developed by collating the health, safety and other risks that need to be considered and managed. The risk treatment element of the process determines how these risks might be managed by suggesting controls that can prevent, mitigate or manage the risks. The output from the risk management process can then be incorporated into a project to help ensure successful implementation and sustainable operation of innovative gold-processing plants by the miners in different Philippine contexts. In doing so, the risk-based model should identify how the plant will impact the ASM miners' health and safety, and inform stakeholders how the risks can be reduced and managed accordingly. The model should also highlight if there are risks that the plant may not directly address or may exacerbate, so these can be addressed as well.

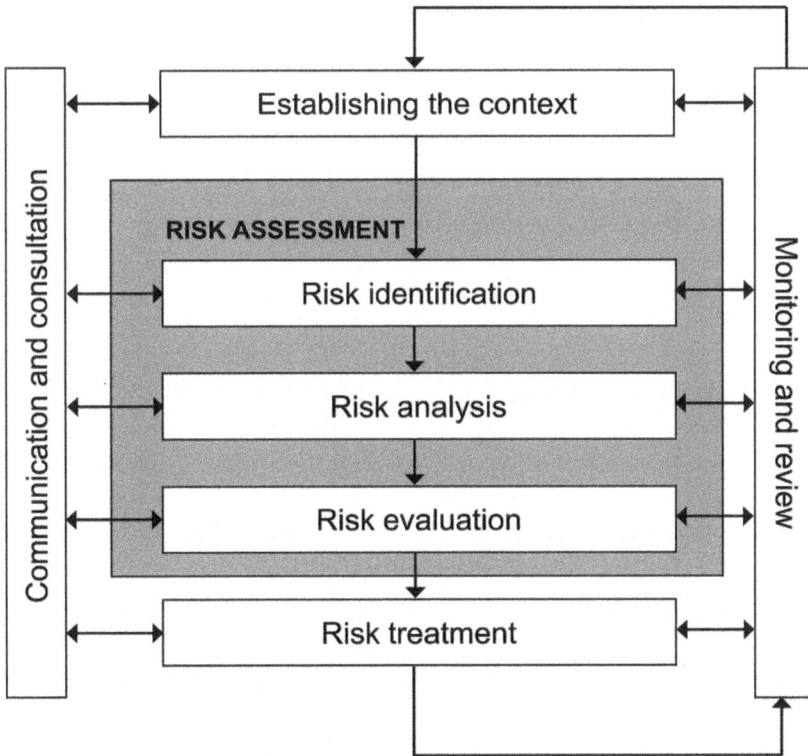

Figure 9.1: The risk management process
Source: AS/NZS ISO 31000:2009.

Methodology trialled

To understand the risks and health and safety considerations associated with introducing mechanisation into ASM, it is necessary to grasp the true ASM context and to understand stakeholders' views about the impacts of the proposed project. In certain cases, it is only through immersion into a community that the full impact and contextual relevance of projects can be understood. This risk-based immersion methodology was used to gain insights into the range of impacts that mechanisation can have on the ASM communities in the Philippines.

Risk identification using field interviews was conducted among representatives from the stakeholder groups. Semi-structured interviews and participant observation were exploratory in nature and aimed to provide baseline information about risks, especially health and safety considerations in gold processing in ASM communities. For the purposes of the case study, semi-structured interviews were conducted in four stakeholder groups, including the technical group (technology providers and other technical experts, external to the processing plant project), the government group (national and local government units), the social group (non-government organisations, civil society organisations, community staff of the processing plant project) and the ASM gold group.

The interview part of the data collection was focused on understanding the Philippine ASM. The interview questions were designed to allow the stakeholders to discuss their work and their perceptions, and to collect information about health and safety. Information was gathered regarding stakeholder perceptions of the impacts that the proposed gold-processing plant might have on the health, safety and other risks associated with ASM activities. These helped reveal the level of acceptance of the proposed technology amongst relevant stakeholders, who will be directly affected by the changes brought about by its actual implementation. The interview questions focused not just on health and safety issues, but also on the equally important dimensions of environment, social and institutional policy and regulation, and governance.

Another important aspect of the risk-based immersion methodology were the insights that resulted from the on-site field observations of the current system of gold processing in different Philippine contexts, which included the two major ASM provinces of the country—namely, Benguet and Compostela Valley. ASM miners were observed going about normal work activities in their natural work settings. These observations were conducted to collect information on the actual conditions at the ASM gold-processing sites, and about the miners' typical workday, including their processes, activities and the equipment they use. Risks and hazards were identified in the working conditions, equipment used and techniques performed. Observations were also made to include the manual and mechanised techniques the ASM miners used to process their gold. A comparison could then be made with the risk perceptions collected, and it helped to understand the details that should not be overlooked during hazard identification by the people designing and implementing mechanisation within ASM contexts.

Observations, including manual and mechanised techniques in the current gold processing of the ASM miners were recorded using a video camera for an hour. This was done to capture as many details as possible about a typical working day in an ASM community, within time constraints, in order to make the hazard identification step effective. Video recording was required for thorough analysis of the entire workplace environment, and it was also a practical way for the researcher to have primary data records to review as often as needed during the data analysis stage of the research. Fieldnotes were also taken to complement video recordings as much as possible.

Key lessons from fieldwork

Important understandings of strengths and limitations emerged when the research methodology was put into practice. Their analysis offers reflective insights in confirming whether the research goals were adequately achieved at this stage of the research study. Additionally, this analysis provides insights into whether it is possible to integrate health and safety risk management into mechanised gold-processing projects.

Strengths

The interview process predominantly used open-ended questions. Initial questions about risk were framed in a general way, so that the participant could openly express which among the risks relating to artisanal and small-scale gold processing was most relevant to them. The open-ended questioning method encouraged participants to raise diversely prolific data relating to the range of threats, opportunities and risks relevant to each stakeholder group. As Creswell (2003) stated on socially constructed knowledge claims, broadly designed questions allow participants to construct the meaning of a situation. Such meaning is a product of actual experiences and a series of interactions. Creswell argues that open-ended interviews are better as they can focus on specific contexts in which people live and work, in order to understand the historical and cultural settings of the participants.

Employing open-ended questions also allowed the participants to express their expectations of the innovative gold-processing plant project. Discussions regarding stakeholder expectations revealed their experiences and perspectives regarding the interventions that attempted to affect change. This created an inclusive space, especially for the ASM miners, in a manner that provided a sense of ownership. For instance, one miner's association in Benguet attributed the reduction in mercury use among its members to a synergy of local efforts and external collaboration. External support alone often furthers the cause only to a certain extent. It was interesting to know that there have been many organisations and government projects that have visited these communities through the years, with similar environmental objectives. Thus, two separate insights were derived: either the environmental outcomes are still yet to be fully realised in the area, or there is simply unnecessary duplication of projects and strategies.

Video recording facilitated an in-depth analysis of health and safety hazards and risks among the artisanal and small-scale mining and processing sites visited. The video recordings also show examples of how risks to health and safety from the gold-processing operations can be potentially increased if mechanisation is applied in the absence of complementary safety measures. It was observed and recorded on video that there was locally fabricated machinery, which was operated to some locally determined standards. It was quite evident how the safety features were suppressed in some cases. With the lack of capital and appropriate technical knowhow, the ASM miners and their families remain exposed to risks in such hazardous working conditions.

There were three sources of mechanised technology apparent in the ASM provinces, captured in the participant observation. First was the application of local ingenuity in order to meet local conditions. The miners find a way amongst themselves and within their community to improve their practices. This is more culturally influenced and can remain within the bounds of traditional practices. This confirms the observations cited from literature (Priester et al. 1993; Hentschel et al. 2002). Another source was the financier's scheme, which can be driven to improve efficiency in order to enhance profit. Financiers support the ASM through the provision of mechanised equipment, under a shared profit scheme. This is a similar arrangement to that found with the small-scale miners in Ghana, where foreign businessmen assist the miners technically and financially in the form of mine support services (Hilson et al. 2014). Finally, there are external sources, such as government projects and foreign investments, which can be driven by both efficiency and risk reduction, especially when they are initiated by government institutions.

The interviews generated rich data. However, it was not always easy to gauge stakeholder perceptions of the significance and prioritisation of the issues and risks, which are important information for those that need to make informed decisions about the development of context-specific technical solutions. It was beneficial to supplement the interview information with a survey instrument that collected participant perceptions on the significance and ranking of different types of risks. To do this, a conjoint rating and categorisation instrument was created. The rating component comprised a six-point scale, where 1 = lowest risk priority and/or slightly relevant to the participant's stakeholder group and 6 = highest risk priority and/or highly relevant to the participant's stakeholder group. The categorisation component allowed the participant to provide feedback for each of the different risks: health and safety, technical, social, environmental and regulatory. The use of two different methods—interview and survey—was beneficial because it helped to neutralise limitations or biases that have been recognised in literature (Creswell 2003). For example, the results from one method can help develop or inform the other method (Greene et al. 1989). Alternatively, one method can be nested within another method to provide insight into different levels or units of analysis (Tashakkori and Teddlie 1998).

Limitations

The responses reveal an overlap associated with perceptions of the different risks (technical, social, environmental, health and safety and regulatory). There seemed to be a lack of clear distinction of each type of risk. For example, with respect to health and safety risks and environmental risks, participants seem to use these terms interchangeably. Similar responses indicating this lack of distinction were obtained from among the ASM miners, the government, partners and even some technical persons. Others use the word 'safety' even if they actually mean security issues arising from conflict. This mismatch of perceptions and expectations among stakeholder groups can introduce confounding findings within results. To overcome this limitation and because the research objectives were focused on health and safety, probing keywords were used, such as sickness, injuries and accidents, to delineate the blurred lines between health and safety and the other different types of risk.

Another limitation that was encountered was the finding that some of the government agencies and relevant stakeholder groups have limited knowledge about the project. Although this finding may be unexpected, it is fairly reasonable, given that the gold-processing plant project was still in its pre-construction stage when the fieldwork was conducted (2015). The limited information at this stage of the project can be acknowledged as a barrier to collecting sufficient data to support claims regarding the project implementation. However, some respondents were conservative in giving information about the project. Reluctance and ambivalence were common responses from stakeholders due to a degree of incompleteness. Indeed, the design and construction stages may still represent the project in its early state. However early, this stage should not be regarded as a premature time to discuss the appropriate framework to explore potential and perceived risks.

Conclusion

ASM is adopting mechanised technology in order to improve efficiency of gold processing and reduce environmental risks. In a given context, the exact purpose is driven by the source of the mechanised technology, which includes local ingenuity, financers' schemes and external sources. With the ongoing efforts of the government to improve the efficiency and mitigate environmental impacts through projects involving mechanised technology, there is an opportunity to formally include health and safety considerations.

Based on the experiences and lessons from the recent fieldwork in two Philippine ASM provinces, important health and safety risks were identified and recognised among the artisanal and small-scale mining and processing sites visited, and these risks should be addressed. However, without appropriate technical and risk management know-how, the miners and their families remain exposed to the risks associated with hazardous ASM working conditions. The risks to health and safety from the gold-processing operations can be potentially increased if mechanisation is applied in the absence of explicit health and safety considerations. There is a lack of clear distinction between environmental risks and health and safety risks, as though these two domains can be used interchangeably. This can result in some unwanted consequences due to mismatch of perceptions and expectations.

To successfully implement sustainable and effective mechanised technology in ASM contexts, it is necessary to understand the range of risks and perceptions from different stakeholder perspectives. One way to understand risk is by immersion-based research. As shown with the case study conducted in the ASM provinces of Benguet and Compostela, rich data on risk and perceptions can be collected from immersion-based research. The data was cross-examined and analysed to identify common themes. These themes can be used to form a risk management framework that helps people make informed decisions on how technical solutions can be best integrated within ASM contexts to minimise health, safety and other risks that can adversely impact the successful implementation and ongoing sustainability of a technology.

References

Alternate Forum for Research in Mindanao, Inc. (AFRIM), 2012. *A Background Study on the Small-Scale Gold Mining Operations in Benguet and South Cotabato and their Impact on the Economy, the Environment and the Community*. Quezon City, Philippines: Bantay Kita/Action for Economic Reforms.

Bugnosen, E., 1998. *A Preliminary Assessment of Small-Scale Mining Legislation and Regulatory Frameworks*. UK: Department of International Development, Intermediate Technology Development Group.

Bugnosen, E., 2002. *Country Case Study on Artisanal and Small-Scale Mining: Philippines*. England: International Institute for Environment and Development.

Buxton, A., 2013. 'Responding to the Challenge of Artisanal and Small-Scale Mining: How can Knowledge Networks Help?' *IIED Sustainable Markets Papers*. London: International Institute for Environment and Development.

Creswell, J.W., 2003. *Research Design: Qualitative, Quantitative, and Mixed Methods Approaches*. 2nd edition. Thousand Oaks: Sage Publications.

Greene, J.C., V.J. Caracelli and W.F. Graham, 1989. 'Toward a Conceptual Framework for Mixed-Method Evaluation Designs.' *Educational Evaluation and Policy Analysis* 11(3): 255–74. doi.org/10.3102/01623737011003255

Hentschel, T., F. Hruschka and M. Priester, 2002. *Global Report on Artisanal & Small-Scale Mining*. Minerals Mining and Sustainable Development, no. 70. International Institute for Environment and Development, World Business Council for Sustainable Development.

Hermanus, M.A., 2007. 'Occupational Health and Safety in Mining Status, New Developments, and Concerns.' *Journal of The South African Institute of Mining and Metallurgy* 107(8): 531–8.

Hilson, G., A. Hilson and E. Adu-Darko, 2014. 'Chinese Participation in Ghana's Informal Gold Mining Economy: Drivers, Implications and Clarifications.' *Journal of Rural Studies* 34: 292–303. doi.org/10.1016/j.jrurstud.2014.03.001

Hruschka, F., 2011. *Swiss Agency for Development and Cooperation (SDC) Experiences with Formalization and Responsible Environmental Practices in Artisanal and Small-Scale Gold Mining in Latin America and Asia (Mongolia)*. Available at doc.rero.ch/record/30885/files/05-GoldMining_LatinAmericaAndAsia.pdf

Lahiri-Dutt, K., 2004. 'Informality in Mineral Resource Management in Asia: Raising Questions Relating to Community Economies and Sustainable Development.' *Natural Resources Forum* 28(2): 123–32. doi.org/10.1111/j.1477-8947.2004.00079.x

Lu, J., 2012. 'Occupational Health and Safety in Small-Scale Mining: Focus on Women Workers in the Philippines.' *Journal of International Women's Studies* 13(3): 103–13.

Marriott, A., 2008. 'Extending Health and Safety Protection to Informal Workers: An Analysis of Small Scale Mining in KwaZulu-Natal.' Research Report No. 76. Durban: School of Development Studies, University of KwaZulu Natal.

Mines and Geosciences Bureau, 2016. 'Mining Industry Statistics.' 5 August. Available at www.mgb.gov.ph/images/MIS_1997-present-5 Aug16.pdf

Mutagwaba, W., R. Mwaipopo-Ako and A. Mlaki, 1997. 'The Impact of Technology on Poverty Alleviation: The Case of Artisanal Mining in Tanzania.' Research Report No 97.2. Dar es Salaam: Inter Press Tanzania Ltd.

Nuwayhid, I.A., 2004. 'Occupational Health Research in Developing Countries: A Partner for Social Justice.' *American Journal of Public Health* 94(11): 1916–21. doi.org/10.2105/AJPH.94.11.1916

Priester, M., T. Hentschel and B. Benthin, 1993. *Tools for Mining: Techniques and Processes for Small-Scale Mining*. Germany: Friedr. Vieweg & Sohn Verlagsgesellschaft mbH (Informatica International, Incorporated).

Rawls, J.J., R.J. Orsi and M. Smith-Baranzini, 1999. *A Golden State: Mining and Economic Development in Gold Rush California*. Berkeley: University of California Press.

Standards Australia/Standards New Zealand Committee, 2009. 'AS/NZS ISO 31000:2009 Risk Management—Principles and Guidelines.' November.

Tashakkori, A. and C. Teddlie, 1998. *Mixed Methodology: Combining Qualitative and Quantitative Approaches*. Applied Social Research Methods Series (Vol. 46). Thousand Oaks, California: Sage Publicaitons.

Verbrugge, B., 2014. 'Capital Interests: A Historical Analysis of the Transformation of Small-Scale Gold Mining in Compostela Valley Province, Southern Philippines.' *The Extractive Industries and Society* 1(1): 86–95. doi.org/10.1016/j.exis.2014.01.004

Verbrugge, B. and B. Besmanos, 2016. 'Formalizing Artisanal and Small-Scale Mining: Whither the Workforce?' *Resources Policy* 47: 134–41. doi.org/10.1016/j.resourpol.2016.01.008

Weber-Fahr, M., 2002. *Treasure or Trouble? Mining in Developing Countries*. Washington, DC: World Bank and International Finance Corporation.

10

Resources and resourcefulness: Gender, human rights and resilience in artisanal mining towns of eastern Congo

Rachel Perks, Jocelyn Kelly, Stacie Constantian
and Phuong Pham

Few things evoke reactions as passionate as issues surrounding gender, conflict and mining in the eastern Democratic Republic of Congo (DRC). At once reviled by international advocacy organisations and celebrated by local communities, mining is viewed as both the scourge and the saviour of a region wrecked by decades of violence. Studies have reported on human rights as well as on the status of women in the DRC, and although some examine the link between mining and sex-based violence, little research explores the gender dimensions of artisanal and small-scale mining (ASM). The research in this chapter was framed by questions such as: Do men and women face similar difficulties when seeking to gain employment in mining? Are they afforded similar opportunities once they have secured access into ASM? What are the most prevalent social, economic and health impacts experienced by individuals. Are these impacts gendered? A human rights–based approach informed the range of issues examined, such as gender, militarisation of the extraction process and free and equal participation in political, judicial and economic systems. By speaking with a wide variety of actors who live and work within these communities, we attempted to identify issues that are common to mining-affected areas. The experiences of both women and men were examined, but a particular focus remained on understanding women's experiences in mining towns. Hence, the research was ultimately guided by the hypothesis that by understanding issues related to safety, security and economic opportunities for women, significant gains in both economic and social development in the eastern DRC could be achieved.

The Harvard Humanitarian Initiative in collaboration with the World Bank conducted two phases of research: Phase 1 (qualitative), followed by Phase 2 (quantitative). Research was conducted in North and South Kivu provinces, two of the most conflict-affected areas of the DRC. Phase 1 took place in 2012, with the aim to elicit experiences of male and female miners, using a human rights framework of inquiry developed specifically for the project. Five key findings resulted from the qualitative research in Phase 1, which then informed the quantitative, population-based survey in Phase 2. These focused on gender-based violence, discrimination, right to health, access to justice and right to participation. Although women were vital actors in mining communities and filled many roles, they were also among the most vulnerable to sexual and economic predation. Sexual exploitation was described as commonplace in mining towns, and many women spoke about engaging in transactional sex out of desperation. It is important to acknowledge the practice of discrimination, because mining jobs are theoretically open to everyone; however, actually acquiring work was dictated by one's ability to pay for access to the mines. Women and other vulnerable populations were generally relegated to marginal support roles in mining towns. Right to health is important because is it crucial to a woman's well-being. Health problems were summarised into three categories: poor labour conditions; poor structure of mining tunnels; and public health problems, such as poor hygiene and high levels of infectious diseases. For the majority of participants, access to justice was a demand arising from the sufferance of human rights violations. Those most vulnerable to human rights abuses, such as orphans, widows, sex workers and displaced persons, were also those least able to access traditional and formal justice mechanisms. Right to participation in a situation of highly restricted access to political participation and widespread discrimination was crucial.

Phase 2 used a quantitative survey to examine the scope of human rights issues emerging from Phase 1 in select mining towns. The survey results affirmed the human rights issues raised in Phase 1, and provided some indications of the frequencies of experience within the total sampled mining populations.[1] The survey was a cross-sectional study from three territories (Kalehe, Mwenga and Walungu) in the South Kivu Province of the eastern DRC. Sites were sampled from a comprehensive list of artisanal mining sites, compiled by the International Peace Information Service. This list documented 800 mining sites and 85 trading centres, with information about armed groups' presence and involvement, and the scale of the mining activity in 2012 and 2013. A total of 998 surveys were collected for this project: 357 individuals were female and 641 individuals were male. The next section outlines selected research findings, combining both qualitative and quantitative data from Phases 1 and 2.

Women, mining and conflict in the DRC

ASM and its relationship to armed conflict in the eastern provinces of the DRC has been has been vividly described, researched and discussed. The wars in the mid-1990s accelerated the disintegration of agricultural economies that had begun during structural adjustment in the early 1980s. Active hostilities stretched into decades of insecurity, and the resulting displacement, crop viruses, threat of violence and danger of looting and predation by armed groups shaped entirely new economic and social systems. ASM assumed a more prominent role in rural economic life. In the DRC, some authors have written of the causes and consequences of 'deagrarianisation'—the transition away from agriculture towards other rural livelihoods—with respect to the proliferation of ASM (Perks 2011; Smith 2011; Geenen 2012; Kelly 2014). Important to note from these prior research findings is that the manner of transformation of the rural economic landscape in the DRC, and the role played by mining in this transformation, mirrors the experiences of other mineral-rich environments. Also important to highlight are the varying degrees of political stability under which 'deagrarianisation' and ASM proliferated in these environments: from politically stable Ghana (Banchirigah 2008), Burkina Faso (Luning 2008) and Tanzania (Bryceson and Jønsson 2010), to outright civil destabilisation in Sierra Leone (Maconachie and Binns 2010).[2] Such varied political landscapes under which ASM has proliferated

---

1   For purposes of brevity, this chapter concentrates on research results that inform the first three findings. Those interested in seeing the full results can refer to the full report from the World Bank (2015a), titled 'Resources and Resourcefulness: Gender, Conflict, and Artisanal Mining Communities in Eastern Democratic Republic of the Congo'.

2   For an overview, see Hilson (2011).

across the subcontinent in the last 30 years beg the question for the DRC as to the extent to which the proliferation of ASM can be best, or perhaps solely, understood from a prism of conflict. Such a reframing of our understanding of ASM might further lead to new methods of inquiry with respect to the social, environmental and economic externalities associated with ASM today. Are human rights issues associated with ASM in the eastern DRC best understood as a product of instability and armed conflict? Or is there a more complex landscape of contributing factors? These are important issues to raise at the outset, for the framing of the cause and consequences of ASM in the eastern DRC has often been associated most vividly with conflict, despite, as suggested, evidence from other environments in sub-Saharan Africa that shows ASM proliferation to have significant roots tied to general economic decline in rural areas since the 1980s, and poor government regulation of ASM equally since that period.

Unpacking myths and exploring realities of human rights in the mines

The nature of ASM in the DRC, given its illegality and the diversity of jobs and types of minerals extracted, has been historically difficult for many in the international community to comprehend, resulting in a wide range of potentially destructive narratives being applied to a remarkably diverse and vibrant local economy.

A prevalent narrative surrounding ASM in the eastern DRC is that the wealth generated from mining fuels the ongoing conflict in the eastern DRC and, as a result, exacerbates the conflict-related abuse of women. Advocacy narratives have named so-called 'conflict minerals' as a main driver of insecurity, and linked it with rape as a weapon of war. However, several academic pieces have refuted this simplistic narrative as follows: this prominent storyline offers an over-simplified version of reality, in which non-state or rebel armies exercise control over mines, and every worker is essentially subject to slavery. Coercion is typified as the manner in which women end up in mines or mining towns; thus, much of the employment is coerced or involuntary. Several scholars have called into question this narrative's simplicity (Geenen and Custers 2010; Autesserre 2012; Geenen 2012; Seay 2012; Bashwira et al. 2014; Perks 2013; Spittaels and Hilgert 2013), the results of which are discussed in the next section.

Results: Why people stopped farming

The fear of sexual violence during conflict was specifically cited as a reason why people ceased farming and entered mines instead. Women in the mining town of Nzibera described how rape in particular affected farming: 'If people went to their farm, they can get raped or even gang raped, because we think if we say "no," they will kill us'. A young woman in Nzibera explained, 'There is lots of hunger here. But if you have been raped on your farm, you will never return, and the farm will just die'.

In Nyabibwe, women described insecurity and violence as an ongoing deterrent to resuming agricultural activities:

> During the war things changed, if you tried to go to the farm they would rape you, take your things, the biggest consequence of war is that it locked people in, there was no freedom to move to different areas to sell your goods or to look for work … After the big war still things remain difficult, people tell you there is peace and that it is OK to go to the farm, but when you go, if they don't kill you they will rape you.

Motivations for entering the mines

Phase 1 results revealed a general decline in the viability of agriculture as a sole source of income due to the conflict and general insecurity following the war. These factors seemed to frame the historical motivation for entry into mining, particularly with older respondents. The quantitative survey examined the drivers of migration to mining towns in more detail. Mining populations participating in the survey were highly migratory: here, 48 per cent of women and 38 per cent of men stated that the town they currently worked in was not their place of origin (n = 412). Women's rate of migration was statistically higher than men's (at $p < 0.05$). Individuals who identified themselves as migrants were asked why they sought work in mining towns. People were asked about their reasons for moving to mining towns, and were allowed to select more than one option. The results show that 55 per cent of respondents stated that there was no work in their communities of origin; almost as many stated that they sought work in mining because there were no services in the areas from where they came. Women were more likely than men to cite lack of money or employment as a driver to migrate, whereas men were more likely to cite lack of services as a motivation. One-third of women and men stated that lack of food drove them to migrate to mining towns.

Violence and insecurity were less commonly cited as drivers to migrate than economic reasons—15.7 per cent of respondents stated these were the factors in their decision-making, with relatively little difference between men and women. One of the least cited reasons for migration was direct displacement by armed groups. These results suggest that the conflict may have created the conditions by which mining became an appealing industry, but direct conflict-related considerations were less important than economic drivers.

Framing the results along the human rights framework

Sexual and gender-based violence: The blurred lines between sex work and sexual exploitation

Many women spoke about engaging in transactional sex out of desperation—a profession that placed them at an increased risk of experiencing rape and other forms of abuse. Sex workers were often migratory and without social or financial support. Sexual predation by armed men was also described as a concern, although it was generally less pressing than the everyday violence and abuse that women suffered as a result of living in mining towns.

The qualitative research illustrated how running a restaurant, engaging in sex work and transporting materials are by no means mutually exclusive roles. Instead, one leads to, or requires, doing another. Many women spoke about having to transact sex simply to survive. Often, this is because they do not have a husband or other male relatives to bring an income home, and are forced to fend for themselves. They went on to describe how trading sexual favours was a prerequisite for gaining access to small jobs in mining towns:

> People also force women to have sex with them by saying that if you don't sleep with me you won't get to keep carrying bags ... Women do prostitution so that they can get other work. You are selling yourself, tiring yourself to get some money for your children.

Women described the close links between economic and sexual exploitation: how they aren't paid or are underpaid for the work they do; are sexually harassed or raped while working; are physically beaten; and are forced to perform sexual favours in order to get clients or employment.

Those women who speak out about their exploitation are forced out of the area under threat. As a young sex worker in Nyabibwe said simply, 'If you refuse them (sexually), they will tell you if you return again to the area they will kill you'.

All actors interviewed in the qualitative phase of this project emphasised that sex work was widespread and commonplace in mining contexts. As one man said, 'The *soko* [market] of the prostitutes is the mining quarry'. Exchanging sex for goods is one of the only ways women can provide for themselves and their families. Poverty was described as a defining driver of prostitution. Even wives of miners spoke frankly about the temptations for their husbands: 'Yes—it [prostitution] exists because of poverty. Around the mines, women prostitute themselves to get money that the miners give them and because otherwise they wouldn't have anything to eat'.

In many sites, sex workers came from other areas and had to build a life without the benefit of social networks or peer support in their new environment. Women may also migrate from mining town to mining town, looking for a better situation. Miners emphasised the transitory nature of the work, saying, 'There are also many prostitutes from all over, Uvira, Bukavu, they here when there is action here. They call their friends, everyone comes, we don't even know their faces'.

Women engaging in sex work may face stigma and social rejection in their home communities when they return, particularly if they have children out of wedlock. This means that once a woman begins to trade sex in mining towns, it can be difficult for her to leave this line of work. Women may find themselves caught in a cycle of travelling to look for work in places with the most productive mines, where men have disposable income.

The 'choice' to engage in sex work is made against a backdrop of desperation, violence and coercion. This reality means that respondents in the research saw sex work and sexual violence as closely linked. Participants described sexual violence as forcing a woman to have sex against her will, but they also described other forms of this abuse, including men refusing to pay for sexual services. As one miner described, 'The meaning of rape is to take someone by force and even if you haven't already agreed to do it you make them have sex with you, this is a common thing in our area'. Women transporters reinforced this, saying, 'When a man is drunk he can also rape a woman without hav[ing] a conversation, and even if it looks like prostitution, it is rape. This happens a lot of time to prostitutes'. One young miner simply stated, 'Ya kila qualite na aina iko hapa nyabibwe [There is sexual violence of every quality and kind here in Nyabibwe]. When God built the mine here he knew that rape would be there as well with the mine'.

Men also said that women could be at fault for their own rape, either by dressing provocatively or by getting themselves into dangerous situations. The responsibility for avoiding violence was put squarely on women's shoulders. As one young miner explained:

> How can we fight rape? It is a personal decision. You [a woman] must protect yourself, if you don't expose yourself in front of a man they cannot rape you … rape is here is because the women expose themselves and comport themselves in a certain way.

These attitudes illustrate the gross inequality that women face. Excluded from power structures and decision-making structures, the only thing they are perceived to be responsible for is the violence that is perpetrated against them. Often, women made differences in sexual violence from civilians versus armed men. Soldiers are described as engaging in forms of militarised rape, which have become common as a result of conflict in the DRC. As miners in Nyabibwe noted, 'Soldiers are the ones who are used to raping women … but it is usually driven by a gun … But it is not that common among us civilians'. As women transporters described, 'Is it here—yes, when there is war. Ntaganda, Nkunda—their soldiers rape IDPs [internally displaced persons], then others keep the practice going'. The leader of the sex workers' association in Nyabibwe said, 'Soldiers they usually don't pay for sex, sometimes demobilised soldiers as well, but even miners themselves, many don't pay for sex afterwards'.

The experiences of sexual and gender-based violence (SGBV) were explored in order to measure the extent of trading sex for basic goods and services. It became clear that money was the commodity that was most often traded for sex: 38.1 per cent of women reported this. Also common was the need for women to trade sex for protection, with one in five women (20.4 per cent) reporting this experience. Protection, as described in Phase 1, could include being granted access to safe working conditions or protection from violence from other actors. In contrast, men's rates of exchanging sex for any good or service (money, protection or work) was quite stable, at roughly 6 per cent. This implies that there is a small but stable population of men among those sampled who engage in transactional sex.

Active sex work in mining towns is, at least in part, influenced by the transient nature of the work, and the fact that men often come to mining areas without a spouse or partner. Women may come to mining towns expecting to engage in other kinds of work, such as trading or selling of small goods, but will find themselves then compelled to engage in regular sex work. One-fifth of all women (20.1 per cent) working in mining towns identified themselves as sex workers, whereas only 1.3 per cent of men did so. Of those who reported engaging in sex work, 93 per cent of women and 100 per cent of men say they work for themselves, rather than for a boss or association. These findings suggest that although sex work is a common profession for women, those engaging in it are not controlled by an outside entity.

Twenty-eight women said they were sex workers, but had not planned on getting into this work when seeking work in mining. Of these women, three-quarters said they were compelled to do sex work because of poverty, and another 18 per cent said that sex work was more profitable than the work they had originally intended to do. One-third of women said they went into sex work because they were not able to get other jobs, and 18 per cent said they did sex work to gain access to other types of work.

Above and beyond the varied forms of SGBV, women surveyed in Phase 2 reported incidences of harassment. For instance, one-third of women stated that they had been harassed by men in the mines. Of these women, 89 per cent said they had experienced harassment in the past 12 months. Sex workers had 10 times greater odds of reporting harassment than non-sex workers (odds ratio = 10.6, $p < 0.001$). Only 7 per cent of women stated that they had ever discussed this harassment or discrimination with others; although, encouragingly, almost 30 per cent of women stated that associations might be able to help with these kinds of problems. When asked what kind of help might be offered, 82 per cent of women said that these organisations could provide advice, and 18 per cent said that these organisations would be able to speak to authorities on their behalf. We return to the pivotal role to be played by forms of social organisation later in this section.

Discrimination

Interestingly, almost five times more men than women reported having ever been denied a job in the mines that they wanted to have. It may be that men aspire to have more lucrative or competitive jobs, whereas women may not seek these positions. When women gave answers about which jobs they were denied, they were more likely (in the open-ended question format) to state that they had been denied access to the mining area as a whole, rather than stating they were denied a specific job. Furthermore, 13 per cent of women surveyed in Phase 2 reported that they had been told, once having secured a job, that they could not earn the same amount of money as men. Finally, 16 per cent of women surveyed in Phase 2 reported that they had to turn down jobs in the mines because of problems at home. The most common problem cited was caring for children (65.5 per cent), followed by caring for others (10.3 per cent).

Outright denial of jobs to women in the mines can, in part, be explained by attitudes towards women's participation in mines, and lack of general knowledge about their rights under the Mining Code, its accompanying regulations and broader laws on women's right to work. Phase 1 highlighted a number of misconceptions about mining laws and regulations. Therefore, Phase 2 incorporated a module dedicated to knowledge about and access to information about mining regulations and laws in the survey instrument. The 'Knowledge, Attitudes and Practices' module attempted to assess knowledge, not only about the existence of mining-related policies, but also about perceptions around these policies. Only 26 per cent of women and 40 per cent of men reported that they knew there was a mining code in the DRC. Seventy per cent of women and 94 per cent of men stated that they would like to know more about it.

One common misconception was that women were not allowed to work directly in mining-related jobs. This may be related to a poor understanding of the policy that bans pregnant women from undertaking hard labour. Data supports the fact that this is a widely held assumption. Only 17 per cent of women and 20 per cent of men thought that women had the right to work as miners. Interestingly, when the survey asked the same question about women's right to work as mineral buyers or traders, 60 per cent of women and 65 per cent of men stated that they did have this right. Therefore, misconceptions about the right to work in mining may, at least in part, be related to ideas about women doing physically taxing work.

The Congolese mining code also allows children over 15 years of age to work in mining contexts, as long as the work is not physically dangerous. However, because of intense international advocacy about the dangers of child labour in mining towns, there is great sensitivity about admitting that children work in these areas. Although the practice is widespread and children often act as porters, washers, vendors and miners, this is often hidden from visitors for fear that mining activities may be suspended if child labour is discovered. This is reflected in the survey responses. Individuals were asked if children were allowed to work in mining towns, if the work was not physically hazardous. Seventeen per cent of women and 20 per cent of men said that children did not have the right to do this kind of work.

Right to health

Working in mining towns brings with it risks for a number of health hazards—from physical injury associated with hazardous physical work, to sexually transmitted infections because of high levels of transactional sex. Health problems may be compounded by limited access to health care in these areas, or limited ability to pay for services that do exist. In the qualitative survey of Phase 1, health problems in mining towns were broken down into three categories by participants: (i) poor labour conditions; (ii) poor structure of mining tunnels; (iii) public health problems such as poor hygiene and high levels of infectious disease. The scarcity of public health resources in mining areas, such as toilets, clean water supply, waste systems and primary care facilities, combined with close-quartered living conditions leads to high prevalence of diarrhoea, tuberculosis, respiratory infections, malnutrition and malaria. These problems are consistent across all the sites visited. Sexually transmitted infections, including HIV/AIDS are also described as significant problems, although people have such poor access to health care that these conditions are often not officially diagnosed. The head nurse in Mulamba describes the biggest health issues facing his community:

> There is a lot of HIV/AIDS; poverty leads women to exchange sex. There are also sanitary problems and many accidents and blunt trauma from landslides and rockfalls. There is tuberculosis in the mines. Close to the mines, we see cholera, amoebas, diarrhea, measles and typhoid.

Lack of access to health care due to financial or geographical obstacles exacerbates these issues. Many participants explain how pursuing health care is not an easy choice. Miners in Mulamba illustrate the sacrifice it often takes to get medical treatment:

> When people need care, they are transported to the nearest health centre. They must sell their fields or goats to pay for health care. They can't make enough to pay for care from the mines.

The head nurse in Mulamba goes even further, saying many must choose between eating and going to the health centre, and that such barriers to health services lead many to put their faith in prayer or magic over health care:

> You will see that in one family, if someone falls sick, the little they make is not enough to pay [for health care]. Many people die because their family decides to eat rather than to pay the clinic fees for one person. This is because of the kind of poverty we live in. People say that health problems are a result of witchcraft. They pray instead of getting care.

Phase 2 used the quantitative survey instrument to delve deeper into frequencies of health issues. Fifty per cent of women and 40 per cent of men reported having a medical issue in the past year. Encouragingly, of those individuals reporting a problem, 78 per cent of women and 70 per cent of men said they accessed care for this problem.

A majority of women (71 per cent) and men (88 per cent) said that the care they received was adequate. However, a notable percentage of women (27 per cent) and men (10 per cent) said that the care they received was not adequate. It is striking that women were three times more likely than their male counterparts to say they received poor care at a health facility (as defined by the participants' perception).

Fourteen per cent of women and 7 per cent of men stated that they had been denied medical care at least once in their lives, with women twice as likely as men to experience this problem. When asked about the reasons why they were denied health care, the vast majority of both women (84 per cent) and men (81 per cent) stated that they could not pay for health care services. The second-most commonly cited reason was that the clinic did not know how to treat their condition; this reason was cited by 10 per cent of women and 7 per cent of men. One in four women and men said that they had taken out a loan to pay for medical care.

When one disaggregates the experience of being denied health care by profession, clear differences emerge. Those in less prestigious jobs are more likely to report denial of care. Miners, food vendors and sex workers are the most likely to report this experience, whereas those in positions of power are much less likely to be denied care. Those who were most likely to report being denied access to health care were those who also reported experiencing a health problem in the past year. These results speak to the fact that those in the least privileged jobs are the most likely to have health issues.

Discussion

The popular advocacy narrative surrounding ASM in the eastern DRC— that the wealth generated from mining fuels the ongoing conflict and, as a result, exacerbates the conflict-related abuse of women—has been questioned by a number of scholars, who argue that over-simplification has led to misguided policies and a lack of understanding about the true issues facing women (and men) in the mining areas of the eastern DRC (Pact Inc. 2010; Hayes and Perks 2012; Seay 2012, 2014; Kelly et al. 2014a, Kelly et al. 2014b). The findings of this project emphasise that the conditions and dynamics in which people work in ASM sites are more complex, but there are also concrete points for positive change.

The respondents in both phases of this project were highly migratory populations with low levels of education. Women cited a lack of money or employment as reasons for migrating, whereas men moved most often because they lacked access to services in the areas from where they came. Women who seek work in mining towns are often the sole income earners in the family, either due to the loss of husbands or male family members through conflict, or due to the failure of their traditional income earning strategies (Hayes and Perks 2012). The higher rates of women's migration to unfamiliar towns can also bring a number of vulnerabilities, including lack of access to social support groups and lack of access to, and knowledge of, goods and services. The least reported reason for migration was displacement by an armed group, which highlights how economic and development opportunities were a greater motivator for the respondents sampled than conflict insecurity—a finding that is consistent with other recent studies (Geenen 2012).

This report also highlights how acts of discrimination and exploitation in mining areas can range from the subtle—such as a lack of representation among mining authorities, political disenfranchisement, exposure to occupational health hazards without proper access to health services and illegal taxation—to the extreme—such as sexual violence, child labour and debt bondage. Although the latter abuses often receive the most attention, the former are perhaps the most pervasive. Consider the findings from a 2014 US Agency for International Development (USAID) survey in artisanal mining towns in northeastern DRC: although labour trafficking, forced marriage, and debt bondage were reported by survey participants, these abuses were relatively rare and many of the perpetrators were described as family members, mining bosses, other miners or government officials, as opposed to armed actors from a particular group (Kelly et al. 2014a). The same report details abuses such as usurious lending, work without pay and predatory taxation. These practices were, however, undertaken mostly by civilian authorities rather than armed actors (ibid.). These types of exploitation also emerge in the findings from the present report, confirming prior research results cited herein.

Although women are vital actors in mining communities and fill a variety of roles, they are also among the most vulnerable to sexual and economic predation. The more profitable jobs are, more often than not, kept for men. Women are paid less than men for the same amount of work or level of effort, and are excluded from high-level conversations that ultimately dictate their circumstances (Hayes and Perks 2012; Kelly et al. 2014a). Unfortunately, such practices are not exclusive to the DRC alone, as described by Eftimie et al. (2012), Lahiri-Dutt and Macintyre (2006), and Hinton et al. (2003) in other ASM contexts across the globe.

This report details how gendered employment patterns emerge in mining towns. The job of miner was almost exclusively filled by men, whereas the vast majority of women worked in support roles, such as selling food or engaging in transactional sex. Past research also found that women are excluded from the highest-paying jobs, with many relegated to occupations such as sorting, gathering, washing, cooking, selling goods and/or engaging in transactional sex (Hayes 2008; Hayes and Perks 2012). No women worked as *chefs d'equipe*, *comptoirs* or PNGs, and only eight men reported being sex workers. Looking at the marital background of men and women who seek work in mining towns highlights a notable difference: whereas most men working in mining towns are single or married, a very high proportion of women therein reported being widowed or divorced. These findings suggest that women who have lost the support of a male relative, either through death or separation, may be forced to seek work in mining towns in order to survive. As noted by Hilson (2009), mining towns can provide opportunities for employment for traditionally marginalised populations. The jobs that are available to these vulnerable groups, however, are often low paying and exploitative.

Results from this work also highlight the overly simplistic nature of the narrative of 'rape as a weapon of war'. Women in ASM communities navigate a complex backdrop of sexual abuse. Sex is frequently a necessary component of employment in ASM, either as a profession in and of itself, or as a requirement in order to access other work opportunities or security. Women deal with myriad challenges, sometimes in the form of men failing to pay after sex or forcing the woman to have sex even if a price has not yet been negotiated. In focus groups, women reported sexual harassment, rape and forced performance of sexual favours, and the USAID report found that 7.1 per cent of the female survey participants had experienced forced sex within the past year (Kelly 2014). Responses to the USAID survey highlight the additional vulnerability implicit in this kind of transactional sex work, as women who reported having engaged in sex work were significantly more likely to also report sexual violence (ibid.). Bashwira et al. (2014) and Mahy (2012) support these findings, reporting on the complex dynamics between sex, employment and power in ASM communities.

Exclusion for political processes and entrenched discrimination were described as the norm in the communities visited for Phase 1. Vulnerable groups, including women, the elderly, the poor and landless, children and the handicapped (to name a few), were absent from political decision-making. Instead, participants stated that power resides in the hands of a few individuals. Marginalised groups described their efforts at self-organising into associations in an attempt to gain a sense of self-agency. However, they lack the financial means, human capacity and political influence to make lasting impacts.

The reasons behind marginalisation of women in particular are myriad. However, they seem to rest principally on three mutually reinforcing axes: cultural norms, altered family structure (resulting from decades of war and destruction) and education. Despite women's rise in the public work domain, many traditional and socially entrenched patriarchal beliefs continue to prevent women from engaging in meaningful, productive employment in the mines. The burden of the female single wage earner becomes even starker, as the most vulnerable women face higher rates of sexual predation than their married female counterparts. This culture has further retained and made remarkable use of common superstitions that perpetuate the belief that women working in mines are a form of bad luck. Hayes and Perks (2012) provide the example of a prevalent belief held among miners in Orientale and Katanga provinces: if a woman enters a mine, the minerals will vanish. These norms and taboos are perpetuated by a lack of education regarding mining laws and codes. Kelly et al. (2014b) provide the example of an article within the 2002 Mining Code for the DRC, dictating that pregnant women cannot engage in hazardous or heavy labour. This specific mining code article has enjoyed sweeping misinterpretation across the mines in the DRC, in order to prevent women from engaging in any mining activity. As shown in the report, people are often ignorant of the rights they have under national and international law. In fact, ignorance and misconceptions about mining laws and regulations are widespread. The continued marginalisation and exploitation of women and children is but one expression of this ignorance.

This work particularly investigates patterns of discrimination and political exclusion. Women were more likely to be denied access to the mining area, overall. Notably, there was a widespread perception that women were not allowed to work as miners or in other physically taxing jobs. These misconceptions do not lead to the total exclusion of women and children from mining areas but, rather, to their marginalisation to low paying support jobs.

The lack of structures to ensure fair practices and to promote justice served to reinforce a culture of predation on women—fostering an ethos of both cultural and judicial impunity. As highlighted by an outside scholar and by an SGBV expert during a feedback session on the report results in Washington, DC, the general erosion of state institutions has contributed enormously to the abuse of women—economically, physically and sexually—both in the mining and non-mining areas of the eastern DRC (see World Bank 2015b). Putting aside the question of whether the culture of impunity is more acute in mining versus non-mining areas, it is important to recall the nature of these extractive environments—by and large remote, isolated geographic spaces where alcohol and drug consumption is prevalent (Pact Inc. 2010). Those most vulnerable to human rights abuses, such as orphans, widows, sex workers and displaced persons, are also those least able to access traditional and formal justice mechanisms. It has emerged clearly from this feedback session that rebuilding state institutions needs to be placed front and centre in the fight to promote women's rights in these settings. In viewing SGBV as an expression of a much wider societal breakdown, one begins to see how multifaceted human rights issues are in mining areas, by consequence requiring a holistic, multisectoral approach towards rebuilding individuals, families and communities at large.

One could consider, for example, the health consequences of SGBV alone to understand the need for a multisectoral approach. Participants in the research project spoke particularly about HIV/AIDS as an emerging threat that the nearly non-existent health infrastructure in most isolated mining areas is not equipped to handle. Close quarters and hard labour mean that other health issues, including physical trauma, diarrhoea, tuberculosis, respiratory infections, malnutrition and malaria, are rampant. It is certainly plausible that HIV/AIDS and other sexually transmitted infections (STIs) may be more prevalent in mining towns in the eastern DRC, as is the case in other settings (Clift et al. 2003; Conro and Walque 2012). Furthermore, superstition often replaces education about the spread of disease, increasing the likelihood of high STI prevalence rates in ASM communities (Pact Inc. 2010). Populations working in and around mines are at risk of a great number of other health risks due to structural issues, exposure to minerals, ergonomic concerns and a lack of sanitation and safety equipment. Health issues include, but are not limited to, mining disasters, mercury exposure, radioactive material exposure, gastrointestinal disease, malnutrition, infection, lung disease, bone and muscle injuries and heat stress (Walle and Jennings 2001; Hayes and Wauwe 2009; Pact Inc. 2010; Kelly et al. 2014a). Pregnancy, reproductive problems and injuries due to SGBV are also issues of concern for women specifically, especially given the general lack of access to adequate health care (Pact Inc. 2010; Hayes and Perks 2012). Drug and alcohol abuse is described as being rampant in mining towns and inside the quarries themselves. Substance abuse has led to an increase in violence and a breakdown in the social fabric of their villages.

Of those who needed health care in the past year, a majority of respondents from Phase 2 said they accessed it and thought the care they received was adequate. Women, however, were three times more likely to say that the care they received was not adequate. This potentially speaks to the health system being ill prepared to provide for the health care needs of women, which includes unintended pregnancy, sexual and physical violence and STIs. Women also reported health facilities as being unable to treat their conditions more than men did. Cost was cited as a barrier to care, with

14 per cent of women and 7 per cent of men saying they had been denied medical care at least once in their lives, often because they could not pay. Those in the most vulnerable and lowest-paying jobs were the most likely to be denied care. This is an especially concerning issue because those most likely to report being denied care are also the most likely to report having had a health problem in the past year. Although those who have health problems may experience more denial of care because they seek care more often, these results speak to the fact that those in the least privileged jobs are the most likely to have health issues. A quarter of the respondents said they had taken loans to pay for medical care. In the USAID study conducted in mining towns in South Kivu and Maniema, paying for health care was cited as the third-most common reason that people took loans (Kelly et al. 2014a).

Against this challenging backdrop, however, the research pointed to promising examples of civic organisations with the potential for providing inclusivity and social support to miners. According to the DRC Mining Code, all artisanal miners are obliged to be part of a cooperative. The experience of cooperative formations in ASM in the DRC is, however, extremely varied. Members pay a range of entrance fees and taxes. Depending on the size and scope of the cooperative, miners may either enjoy profit-sharing arrangements and protection, or receive no such benefits (Pact Inc. 2010). Although hailed by the government as critical for better organising ASM, the perception of cooperatives' benefits for miners varies. For instance, Kelly (2014) reported that certain mining communities perceive cooperatives as being embedded within the hierarchy and failing to represent many within the community. Geenen (2014) expresses a similar concern, specifically in the gold mines of South Kivu.

Yet, the power of a collective organisation as a means to represent miners' concerns and respond to immediate issues cannot be altogether dismissed. The research results from this project suggest that smaller, trade-specific groups known as associations may be more beneficial for marginalised communities, thereby supporting prior research in specific mining areas. For instance, a USAID survey found that less than 20 per cent of workers in the lowest-paying trades, such as potters, food venders and sex workers, reported belonging to a community group, compared to 50 per cent of mining team leaders and 25 per cent of miners (Kelly et al. 2014a). Additionally, 25 per cent of men reported belonging to groups, compared with 17 per cent of women (ibid.). For instance, in select areas of northern Orientale, associations that combine literacy and savings objectives for miners have proven to be effective vehicles for self-organisation (Hayes and Wauwe 2009). Meanwhile in South Kivu (Kelly et al. 2014a), there is the example of the Association of Free Women, representing women involved in sex work. Members of the group expressed the critical role it has played in their access to health care, financial stability and social support. In Phase 2, women surveyed reported greater access to savings and loan programs than men, possibly because non-government organisations target women for these programs. Utilisation of these services, however, was extremely low (9 per cent for women and 6 per cent for men), despite the fact that half of all women and three-quarters of men stated they wanted access to these programs, indicating a large unmet need. Such forms of self-organisation described earlier could be better expanded.

Indeed, in the absence of courts and tribunals, as well as capable national human rights institutions or ombudspersons, participants in the research said they relied on each other and small associations as the most neutral and capable parties for bringing some level of justice in their communities. Active systems of association were in fact present at every site, speaking of an enterprising spirit and a desire for more inclusive political engagement. People often self-organised to promote a common profession, cause or interest. However, these groups lack the financial means, human capacity and political influence to make sustainable change. Organising to create more effective unions and associations, while promoting grassroots inclusive economic cooperatives, offers a potentially more meaningful opportunity for improving life in mining towns.

The implementation of laws and policies is a critical part of the formalisation process for artisanal mining in the DRC. Knowledge of and adherence to these laws at the local level is the only avenue for successful implementation. A minority of men and women surveyed, however, stated that they were aware of the existence of the mining code and of related laws. Women were far less likely than men to report knowledge of the mining code. Community organisations were described as an effective avenue for education for all members of the community. There were, however, gendered differences in ways that people gained knowledge. Men were most likely to receive information from radio messages, whereas women stated they received information more readily from their peers.

Findings from this work bring to light several new pathways for understanding the challenges facing both men and women in the mines of the eastern DRC. These pathways also emphasise the need for holistic responses to ensure sufficient change in behaviours and practices in the mines of the DRC. These programmatic and policy responses will need to be fundamentally linked to the long-term project of rebuilding the state in the DRC. Most importantly, reforms should be informed by, and responsive to, the needs of those working in mining, so that all of those seeking fair and equitable work opportunities can find them.

References

Autesserre, S., 2012. 'Dangerous Tales: Dominant Narratives on the Congo and their Unintended Consequences.' *African Affairs* 111(443): 202–22. doi.org/10.1093/afraf/adr080

Banchirigah, S.M., 2008. 'Challenges with Eradicating Illegal Mining in Ghana: A Perspective from the Grassroots.' *Resources Policy*, 33(1), 29–38. doi.org/10.1016/j.resourpol.2007.11.001

Bashwira, M.R., J. Cuvelier, D. Hilhorst and G. van der Harr, 2014. 'Not Only a Man's World: Women's Involvement in Artisanal Mining in Eastern DRC.' *Resources Policy* 40: 109–16. doi.org/10.1016/j.resourpol.2013.11.002

Bryceson, D.F. and J.B. Jønsson, 2010. 'Gold-Digging Careers in Rural East Africa: Small-Scale Miners Livelihood Choices.' *World Development* 38(3): 379–92. doi.org/10.1016/j.worlddev.2009.09.003

Clift, S., A. Anemona, D. Watson-Jones, Z. Kanga, L. Ndeki, J. Changalunch, A. Gavyole and D.A. Ross, 2003. 'Variations of HIV and STI Prevalences within Communities Neighbouring New Goldmines in Tanzania: Importance for Intervention Design.' *Sexually Transmitted Infections* 79: 307–12. doi.org/10.1136/sti.79.4.307

Conro, L. and D. Walque, 2012. 'Mines, Migration and HIV/AIDS in South Africa.' *Journal of African Economics* 21(3): 465–98. doi.org/10.1093/jae/ejs005

Eftimie, A., K. Heller, J. Strongman, J. Hinton, K. Lahiri-Dutt, N. Mutemeri, C. Insouvanh, M. Godet Sambo and S. Wagner, 2012. *Gender Dimensions of Artisanal and Small-Scale Mining:*

A Rapid Assessment Tool. Washington, DC: World Bank Group's Oil, Gas and Mining Unit. Available at siteresources.worldbank.org/INTEXTINDWOM/Resources/Gender_and_ASM_Toolkit.pdf

Geenen, S., 2012. 'A Dangerous Bet: The Challenges of Formalizing Artisanal Mining in the Democratic Republic of Congo.' *Resources Policy* 37(3): 322–30. doi.org/10.1016/j.resourpol.2012.02.004

Geenen, S., 2014. 'Dispossession, Displacement and Resistance: Artisanal Miners in a Gold Concession in South Kivu, Democratic Republic of Congo.' *Resources Policy* 40: 90–9. doi.org/10.1016/j.resourpol.2013.03.004

Geenen, S. and S.R. Custers, 2010. 'Tiraillements Autour du Secteur Minier de l'Est de la RDC.S.' *L'Afriique Des Grands Lacs. Annuaire 2009–2010*, Paris, pp. 231–58.

Hayes, K. (ed.), 2008. '2008 Regional Workshop: Small-Scale Mining in Africa: A Case for Sustainable Livelihood.' Amsterdam: Commonwealth Fund for Commodities.

Hayes, K. and R. Perks, 2012. 'Women in the Artisanal and Small-Scale Mining Sector of the Democratic Republic of the Congo.' In P. Lujala and S.A. Rustad (eds), *High-Value Natural Resources and Peacebuilding*. Oxon and New York: Earthscan.

Hayes, K. and V. Wauwe, 2009. Microfinance in Artisanal and Small-Scale Mining.' Background Papers for 9th Annual CASM Meeting. Maputo, MZ: Communities and Small-Scale Mining.

Hilson, G., 2009. 'Small-Scale Mining, Poverty and Economic Development in Sub-Saharan Africa: An Overview.' *Resources Policy* 34(1–2): 1–5. doi.org/10.1016/j.resourpol.2008.12.001

Hilson, G., 2011. 'Artisanal Mining, Smallholder Farming and Livelihood Diversification in Rural Sub-Saharan Africa: An Introduction.' In G. Hilson and M. Hirons (eds), *Artisanal Mining, Smallholder Farming and Economic Development in Rural Sub-Saharan Africa*. Special Issue of *Journal of International Development* 23(8): 1031–41. doi.org/10.1002/jid.1829

Hinton, J., M. Veiga and C. Beinhoff, 2003. 'Women and Artisanal Mining: Gender Roles and the Road Ahead.' In G. Hilson (ed.), *Socioeconomic Impacts of Artisanal and Small-Scale Mining in Developing Countries.* Netherlands: Swets & Zeitlinger B.V. Publishers. doi. org/10.1201/9780203971284.ch11

Kelly, J., 2014. '"This Mine has become our Farmland": Critical Perspectives on the Coevolution of Artisanal Mining and Conflict in the Democratic Republic of the Congo.' *Resources Policy* 40: 100–8. doi.org/10.1016/j.resourpol.2013.12.003

Kelly, J., N. Greenberg, D. Sabet and J. Fulp (eds), 2014a. 'Assessment of Human Trafficking on Artisanal Mining Towns in Eastern Democratic Republic of the Congo.' Arlington: US Agency for International Development and Social Impact, Inc.

Kelly, J., T. King-Close and R. Perks, 2014b. 'Resources and Resourcefulness: Roles, Opportunities and Risks for Women Working at Artisanal Mines in South Kivu, Democratic Republic of the Congo.' *Futures* 62A: 95–105. doi.org/10.1016/j.futures.2014.04.003

Lahiri-Dutt, K. and M. Macintyre, 2006. *Women Miners in Developing Countries: Pit Women and Others.* London: Ashgate.

Luning, S., 2008. 'Liberalisation of the Gold Mining Sector in Burkina Faso.' *Review of African Political Economy* 35(117): 387–401. doi. org/10.1080/03056240802411016

Maconachie, R. and J.A. Binns, 2010. 'Farming Miners or Mining Farmers? Diamond Mining and Rural Development in Post-Conflict Sierra Leone.' *Journal of Rural Studies* 23(3): 367–80. doi. org/10.1016/j.jrurstud.2007.01.003

Mahy, P.K., 2012. *Gender Equality and Corporate Social Responsibility in Mining: An investigation of the Potential for Change at Kaltim Prima Coal, Indonesia.* The Australian National University (PhD thesis).

Pact Inc., 2010. *PROMINES Study: Artisanal Mining in the Democratic Republic of Congo.* Washington, Congo and Kinshasa: Pact Inc.

Perks, R., 2011. 'Can I Go? Exiting the Artisanal Mining Sector in the Democratic Republic of the Congo.' *Journal of International Development* 23(8): 115–27. doi.org/10.1002/jid.1835

Perks, R., 2013. 'Digging into the Past: Critical Reflections on Rwanda's Pursuit for a Domestic Mineral Economy.' *Journal of East African Studies* 7(4): 732–50. doi.org/10.1080/17531055.2013.841025

Seay, L., 2012. 'What's Wrong with Dodd-Frank 1502? Conflict Minerals, Civilian Livelihoods, and the Unintended Consequences of Western Advocacy.' Working Paper 284. Center for Global Development.

Seay, L., 2014. 'Did Cutting Access to Mineral Wealth Reduce Violence in the DRC?' *The Washington Post*, 25 March. Available at www. washingtonpost.com/news/monkey-cage/wp/2014/03/25/did-cutting-access-to-mineral-wealth-reduce-violence-in-the-drc/

Smith, J., 2011. 'Tantalus in the Digital Age: Coltan, Ore, Temporal Dispossession, and Movement in the Eastern Democratic Republic of the Congo.' *Journal of the American Ethnological Society* 38(1): 17–35. doi.org/10.1111/j.1548-1425.2010.01289.x

Spittaels, H. and F. Hilgert, 2013. 'Analysis of the Interactive Map of Artisanal Mining Areas in Eastern DR Congo.' Viewed at ipisresearch. be/publication/analysis-interactive-map-artisanal-mining-areas-eastern-dr-congo/

Walle, M. and N. Jennings, 2001. 'Safety and Health in Small-Scale Surface Mines: A Handbook.' Working Paper No. 168. Sectoral Activities Program. Geneva: International Labour Organization.

World Bank, 2015a. 'Resources and Resourcefulness: Gender, Conflict, and Artisanal Mining Communities in Eastern Democratic Republic of the Congo.' Working Paper 95971. Washington, DC: World Bank Group. Available at documents.worldbank.org/curated/en/2015/04/24418255/resources-resourcefulness-gender-conflict-artisanal-mining-communities-eastern-democratic-republic-congo

World Bank, 2015b. 'Empowering Women in the Mines of the Eastern Democratic Republic of the Congo.' 30 April. Washington, DC: World Bank. Available at www.worldbank.org/en/news/feature/2015/05/04/empowering-women-in-the-mines-of-the-eastern-democratic-republic-of-the-congo?cid=ISG_E_WBWeeklyUpdate_NL

Section Three: Conflicts and governance

11

Historical trajectory of gold-mining in the Nilgiri–Wayanad region of India

Amalendu Jyotishi, Kuntala Lahiri-Dutt
and Sashi Sivramkrishna

This chapter builds on the earlier work of Lahiri-Dutt (2004) to interrogate categories of 'formal' and 'informal' as they are applied to mining, both historically and in the contemporary era. This historical perspective reveals the many difficulties that come with consistently applying ideas of formality to diverse forms and institutions of gold-mining throughout time. The difficulties of constructing such a multivariate definition of informal mining renders such a project ambiguous at best. We, therefore, propose that any suitable uni- or bi-variate definition must be recognised as contextual and subjective, dependent on the specific concern that is being raised about mining. In particular, this definitional complexity has important implications for policy, which suggest the formalisation of informal (gold) mining as a solution.

We draw these conclusions from an in-depth study of gold-mining in the Nilgiri–Wayanad region of southern India between the early nineteenth century and the present. We build a case that challenges the popular notion that institutions evolve from informal to formal, and additionally suggests the role of external rather than internal factors in precipitating any institutional changes. Given the particular evolutionary trajectory of gold-mining in the Nilgiri–Wayanad region, it is pertinent to reflect on these changes to suggest implications for the future of these institutions and the livelihoods of miners who work within them.

Institutions: Formality, informality and property rights

Institutions are 'the rules, enforcement characteristics of rules, and norms of behaviour that structure repeated human interaction' (North 1990). The conventional view on the emergence and evolution of institutions tracks a linear path from informal to formal (Zenger et al. 2001). The evolution of institutions happens through repeated interactions between individuals, groups and organisations. In this process, norms and behaviours are established through various processes, such as collective action (Ostrom 2005) in responses to uncertainty, risk and other transaction costs (Williamson 1981, 2000), or through clearly defined property rights (Coase 1937, 1959; Demsetz 1967; Libecap 1978, 1989; Barzel 1997). Implicitly, the literature in institutional economics argues that internal factors play a more critical role than the external factors in the evolution of institutions. The resistance to change, known as 'path dependency',[1] is an internal factor to the institutional continuity. Another implicit contention of this literature is that institutions become more efficient over a period of time through repeated transactions, by minimising transaction costs or setting the rules of the game through well-defined property rights. The bundle of rights that comprise property rights on the goods or resources in question includes the right to use, right to earn incomes and the right to transfer. When these set of rights exist upon a good or a resource, and are enforceable, the property rights are then said to be well defined. Well-defined property rights play important roles in the efficient allocation of resources. Whatever may be the origin of the institution, it is assumed that the more widely it is recognised, the better it will function. Full recognition is achieved when the norm is endorsed by the state as legally binding. However, it is important to note that not all institutions require the support of governments. The political economy of institutional change is therefore important in that they may evolve to confer privileges on particular groups (Sikor and Lund 2009).

---

1   Path dependency can be defined as a decision-making process that is not only dependent on present circumstances but also on the past and, hence, acts as a constraint for efficient allocation of resources (Liebowitz and Margolis 2000).

In this chapter, we use the standard classification of economic goods based on the property rights attributes of 'rivalry' and 'excludability'. Based on the relative strength of these attributes and their combinations, we can identify four major types of institutional structures: private, commons, club and open access (see Figure 11.1). This simple classification not only provides an understanding of the different types of institutions based on the combination of these attributes, but also does not separate formal and informal institutions explicitly. Yet, of these institutions, those with high excludability are often based on formal structures, whereas institutions operating with low excludability are guided by informal norms. As one moves from a low to a high degree of rivalry and excludability, the institutional process also moves towards more formal and structured institutions. Thus, low rivalry and excludability often lead to informal arrangements for the institutions. The 'efficiency' in resource use parameter predominates in the degree of excludability, and the 'equity' in resource distribution parameter predominates in the degree of rivalry.

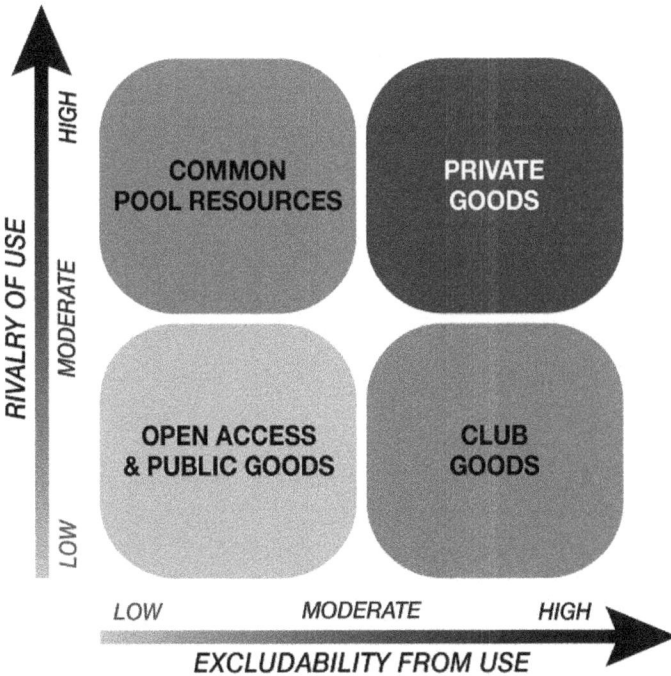

Figure 11.1: Excludability and rivalry principles of property rights
Source: The Australian National University CartoGIS.

Using this framework and understanding, we chronicle gold-mining in the Nilgiri–Wayanad region, and evaluate how the institutions have evolved and been classified over a long period of time. In the following section, we engage more deeply with the debates around nomenclature and taxonomy.

Informal, artisanal or something else: Dynamism in gold-mining

Debates about formality and informality of mining institutions are important because of the politics that lie behind these taxonomic assumptions (Lahiri-Dutt 2004). Yet, mining that is often small in scale—bordering on illegal, using artisanal technology along with entrepreneurial labour, that itself is often an expansion of a peasantry system—is difficult to slot into either category of formal or informal. The widely used term, artisanal and small-scale mining (ASM), was popularised by *Breaking New Ground: Mining, Minerals and Sustainable Development* (International Institute for Environment and Development (IIED)/ World Business Council for Sustainable Development (WBCSD) 2002); it combines two pre-existing terms, 'artisanal' (United Nations Development Program (UNDP) 1999) and 'small-scale', used by the International Labour Organisation. ASM, like other attempts to categorise mining as either 'formal' or 'informal', loses explanatory potential by focusing on a single aspect at the expense of others. By emphasising one aspect of the mining activity, such as artisanal technology, small-scale (dis)economies of scale or informal rules and norms governing the mining activity, these definitions neglect the ways in which these features not only coexist but also change their relation to one another over time. When considering mining in a single region from a historical perspective, it becomes important to consider the flexibility of these definitions to changes in mining activities or the institutions governing them. As such, one of the major shortcomings of these definitional practices is the assumption of 'staticness' of the form or practice of mining. However, a robust historical perspective can illuminate considerable changes in the form and the institutions that govern mining activities, even within a single region.

Table 11.1 offers a broad overview of how, based on these multiple characteristics, formal and informal mining is conceived. Strictly speaking, perhaps the most critical distinguishing features between formal and informal mining are that the latter is non-legal—based on customary law and/or illegal. This highlights the role of governance in the classification of formality and informality. Important differences between formal and informal mining may occur at the level of governance—the keeping of records and the monitoring of informal activities by the state, making it (im)possible to collect rent or tax.

Table 11.1: Multiple characteristics of 'formal' and 'informal' mining

	Size	Technology	Legality/ Institutional Structure	Economic Rationale	Tax	Economic Structure	State Monitoring/ Records
Formal	Large/ Small	Mechanised/ Modern/ Capital intensive/ Traditional/ Artisanal/ Labour intensive	Licensed/ Statutory	Profit	Yes	Capitalist/ Cooperative	Yes
Informal	Medium/ Small	Rudimentary/ Traditional/ Artisanal/ Labour intensive/ Low level of mechanisation	Licensed/ Non-legal/ Customary/ Illegal	Profit/ Livelihood/ Windfall	No	Individual/ Family/ Cooperative	No

Source: Authors' work.

With respect to institutional change, one of the key challenges in studying mining historically in a single site is to find an appropriate setting to empirically substantiate any resulting hypothesis, not simply quantitatively but qualitatively. Gold-mining in the Nilgiri–Wayanad region of southern India provides an interesting context for the study of institutional evolution for two reasons: first, the fundamental nature and purpose of the activity have remained unchanged over generations; second, the technological changes have been either manual or mechanical, having no major impact on the final product per se. Despite this 'staticness' of the operation, the institutions involved in such practices and their relationship to structures of governance have undergone extensive changes during this period.

In the context of mining, the establishment (and protection) of property rights is perhaps the most basic institution that performs a crucial economic function: that of presiding over competing claims of 'ownership' over both the land from which minerals are extracted and the commodities that are extracted from it. Property rights over mines can be accorded formally through state agencies and/or informal regulating agencies that include local customs and local political authorities. Historical property rights are thus recognisable in a range of forms, varying from written documents to oral histories. While property rights can be relatively easily identified at specific points of time, it is more difficult to understand how and why these institutions evolve and change over time, specifically over long periods of time. This particular empirical study of gold-mining in the Nilgiri–Wayanad region helps us address several questions that arise when considering institutional transitions over the longue durée.

Chronicles of gold-mining in Nilgiri–Wayanad

The Nilgiri–Wayanad region (Figure 11.2) lies at the southern end of the Western Ghats—the ranges lining the western coast of peninsular India. It is located in the western part of Tamil Nadu, bordering the states of Karnataka and Kerala. During the nineteenth and twentieth centuries, this area was governed by a complex network of actors that included the East India Company, local princes, *zamindars,* and the British, and from 1947 onwards, the state of India. For the larger part of history, the *Rajah* (*zamindar*) of Neelambur (which was a vassal state of the princely state of Calicut) was the ruler. Therefore, all the land transactions in the region for agriculture, mining and plantations were conducted by the various *zamindars* of Neelambur (or Nilambur). From the early 1800s to early 1900s, the Neelambur *zamindars* leased land for coffee plantation, gold-mining and tea plantation.

In view of the dearth of sources on the extent and magnitude of informal gold-mining in India, we have collected the material from a combination of historical documents (Jennings 1881; Francis 1908), followed by a primary survey. Previously, Deb et al. (2008) documented the existence and exploration of gold in the region by various companies. Initially, we obtained information on the existence of gold-mining from the Pandular website.[2] We contacted the website owner and formed a strong working relationship. In the initial phase, his association and assistance allowed us to visit the informal gold-mining sites in the region. Subsequently, our research led us to explore the documented history of gold-mining over the last two centuries in its continuity, and more extensive socio-economic and focus group surveys.

---

2   www.pandalur.com.

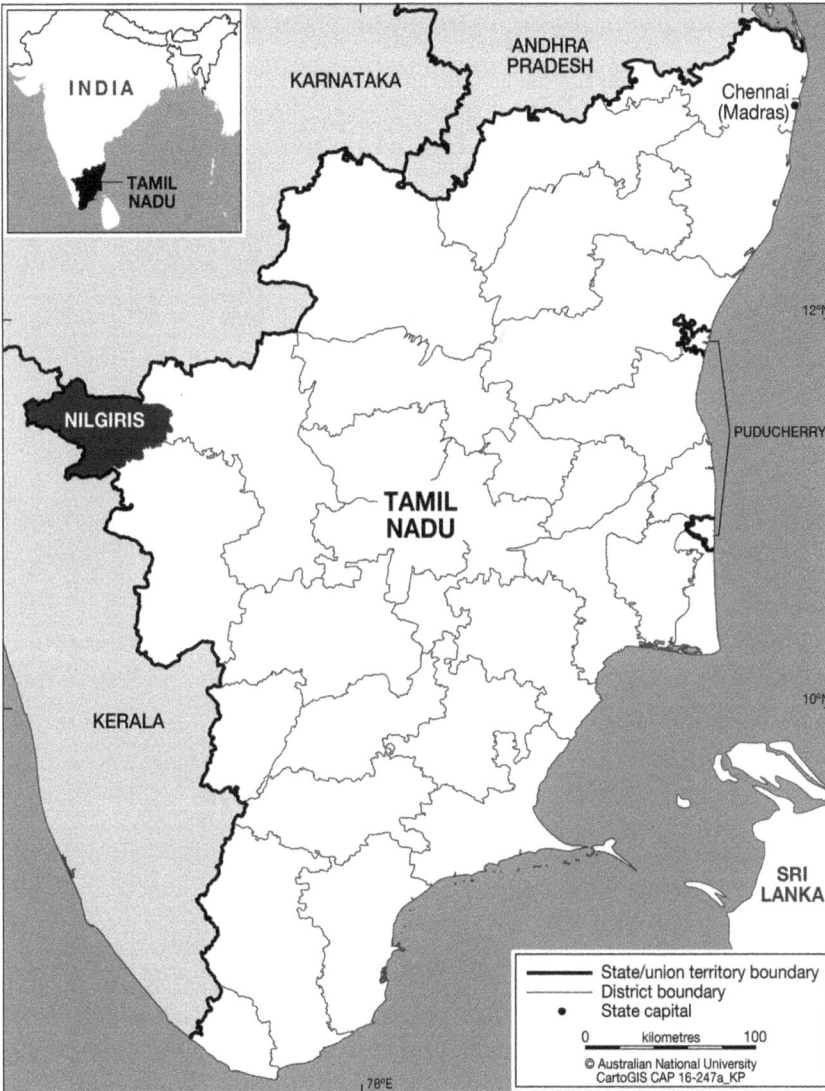

Figure 11.2: The study region

Note: Map not to scale.

Source: The Australian National University CartoGIS.

The following analysis is divided into four historical phases, each constituted by the predominant institutional structure that governed the gold-mining activities in the region at that time.

Institutional changes over two centuries of gold-mining

Phase I: Up to the 1860s

In one of the earliest colonial records on this region, Thomas Barber, an official in the East India Company, recorded in his 'Journal of a Route of the Neelghurries from Calicut' (1830) the low returns on gold-mining and payment of a tax by miners in this region (p. 314):

> Two persons are employed to each patty [tray], one to dig the earth, the other to hold the patty, wash the earth away, and extract and unite, by means of quicksilver, the golden particles. Each patty pays a tax to government of 3 rupees per month, which my informers added, absorbed about two-thirds of the net profits; and from the wretched appearance of the person employed in working the patties, it is evident they are miserably paid.

The East India Company faced strong resistance to their attempts to assess gold mines in Wayanad. An interesting letter published by Bergein (1834: 124), a correspondent of the *Mechanics' Magazine*, states:

> The natives, one and all, are very averse to our having anything to do with the mines, for obvious reasons: several hundred families are at present supported by their produce; the zemindars of the districts obtain from them a moderate revenue, which would be lost to them were these mines to come into the hands of government …[3]

Similarly, Bergein (1834) records how the *Rajah* of Neelambur and his people resisted the British—the *Rajah* tried to starve a British mining contingent by prohibiting the sale of food and even poisoning their cattle. Additionally, a number of British shafts were also filled out by the *Mapillahs*,[4] and then covered with trees and shrubs (ibid.). These conflicts aside, it appears that gold-mining, although artisanal in technology, was in this period formal in the sense that reasonably good records were being maintained and taxes were being collected. As far as institutional framework was concerned, what particularly interested us here was the

---

3　'Government' here refers to the East India Company.
4　One of the largest Muslim communities from Kerala. This community is a local source for labour in the region.

presence of a formal tax structure implemented by the *Rajah* of Neelambur. The formal tax structure was built on the 'excludability' principle for those who did not pay the tax.

An 1843 report titled, 'Mineral Resources of Southern India, no. 4 Gold, Tracts', by Lieutenant Newbold of the East India Company, recorded that, in the South Maratha country (now Gadag District in the state of Karnataka), mining was a seasonal occupation and carried out predominantly by a caste called *Jalgars*,[5] who had been acquainted with the art of gold-mining 'from time immemorial' (Newbold 1843: 207).

The report also notes that 'the rivulets of Hurti and Soltoor … are more productive, though I did not see many persons at work at Hurti' (Newbold 1843: 206). Moreover, the overall production of gold from the three rivulets, Hurti, Soltoor, and Doni, was not insignificant—some 200 ounces (5.7 kg) of gold was mined annually after the monsoons. These remarks point towards the relative abundance of gold and the non-competitive nature of artisanal mining during this phase in history. Newbold (1843: 208) also illuminates the governance structure of mining:

> A Mussulman, who accompanied me to the spot, informed me that he had obtained four rupees' worth of gold in two days. The hire of the three Jalgars, &c., whom he employed amounted to half that sum, leaving two rupees clear profits. The ancient lords of the soil, the Dessayes of Dummul, formerly levied a toll from the gold-washers, which ceased with the authority of the last chief, who was hanged, by the order of the Duke of Wellington, over the gate of his own fortress, for firing on a flag of truce at the siege of that place.

A careful reading of this statement reveals a lot of information. In earlier times, under the *Dessayes*,[6] it seemed that the *Jalgars* mined for gold directly upon paying a toll to the rulers. With the end of their rule, the *Jalgars* worked more as labourers for a Mussulman who bore the risk of fluctuating returns, yet retained a relatively significant portion of the revenue. It is likely that the Mussulman was a trader or capitalist and was an integral part of the supply chain of gold.

---

5   The term *Jalgar*, frequently used in colonial documents related to gold-mining, is possibly associated with gold-mining and panning communities. Even today, the community with the surname Jalgar exists in the Northern Karnataka region.

6   Also know as *Desai*, these are the village heads.

While Newbold (1843) reported on gold-mining in various parts of India, of particular interest here are his remarks on gold-mining at the base of the Nilgiris, at the frontier with Mysore state. He writes, 'the mines of Wayanad district then worked were those of Cherankode Devala, Nelyalam, Ponery and Pulyode' (Newbold 1843: 208). Some 750 ounces (21 kg) of gold was extracted annually in Nilambur, where 'the golden region is rented from Government by a native chief'[7] (ibid.: 209). In the report, 'Preliminary Notes on the Gold Fields of South East Wayanad', King (1875: 45) described the land tenure in Wayanad as follows:

> Hitherto the land in Wayanad has been principally parceled out in coffee gardens, either free-hold, or paying an annual rent to the Rajahs who hold a great quantity of the ground, or direct to the Government. At the same time, after a certain period, a revenue is derived from all the gardens by the Government, whether it be Rajah's land, or not.

Ball (1869) also lends credence to our hypothesis of 'excludability'; people rarely crossed boundaries in their search for gold. This happened elsewhere in India, for example, in Singbhum district (now in the state of Jharkhand) he records (ibid.: 12):

> On my arrival in Dulmi (which is situated on the faulted boundary of these two groups of rocks) when marching northwards from the lower part of Pattrum, the gold-washer asked to be allowed to return to his own country (Dalbhum), stating that none of his race ever went north of Dulmi.

He further reports on the governance structures of gold-mining in this region, which he defines as categorised by caste exclusion and low standards of living for workers—substantiating our claim that gold-mining was essentially a 'club' good in the pre- and early colonial period (ibid.: 13).

> The gold-washers belong to the lowest and poorest races in the country, Gassees[8] according to Colonel Haughton, but some of those which I met with were a race of kumars, called Dokras. Their numbers have been greatly reduced by the famine; without exception they are all in the power

---

7   Here, the government means the East India Company, and the *Rajah* (*zamindar*) of Neelambur is the native chief. The Neelambur *Rajah* also leased out land to the East India Company and others for mining purposes.

8   The name *gassees* or *ghasi* refers to the occupation of cutting grass, especially for horses. In addition to collecting and selling grass, the *ghasis* are employed in scavenging work.

of the Mahajuns,[9] for whom they work at a low rate, and are never able to free themselves of the claims which the Mahajuns make on account of the advances. The daily earnings of the gold-washers are small, but might no doubt be increased, if it were not that they are always satisfied when enough gold has been found for procuring the day's subsistence.

In 1875, Dr W. King of the Geological Survey was appointed by the government to survey the 'country'. He reported that gold-mining in this region was carried out by two tribes, the Paniyas[10] and the Kurumba[11] (also referred to as Pannirs and Korumbars), each possessing their own specific domain of labour. 'The Pannirs wash for it in the alluvium, surface soils, and river sands. The Korumbars dug down to and excavated the quartz leaders' (King 1875: 30). This description reinforces our understanding that caste or tribe provided some excludability to gold-mining. Apart from a tax payable to the 'owner' of the land, there does not seem to be any rivalry between the members within the two tribes. However, there are indications that one group (Paniyas) were employed as labour, rather than working as independent miners; they earned no more than a quarter of a rupee per day, thereby engaging in mining only during the lean season, when employment was sparse on the coffee gardens. It is also mentioned in Ball (1869: 33) that the Paniyas were 'driven by land-owners to search for gold, the land not being so well adapted for agricultural work'. On the other hand, it seems that the Kurumbas were more independent, although they may have been forced to share their profits with the *Rajahs*. Moreover, the Kurumbas were capital constrained, as mentioned in King (1875: 42):

> The big reefs were not worked by these men on account of the difficulty in breaking up the stones, and because the gold is distributed too finely through it to have paid hand labor. With machinery and modern appliances, the reefs should pay even if only 3 dwts. of gold are got always from the ton of quartz.

---

9   *Mahajuns* are the local moneylenders or usurers.

10   The Paniya, also known as Paniyar and Paniyan, are an ethnic group primarily inhabiting the Wayanad region. The Paniya have historically worked as agricultural labourers and serfs.

11   Traditionally, the Kurumbas or Kurubas have subsisted as hunters and gatherers. They largely reside in the Nilgiri region of Tamil Nadu. There are different groups within this tribe; some among them are Alu-Kurumbas, Betta-Kurumbas, Jenu-Kurumbas, Kurubas, Mudugas, Mulla-Kurumbas, Palu-Kurumbas, and Urali-Kurumbas.

It is also mentioned that the Kurumba sometimes found large amounts of gold, but probably had given up this work by 1875 because, as King records, the pits 'would be too deep for their style of work, water being the great obstacle likely to be met with' (ibid.: 42).

From these historical sources, we can conclude the existence of gold-mining in the region prior to 1830. Use of wage labour, as well as the presence of independent miners is likewise identifiable during this period. The existence of tax structure, payment to the *zamindars*, and the presence of intermediaries and financers (*Mahajans*) are also reported. The tax on panning trays also suggests formal recognition of artisanal forms of mining. Most importantly, it has been identified that select castes and tribes were engaged in gold-mining activities. Based on this assessment of archival records, gold-mining in Phase I may be characterised as formal, although the technology was artisanal and caste-based. Entrepreneurial labour and rudimentary technology are not by themselves sufficient conditions to make an activity informal. 'In 1879 the government of India employed Mr. Brough Smyth (for many years the Secretary for Mines in Victoria and held to be the greatest authority on the subject in Australia to examine the Wayanad reefs)' (Francis 1908: 15). It is this report that provided the basis for the rapid escalation of British interest and development of the region that we discuss next.

Phase II: From the 1870s to the 1890s

In his study of the relevance of the gold standard to India, Sivramkrishna (2015) elaborates that it was at the International Monetary Conference of 1867 that a move towards a common international currency was first propagated. Gold, which had become relatively abundant by then on account of its discoveries (and the gold rushes in America and Australia), was to be the basis of this international currency. The danger of a shortage in gold, raising its price and leading to deflation, had to be countered through an increase in the supply of gold—this led to intensification in the search for gold across the world. It was in the context of these changes in the global monetary system that India, and more specifically the Nilgiri–Wayanad region, was drawn into an international gold rush. The desperation of this search for gold proved a sufficient trigger to induce the British to explore the possibility of setting up industrial-scale gold-mining operations in the Wayanad region, in the 1870s. To do so, their immediate concern was getting access to the land. The following extract from King (1875: 45) clearly exemplifies this major issue:

Now that gold-mining is likely to become an industry, a new set of land interests are being developed. The Rajahs, of course, retain their right to all minerals and can sell these as they like. The Government of Madras has not yet, I believe, decided as to how they are to act in the matter, except that applications for land for gold-mining and for agricultural purposes on which quartz reefs are supposed to exist, are being reserved for consideration until the question of mining is settled.

In the meantime the Rajah of Nellamborhas (according to their prospectus) leased a block of 15 acres of land near Dayvallah to the projectors of the Alpha Gold Company for twelve years at an annual rent of Rs.225. Since then it is reported that the Rajah in recent applications demands 10 per cent on the outturn of any gold-mining which may be carried on; and it is very probable that he may change this rate. Nearly all the land in the Nambalycode Amsham is owned by the Rajah of Nellambor. Equally, as with the revenue derived from estates on Rajah's lands, it may be found advisable that the gold from these reefs should pay a royalty to Government.

This statement underlines two points: the growing concern of the industrialists on the ownership of land by local feudatories, and the extraction of high lease rents; and the role played by the British Government in Madras in settling these issues.

During this period, enormous amounts of investment flowed into Wayanad. Some 41 companies with a paid-up capital of more than GBP5 million were set up. Domestic companies invested some GBP250,000 and began mining operations in Wayanad. Writing in the *Madras District Gazetteers: The Nilgiri* in 1908, Francis reflects on this period and suggests that the mood was upbeat and that 'the whole mountain is worth putting through the stamps' (Francis 1908: 17). Most of the capital invested went into procuring land through mining leases; land prices varied between GBP70 and GBP2,500 per acre. These reflections by Francis provide a vivid picture of the magnitude of corporate activities in gold extraction that began in Wayanad in the late nineteenth century. The quote given below from Francis (1908: 27–9), suggests the magnitude of corporate activities in the Nilgiri–Wayanad gold extraction in the late nineteenth century:

In 1879, however, the Government of India employed Mr. Brough Smyth (for many years Secretary for Mines in Victoria and held to be the greatest authority on the subject in Australia) to examine the Wynaad reefs; and his report, written in October 1879, was, to say the least, distinctly encouraging. He discussed in detail the value of a great number of the known reefs, most of which crop out in the country traversed by the road from Nadgani (same as Nadugani) to Cherambadi; gave the results of assays made by himself and others which ranged from nil to no less than 204 oz. of gold to the ton of ore; considered that low-grade ores, running even as low as 3 dwts.[12] to the ton, could be worked at a profit.

The result of this sanguine report was the artificial boom of 1880. The stock markets were ripe for any speculation, however wild. Low rates of return on British Government stocks, a paucity of foreign loans, flourishing trade and an unusual scarcity of gold all contributed to make miscellaneous enterprises more attractive; while coffee-planting' was already on the down-grade and owners of estates containing reefs were only too glad to seize a chance of disposing of them at a profit. The mania began in December 1879, when a company with a capital of 100,000 pound was launched; and in the next nineteen months the number of companies floated in England amounted to no less than 41 with a capital of over five million sterling, while during the same period six companies with a total capital of £261, 000 were also started in India itself. Of the English companies, 33 went to allotment and the sum obtained by them for investment in the industry amounted nominally to £4,050,000. Of this, however £2,375,500 was allotted for payment for the land in which the supposed mines were located—the prices of this ranged from £70 to no less than £2,600 per acre—so that the sum left for working expenses was not more than £1,674,500. Mr. Brough Smyth himself was appointed manager of two of the companies, but retired on the ground of ill-health in 1882, when the tide had begun to turn.

Phase II was a marked by intense capital investment, employment of wage labour and setting up of modern public limited companies listed on stock markets, which in turn lead to a high magnitude of speculation. However, by the early 1880s, the edifice of gold-mining in the region collapsed; the concentration of gold was too low to make large-scale mining feasible. Most of the companies had to shut down and investors lost large sums of money. Tremors from the bust even rattled the London stock market. The short-lived formal corporatisation phase seemed to be merely speculation driven. However, the repercussion of this phase had significant influence on the subsequent phases of informalisation of gold-mining in the region. However, it is clear from a study of colonial records that although gold-mining remained formal, the technology employed had changed from artisanal to modern—the latter including a higher degree of capital-intensive mechanisation as well as the use of wage labour.

12 dwts is also known as penny weight. 20 dwts=1 troy ounce (also written as oz.)= 31.104 g.

Phase III: From the 1900s to the 1970s

There is little known to us of this post-corporatisation phase, apart from the gloomy picture portrayed by Francis (1908). He describes how Pandalur and Devala, by the turn of the twentieth century, had turned into nothing but ghost towns (ibid.: 31):

> At Pandalur three or four houses, the old store, and traces of the race-course survive; at Devala are a grave or two; topping many of the little hills are derelict bungalows and along their contours run grass-grown roads; hidden under thick jungle are heaps of spoil, long-forgotten tunnels used only by she-bears and panthers expecting an addition to their families, and lakhs' worth of rusting machinery which was never erected; while along the great road to Yayitri, which now, except for the two white rats worn by the infrequent carts, is often overgrown with grass, lies more machinery which never even reached its destination. Moreover, most of the numerous coffee- estates which formerly bordered this road all the way from Gudalur to Cherambaidi were acquired by the gold companies and thenceforth utterly neglected; and now not a single one of them all is kept up. They have all gone back to jungle and are covered with such a tangle of lantana and forest that it is hardly possible to make out their former boundaries. Thus the coffee industry is dead and the mining industry which killed it is dead also; and this side of the Wynaad is now perhaps the most mournful scene of disappointed hopes in all the Presidency.

What happened to gold-mining in Nilgiri–Wayanad after that is less known. This period also saw major changes taking place in monetary systems across the world that would have naturally impacted gold-mining in the Nilgiri–Wayanad region, and industrial mining in particular. Talking to a few individuals in the region and their recollections of their grandparents suggests that the mining activities continued, albeit informally, in the ruins and remains of the tunnels abandoned during the earlier phase. It would certainly be of interest to find out more about who these miners were. In materials gathered by the family members of the Neelambur *Rajah*, we found at least two expressions of interest by corporate entities to extract gold and mica, dated 1924 and 1932. However, these letters remained at the expression of the interest stage and the projects never materialised. As such, we want to suggest that this phase of mining was an 'open-access' phase without any excludability or rivalry, where a range of individuals or groups could freely ride on the already excavated tunnels. The end of the gold rush, which left gold-mining in the region unattractive for industrial-scale mining, opened up the region for open-access mining. A large tract of the land, known as section 17 or Janmam land, mired into litigation in the court of law, leaving no clear ownership of land (Menon 2014). Multiple stakes existed, including the forest department of the Government of Tamil Nadu, the erstwhile Rajah of Neelambur and tea estate and other plantation owners, all of whom left the land ungoverned and an open-access space for later entrants.

Phase IV: From the 1980s onwards

Our several visits to the field between September 2014 and May 2016 revealed the contemporary narrative of gold-mining in the region. We travelled through Nadugani, Pandalur and Devala in Gudalur *taluka*[13] of Nilgiri district and identified many individuals engaged in gold-mining. The current practice of gold-mining is informal and artisanal. The number of people engaged in this activity seems to have surged in recent decades, possibly due to the increase in global gold prices. Our attempt to identify the genesis of persistence and the regrowth of gold-mining in this region led us to understand demographic changes that occurred here during the early 1970s, when Tamil workers, previously employed on Sri Lankan tea plantations, were resettled in this region. This resettlement was the result of a pact between the then Indian prime minister, Lal Bahadur Shastri, and his Sri Lankan counterpart, Sirimavo Bandaranaike, known as the Shastri–Bandaranaike Pact (Pillai 2012). This led to a number of Tamil migrants working in Sri Lankan plantations being brought back to India in the 1970s and settled in various places, including Gudalur *taluka*. The Sri Lankan diplomat K. Godage (n.d.) wrote:

> Indentured labour from South India was brought to work on our tea and rubber plantations.[14] The issue of Indian immigrants became an intractable problem between the two countries in 1953 when India resiled from its position that these immigrants were Indian nationals. It was only in 1964 that an agreement was reached between Prime Ministers Sirimavo Bandaranaike and Shastri. This agreement was supplemented by another between Prime Minister Indira Gandhi and Bandaranaike in 1974 with India agreeing to take back 600,000 and Lanka agreeing to grant citizenship to 373,000.

---

13 *Taluka* is a sub-district.

14 The arrival of labourers from South India to Sri Lanka has a long history. After Ceylone was conquered by East India Company, a large-scale temporary migration happened in the 1840s to the coffee plantations, especially during harvest season. Permanent migration and settlement gained momentum since the 1870s, with the initiation of tea plantations that required regular labour.

Our observations during recent field visits show that large numbers of informal gold miners belong to this Ceylon[15] repatriate community. To provide employment to the settlers, the Government of Tamil Nadu created TANTEA, a tea plantation company to absorb the repatriates who were used to a similar kind of work environment in Sri Lanka. The state provided a financial incentive to private estates to absorb repatriates (Menon 2014). TANTEA plantations established between 1968 and 1980 comprised about 2,500 ha, created by clear felling of forests (ibid.). However, the time lag between the arrival of repatriates and absorbing them into plantations would have been one of the prime causes of a spillover into informal mining activities by this community. This also suggests that the gold-mining fields were open-access grounds in the early twentieth century for the repatriates and others to work in and earn their livelihood. Added to the time lag in absorbing the workforce, the downfall in the tea market would also have played a critical role in the proliferation of 'informal gold-mining', where neither formal records were kept nor taxes collected. The technology employed in these mines is rudimentary while the economic structure is more cooperative and family based.

In Pandalur, we found mines located within private tea estates, whereas in Devela, most of these are inside the forest reserve areas, and in Nadugani, there was a mix of both. In Pandalur, Nadugani and Devala, with the help of local people, we visited several tunnels where mining activity currently occurs. These tunnels can be 500–3,000 m in length. Some tunnels are horizontal throughout, whereas some are vertical over a considerable length before becoming horizontal. In the tunnels are several types of quartz stones with tinges of gold, mixed with varieties of other materials (especially mica). These stones are collected by these informal miners. The collection process involves digging with shovels, hammers, etc. in the tunnel, and often blasting with low-intensity explosives. In a day, a miner collects about 10 kg of such stones, which may translate into less than a gram of gold. One of the miners told the researcher, 'our target is to make 500 rupees in a day'. In today's context, this translates into a third of a gram of gold (excluding other expenses of crushing, etc.). This particular miner was young, just about 25 years of age, belonging to one of the repatriate communities from Sri Lanka resettled here.

---

15 Ceylon was renamed Sri Lanka in 1972. We use Ceylon and Sri Lanka alternatively, the first representing an historical sentiment and the latter a contemporary thought.

The selected quartz ores are taken to a crushing mill, in a few cases called 'rice mills'. The reason behind this obviously lies in how statutory law perceives informal mining. The crushing mill owners are often different from the miners and they charge around INR4–10 per kilogram of stone. However, it is likely that some of the crushing mills, apart from crushing the stones brought in by miners, would crush their own collected stones. Stone-crushing mills are important factors contributing to efficient production of gold. After all, a significant portion of investments of the colonial industrial-mining operations was in large-scale stone-crushing equipment. Once the stones are crushed, it is panned using mercury and gold particles are collected. These gold particles are then sold to jewellers or local finance companies in the region. With a higher level of awareness among the informal miners, the price at which the gold is sold moves with the change in market price, although the rates are lower than the market price.

Synthesising the phases

From the available records and documents in the early colonial period, we identify that a formal tax structure (in the form of trey tax) was used by the local *Rajahs* and *zamindars* to provide rights to the gold miners till the 1860s. The tax structure is a clear form of exclusion for those who do not pay the tax. In addition, like most other occupations in India, there were specific castes and tribes that engaged in gold-mining. Put together, we have interpreted one element in the then existing property rights regime as 'excludability'. The wide dispersion in rather small quantities of gold across a region, as well as the lack of capital, meant that several persons from the castes and tribes were engaged in the occupation; we therefore also impute an element of 'non-rivalry' amongst the miners, as long as they were able to pay the stipulated fee to the feudatory. The characteristics of 'excludability' and 'non-rivalry' are the essence of 'club' goods, which was considered to be the dominant form of property rights until the 1860s. There are also several instances of large tracts of land directly occupied by feudatories or *Rajahs*, who mined gold using 'slave labour' at very low wage rates. In these cases, it becomes more appropriate to characterise the property rights regime as 'private goods'. Therefore, Phase I has hues of both: club goods as well as private goods. More importantly, we are clearly able to establish that gold-mining was not an 'open access/public good' (non-excludability, non-rivalry) or a 'common pool' (non-excludability, rivalry) resource in the late medieval or early colonial period.

Gold-mining during the Wayanad gold rush (Phase II) was clearly a private good, exhibiting elements of excludability and rivalry amongst miners. Not only that, this period was also marked with intensive speculation and making short-term gain out of the speculation. The elements of speculation were in the geo-political and economic order of that time. However, Nilgiri–Wayanad witnessed it in the form of intensive deployment of capital and wage labour.

It would not be wrong to say that Phase III of mining was an 'open-access' phase, without any excludability, rivalry or free riding on the opened tunnels; though it would be worthwhile to determine who the miners were during this phase. Very little information is available on this phase. In the later phase, the existence and surge of gold-mining by Sri Lankan repatriates (who happened to be late entrants to the region, between the 1970s and 1980s) suggest that the gold-mining areas were open access, to be appropriated by anyone interested in this activity.

The fourth phase of informal mining has possibly taken the shape of commons, where creating new mining areas is not as easy. A large tract of land is under litigation due to multiple claims by the erstwhile *zamindars*, the state and the estate owners, as well as later settlers. This tract of land is known as Janmam land or section 17 land (Menon 2014). However, in recent times, the forest department is believed to have a de facto claim over the land, especially in the region where there is forest cover. In these lands, the forest department does not allow any expansion of new mining activities. This suggests that the mining activities happen in the defined and already opened areas, leading to rivalry in natural resource exploitation. The groups of people involved have shifted and changed over time. Paniyas remain active and Sri Lankan repatriates are being assimilated; however, one does not find as many Korumbas (or Kurumbas) as evident in the first two phases. If we identify the informal rules and norms governing access to the mining areas, as well as the conflicts between groups in accessing the mines for extraction of quartz stones, it would be more evident that there is rivalry in appropriation.

Governance structure over the centuries

The two centuries of the history of gold-mining in the region show the changes in its institutional and governance structure. These changes are not necessarily due to internal factors. Rather, external circumstances have played a critical role in shaping and reshaping the institutions involved in gold-mining.

We attempted to analyse these phases of changes using the concepts of property rights. The principle of 'excludability' and 'rivalry' provided us a framework to analyse the institutions and governance structure in the four different phases. Table 11.2 summarises the four phases in terms of property rights–based institutional structure.

Table 11.2: Institutions of gold-mining over the past two centuries

Phase	Excludability	Rivalry	Legal Status and Scale	Communities
Early Colonial Phase (up to the 1860s)	Yes (with tax structure)	Minimal	Formal–Artisanal	Kurubas, Paniyas, Mapillahs (engaging labour), landlords and feudatories
Corporatisation Phase (1870s to 1900)	Yes	Yes	Formal–Industrial	Privatisation of property with corporate mining initiatives
Third Phase (1900 to the 1970s)	Minimal	Minimal	Informal–Artisanal	Little is known of this phase. But mostly by Kurumbas, Paniyas and Mapillahs (engaging labour)
Ceylon Repatriate Phase (1980s onward)	Minimal	Several instances	Informal–Artisanal	Interestingly, Kurumbas are not much involved in the activity; Paniyas continue, Ceylon repatriates are in good numbers

Source: Authors' work.

The property rights–based analysis suggests that gold-mining in the region has undergone institutional changes. The first two phases show a formal mining setup, whereas the next two phases are informal. Within this transition in institutions, external factors (like the gold rush and the arrival of repatriate labour) played a critical role.

In contemporary literature, one finds classifications based on formal and informal mining, as well as in terms of artisanal and industrial-scale mining. Artisanal mining is often classified with small-scale mining as ASM. Based on this classification, approaches to understanding and categorising mining activities have been varied. In this chapter, we have added another dimension to the debates about classification by not only considering property rights, but also the way these shift and change over time, influencing the forms institutions can take. Contemporary approaches to informal mining emphasise its operation in an ungovernable space, and ignore its role as a source of livelihood for local people. Approaches to artisanal mining emphasise the scale and forms of technology used in the activity—discussions that tend to beget debates about efficiency. Property rights–based classifications, on the other hand, combine institutional and economic approaches useful for understanding governance, from both the perspective of livelihood and efficiency.

Conclusion

A number of factors worked together to create a situation in which informal mining came to prevail in the Pandalur, Nadugani and Devala settlements of the Nilgiri–Wayanad region. Mining activities in the region date back to the 1830s in the available documented history at which time, as we have argued, it was formal–artisanal; however, contemporary mining was largely shaped by the unique history of the local context. Currently, a large proportion of informal miners belong to the Sri Lankan repatriate communities. Additionally, a smaller proportion of the miners belong to the native population.

The historical, social, economic, geographical and geological factors considered together show the complexity associated with mining that transited from a formal to informal process. A long history of the profession, position of statutory law and the prevalence of customary practice add to this complexity. However, a few important aspects that come out clearly from this explorative study are worth taking note of. First, existence and prevalence of informal gold-mining in India is a fact. Second, people engaged in this mining activity are poor and belong to vulnerable social groups, and informal mining contributes substantially to their livelihood. Third, the long history of gold-mining shows the transition in the institutions—these are not only internal to the economy, but are also impacted by external, global, political and economic factors that play a critical role in shaping the institutions.

In the process, this study defies a few theoretical positions. First, a body of literature suggests that changes in institutions and the governance structure are largely due to the internal dynamics, as suggested in the evolutionary strand of institutional economics. However, our study shows that external factors like the gold rush, collapse of the gold rush and settlement of Sri Lankan repatriates in the region did play a critical role in the evolution and changes in the institutions relating to gold-mining in the region. Second, institutional economics approaches tend to emphasise either property rights or transaction cost economics, which suggest that institutions evolve over a period of time, becoming more efficient (in terms of minimising transaction costs) and well defined (in terms of property rights). However, in our study, we identify that the institutions have moved from a formal to an informal structure, or from a high-excludability to a low-excludability structure. In the pre– and post–gold rush phase, the institutions have shown a tendency towards greater rivalry in use, over a period of time.

Acknowledgements

This study was conducted with financial and technical support from the South Asian Network for Development and Environmental Economics (SANDEE). SANDEE also provided academic support from experts through regular meetings. It also received financial support from the Australian Research Council through its ARC LP130100942, 'Going for Gold'. Earlier versions of the chapter were presented in the Development Convention, Gujarat Institute of Development Research, Ahmedabad, India; International Association for the Study of the Commons (IASC) 2015 at University of Edmonton, Alberta, Canada; the School of Spatial Sciences, University of Groningen, The Netherlands; Informal Mining in the contemporary world at ANU College of Asia and Pacific, The Australian National University, Canberra; and Commission on Legal Pluralism conference at Indian Institute of Technology (IIT), Bombay. The authors would like to thank the audience of these conferences for their comments.

References

Baber, T.H., 1830. 'Journal of a Route of the Neelghurries from Calicut.' *The Asiatic Journal and Monthly Register of British and Foreign India, China, and Australasia,* Vol. III, New Series, Sept–Dec.: 310–16. London: Parbury, Allen & Co.

Ball, V., 1869. 'On the Occurrence of Gold in the District of Singhbhum.' *Records of the Geological Survey of India* II(1): 11–14.

Barzel, Yoram, 1997. *Economic Analysis of Property Rights.* 2nd edition. Cambridge: Cambridge University Press. doi.org/10.1017/CBO9780511609398

Bergein, 1834. 'Gold Mines of Malabar.' *The Mechanics' Magazine, Museum, Register, Journal, and Gazette,* Vol. XX, 5 October 1833 – 29 March 1834: 123–4. London: M. Salmon, Mechanics Magazine Office.

Coase, R.H., 1937. 'The Nature of the Firm.' *Economica* 4(16): 386–405. doi.org/10.1111/j.1468-0335.1937.tb00002.x

Coase, R.H., 1959. 'The Federal Communications Commission.' *Journal of Law and Economics* 2(October): 1–40. doi.org/10.1086/466549

Deb, M., G. Tiwari and K. Lahiri-Dutt, 2008. 'Artisanal and Small-Scale Mining in India: Selected Studies and an Overview of the Issues.' *International Journal of Mining, Reclamation and Environment* 22(3): 194–209. doi.org/10.1080/17480930701679574

Demsetz, H., 1967. 'Towards a Theory of Property Rights.' *The American Economic Review* 57(2): 347–59.

Francis, W., 1908. *Madras District Gazetteers: The Nilgiris.* Madras: Government Press.

Godage, K., n.d. 'Historical Continuities.' India Seminar. Available at www.india-seminar.com/2002/517/517%20k.%20godage.htm

International Institute of Environment and Development (IIED) and World Business Council for Sustainable Development (WBCSD), 2002. *Breaking New Ground; Mining, Minerals and Sustainable Development.* London: Earthscan Publications Ltd.

Jennings, S., 1881. *My Visit to the Goldfields in the South-East Wynaad.* London: Chapman and Hall Limited.

King, W., 1875. 'Preliminary Notes on the Gold Fields of South East Wayanad, Madras Presidency.' *Records of the Geological Survey of India* VIII: 29–45. London: Trubner & Co.

Lahiri-Dutt, K., 2004. 'Informality in Mineral Resource Management in Asia: Raising Questions Relating to Community Economies and Sustainable Development.' *Natural Resource Forum* 28: 123–32. doi. org/10.1111/j.1477-8947.2004.00079.x

Libecap, G.D., 1978. 'Economic Variables and the Development of the Law: The Case of Western Mineral Rights.' *The Journal of Economic History* 38(2): 338–62. doi.org/10.1017/S0022050700105121

Libecap, G.D., 1989. 'Distributional Issues in Contracting for Property Rights.' *Journal of Institutional and Theoretical Economics* 145(1): 6–24.

Liebowitz, S. and S. Margolis, 2000. 'Path Dependence.' In B. Bouckaert and G. De Geest (eds), *Encyclopedia of Law and Economics, Volume I. The History and Methodology of Law and Economics.* Cheltenham: Edward Elgar.

Menon, A., 2014. 'The Godavarman Judgment: Erasing the plurality of land use in Gudalur, Nilagiri.' In M. Bavinck and A. Jyotishi (eds), *Conflict, Negotiations and Natural Resource Management: A Legal Pluralism Perspective from India.* London: Routledge.

Newbold, Lieut., 1843. 'Mineral Resources of Southern India, No. 4, Gold Tracts.' *The Journal of the Royal Asiatic Society of Great Britain and Ireland* 7: 203–11. doi.org/10.1017/S0035869X00155856

North, D.C., 1990. *Institutions, Institutional Change and Economic Performance.* New York: Cambridge University Press. doi.org/10.1017/ CBO9780511808678

Ostrom, E., 2005. *Understanding Institutional Diversity.* Princeton: Princeton University Press.

Pillai, R.S., 2012. 'Indo–Sri Lankan Pact of 1964 and the Problem of Statelessness: A Critique.' *Afro Asian Journal of Social Sciences* 3(1): 1–14.

Sikor, T., and C. Lund, 2009. 'Access and Property: A Question of Power and Authority.' *Development and Change* 40(1): 1–22. doi. org/10.1111/j.1467-7660.2009.01503.x

Sivramkrishna, S., 2015. *In Search of Stability: Economics of Money, History of the Rupee*. New Delhi: Manohar.

United Nations Development Programme (UNDP), 1999. *Artisanal Mining and Sustainable Livelihoods*. New York: United Nations Development Programme.

Williamson, O.E., 1981. 'The Economics of Organization: The Transaction Cost Approach.' *The American Journal of Sociology* 87(3): 548–77. doi. org/10.1086/227496

Williamson, O.E., 2000. 'The New Institutional Economics: Taking Stock, Looking Ahead.' *Journal of Economic Literature* 38(3): 595–613. doi.org/10.1257/jel.38.3.595

Zenger R.D., S.G. Lazzarini and L. Poppo, 2001. 'Informal and Formal Organization in New Institutional Economics.' Viewed at apps.olin. wustl.edu/faculty/zenger/advances6u.pdf (site discontinued)

12

Conflicts in marginal locations: Small-scale gold-mining in the Amazon

Marjo de Theije and Ton Salman

Researchers of the GOMIAM consortium[1] have worked on conflicts surrounding small-scale gold-mining in the Amazon since 2010. We understand conflict in a broad sense and have studied many different forms of it, ranging from armed conflict and environmental conflict, to conflicts of interest between local populations and migrant miners, between large-scale mining and small-scale mining, between local, regional and national governments on issues of small-scale mining, and even between small-scale miners and other small-scale miners, like between equipment owners and operators. GOMIAM researchers work in Bolivia, Brazil, Colombia, French Guiana, Peru and Suriname.

All the various types of conflicts we mentioned have increased dramatically because of the tremendous growth of small-scale gold-mining, but are also seen in medium- and large-scale gold-mining. Environmental issues are among the most prominent causes of conflict. The environmental threats connected to gold-mining include deforestation and air and water pollution from cyanide and mercury contamination. Polluting effects

---

1   GOMIAM is a comprehensive research project on small-scale gold-mining and social conflict in the Amazon region, comparing states, environments, local populations and miners in Bolivia, Brazil, Colombia, Peru and Suriname. See www.gomiam.org.

of small-scale gold-mining often threaten the livelihoods of indigenous peoples or peasants in the vicinity of the mining operations. Small-scale mining activities are often relatively low-tech, organised with informal labour and family networks and, although limited in geographical scale, are still very harmful because of the uncontrolled expansion of (albeit often small) operation sites, and because of the frequent careless use of mercury and other chemicals, with little awareness about the effects of inorganic waste.

Not only does the environment cause conflicts, the land does too. Land conflicts between miners and other collective or private concession holders or local inhabitants, like indigenous communities, peasants, loggers, Amazon nut-collectors or those involved in other forest extraction activities, abound. Quite often, due to deficiencies of government institutions, overlapping concessions are given out to various local players, like miners, agro-business or eco-tourism entrepreneurs, and indigenous communities. Conflicts between the various state ministries responsible for these contradictory policies are often minor in comparison to the conflicts that erupt on the ground because of what was granted to such players.

Such effects obviously bring up critical questions about concessions given out by national or local authorities, about the lack of effective control and protection for the local population, in case the concessions were not officially granted, or if the miners or mining companies do not respect the conditions that were attached to it.

These are only some of the fields of conflict to be found. In our analysis of the gold conflicts, we have found many similarities between the countries studied. Next to the elements we mentioned above, there are many other parallels; for instance, in all countries the state makes efforts to control or profit from the gold-mining, usually initiating processes of 'formalisation' that include licensing, taxation and environmental requirements. The success of such efforts varies. In all countries, local people and migrants compete and/or collaborate in artisanal and small-scale mining (ASM), making a living from it in situations where they often have very few alternatives. Also, everywhere, small-scale gold-mining is associated with negative qualifications of illegality, criminality, prostitution and violence. The reactions researchers get when announcing a visit to the mining areas are always full of warnings and calls to be careful. In the imagination of many, small-scale gold-mining is full of corruption, vigilantism, an eye for an eye and a tooth for a tooth. Finally, and in spite of much heterogeneity in this regard, we found that it is not anarchy or sheer ad hocism that rules in the mining areas. Everywhere there is some form of association, organisation and rules.

Our analyses of the conflicts have focused on different forms of legal systems (Theije et al. 2014), on formalisation and inefficient state capacity (Damonte Valencia 2013; Heemskerk et al. 2015), on migration (Theije and Heemskerk 2009), on organisation of miners (Carrillo et al. 2013) and more. However, the conflicts are complex and multilayered, and we need theoretical tools that might help to unpack and analyse the nature, layers, time frames and possible ways forward to address these conflicts. Elsewhere, (Salman and Theije 2017) we have suggested a multitemporal model to understand the structural backgrounds, the actors' dispositions and capacities, and the strategic and tactical manoeuvring of different stakeholders in conflicts. Here, we focus on a feature of another nature: the spatial characteristics of most gold-mining operations.

In this chapter, we will take a small step in the development of such a framework by attempting to shed light on one specific characteristic of mining that became apparent in all our case studies—in all countries, the mining takes place in remote parts, distant from urban centres, distant from regulating state apparatuses, distant from services and legal markets. What is the role of this physical distance, this far away situation and this isolation from urban centres in the conflicts surrounding mining, and how can our attention on this fact lead to insights that might contribute to a more thorough understanding and possibly to a solution for the conflicts?

We therefore explore the multilayered nature of conflict using the notion of marginal territory, to shed light on some of the most notable features of conflict and the processes underlying the conflicts in the small-scale gold-mining in the Amazon. The location and marginality of the mining regions is part of the structural dimension of the conflicts related to small-scale gold-mining. What is the role for the actual interactions and encounters, the confrontations between actors in the field and the choices people make to join in, be part of the conflict, or stay out?

We will distinguish four different takes on the remote, isolated, far-away situation of many of the small-scale gold-mining operations in the Amazon, and discuss each of these on the basis of a short case. Based on the results, we come to some general conclusions about the role of geographical and political distance as a structural feature in the conflicts around gold-mining.

Mining in marginal territory (outside state control)

The mining activities we study usually take place in remote parts of nation states, in the dense and often difficult to access remote areas in the tropical rainforest. The physical and organisational distance from state control affects conflicts between state representatives and groups of the population, and miners who have different accesses to state backing and authorisation. Remoteness means that activities take place in areas that are marginal from the perspective of the political and economic centres of that region of the country. For the parties involved, it means exclusion from locations of decision-making, as well as a lack of representation and presence in public spheres.

A good case in point is the Peruvian region of Madre de Dios, which is the scene of an enduring dispute between small-scale miners and the authorities. In part, the dispute can be explained by knowledge hiatuses, miscommunication and refusal to understand the other party's interests. Damonte (2016) explains how the Peruvian Government attempted to formalise small-scale mining as a measure in a larger plan to counter the environmental impact of illegal mining. The formalisation, however, included declaring all small-scale mining in the protected non-mining zone areas as illegal, thus creating more illegality. From the perspective of the miners, this was a threat to their livelihoods and a violation of agreements they had about the use of the land. So far, no workable solution has been found to resolve the conflicting understandings of the situation.

The central government in Lima is not really interested in the miners, or in the region, and gives no priority to combating illegal small-scale mining. But at the same time, the government is whimsical in this respect. Pressure to intervene comes from the international community and Peru's own (weak and inefficient) environmental department. The benefits of the small-scale mining in economic terms are marginal for the state, as it makes no significant contribution to Peru's gross domestic product. What remains is that for the Peruvian state, the formalisation process is a governance instrument for a remote region that it still is unable to govern.[2] The central government fails to acknowledge the complex social dynamics that rule the remote Amazonian region.

Historically, the Peruvian state followed a 'conquest' approach that solely focused on extraction of resources from the region. In Madre de Dios, the state never developed a program of social and economic development. As a consequence, the state can hardly 'read' contemporary local society (Damonte 2016) and resorts to punitive actions against the miners on an ad hoc basis, most often without success. The central government has proved incapable of collaborating with regional authorities, for whom the small-scale miners represent an economically important factor, and a politically dominated area. As a result, the miners refuse to collaborate, and instead contest, delay and sabotage the government's plans to reorganise their mining activities. The authorities are insufficiently equipped to enforce the law, because they are unable to grasp the complex societal dynamics underlying small-scale mining. Specifically, the failure of the state to integrate local people and local agendas in the process of formalisation hinders current attempts to intervene with any real effect. For the state, operating 'from Lima', Madre de Dios continues to be a marginal, unknown and unruly area.

Mining in marginal territory (delegated state control)

The small-scale miners in Peru worked and organised the territory in the absence of the state for a long time. Now that the state wants to interfere, they ignore, resist and subvert the actions of the state. This also happens in other parts of the Amazon. Others may step into the void to 'represent' the state or assume the tasks it is unable to fulfil. Who are the actors that can do that, and what are the means they apply to do so?

---

2   Damonte refers to Foucault's (2009) discussion of the concepts of sovereignty and governmentality, as the different and successive ways states can establish governance. A state achieves territorial sovereignty when it is capable of controlling the territory through territorial knowledge (territorial data) and administrative power. 'Governmentality' refers to the 'institutions, procedures … that allow the exercise of … power that has the population as its target' (Foucault 2009: 108), and to the government apparatuses and related knowledge used to govern the population that lives in its sovereign territory.

To elucidate this question, Suriname offers a good example. Suriname is a small country in the northeastern part of the Amazon, which witnessed a substantial increase in small-scale gold-mining activities in the 1990s. The gold-bearing lands are located in the sparsely populated interior, where only Maroons and Amerindians live. In this part of the country, the state has historically been completely absent, and still there are hardly any services provided to the people: no schools, no health services and no police, to name just some basics. The local inhabitants were totally overlooked when mining began to develop. The government issued several large concession areas for prospecting to some businessmen and to the national mining company. In and around these 'official' concessions, the local tribal people in whose traditional territory the concessions were located also started mining more actively than they had done in previous decades. But land rights of the tribal people are not recognised in Suriname, and the rights to mine were never assigned to them either. This resulted in colliding claims and overlapping operations.

In 1991, two concessions were issued to people from the capital, Paramaribo, at the Lawa River—traditional territory of the Aluku (or Boni) Maroons and Wayana indigenous peoples. For the indigenous people, the new concession holders, with official documents in their hands, were complete strangers, even aliens—as if Columbus had landed again on the (allegedly uninhabited) territory of the continent and, without even blinking, declared, 'In the name of the King of Spain I take possession of this land and everything that belongs to it …'. Now, new territorial names and rights were created. The Antino concession came into the hands of Nana Resources, a family company of the Naarendorps; the Benzdorp concession went to Grassalco, the state business company. The two concessions each comprise about 140,000 hectares. Although, strictly speaking, the rights conferred to these outsiders were exploration rights and not exploitation rights, the companies soon started to effectively mine on the concessions. They did so by inviting others—migrant miners and local Maroons. Instead of initiating operations themselves, they gave access to migrant small-scale miners, most of them Brazilians, with a lot of knowledge and experience as small-scale gold miners (usually called *garimpeiros*), who brought their own equipment and paid 10 per cent of the production to the concession holders.

Subletting is prohibited in the national mining law, but in the miners' culture it is an accepted rule that the 'owner of the mining land' has the right to collect a percentage of the gold. For the Brazilian *garimpeiros,* this was a known practice. However, the indigenous Aluku and the Wayana felt that this completely ignored and overruled their (customary) rights and they resisted complying with these rules, which were now claimed and implemented by the new 'owners of the land'. In this struggle, the state remained absent on the ground. In Paramaribo, a minister or another official would occasionally explain on television that all mineral resources belong to the nation, suggesting that these firms (having their base in the city of Paramaribo) were representing the state and operating on the basis of its ruling and, therefore, had more right to mine than local inhabitants or migrants, who in actual fact were doing all the hard work. This was all in the midst of internal frictions between the *garimpeiros* and the Maroons, and with the 'rights dispute' remaining a moot point.

The situation became even more complex when the two firms that were the concession holders started negotiating with 'big fish' from elsewhere: Canada- and USA-based junior companies; exploration companies looking for gold deposits that have potential for large-scale exploitation; and large-scale mining companies. The two firms, Nana Resources and Grassalco, acted as gatekeepers and facilitators. They had probably taken on the concessions because they hoped to attract foreign investors, who would have the knowledge and capital to undertake serious prospecting, and would eventually build and exploit large-scale mines on their concessions. In the early 1990s, such companies initiated professional exploration activities: Golden Star at the Antino concession, and Canarc at the Grassalco concession. In both cases, it did not result in large-scale exploitation projects. The gold reserves were not big enough to justify the scale of investment that would be necessary. It was unlikely that costs and yields would bring a beneficial outcome.

Whereas the local gold miners—the Maroons—refused to pay 10 per cent of the produce to the concession holders, the migrant miners adapted to the rules set by the firms. The small-scale miners were allowed to work wherever they wanted on the large concessions. They had to prospect for themselves, and could then talk to the 'security man' of the firm to obtain permission for opening a mine on the location of their choice. Doing so in large numbers, the *garimpeiros* became the prospectors for the international firm, while they also paid 10 per cent of their raw production to Nana Resources (Theije and Bal 2010). The concession owners collected kilograms of gold every week. In the heydays of the Antino concession, at the beginning of 2006, more than 60 small-scale mining operations were producing gold. Although later, the number of producing miners went down—amongst other reasons, because they were sent away when Nana Resources eventually started a mining operation on part of the concession—the subletting system continues to be an important part of the local organisation of the mining business.

So in the absence of, or under ineffective control by, state representatives, gold production is completely managed according to the rules set by the concession owners. On the margins of Suriname society, far away from the central state, but deriving authority from this same centre, delegated through the concession system, the concession owners use this authority to guarantee and monopolise for themselves, and subsequently distribute to others, the access to profits (Ribot and Peluso 2003) from the gold-mining activities. Their power, however, has not been effective enough to make the Maroons obey.

Marginal territory creates local control

In the absence of central authority, or in the presence of ineffective authority, conflict resolution and organisation at the mining sites is often self-administered and self-regulated. The self-administration is, of course, more flexible than bureaucracies or other institutionalised forces imposing the rules. But it often comes at the cost of imbalances in terms of abilities to promote one's interests. Yet, there is a common interest in security about the ways to act, to obey and to comply. Even in makeshift circumstances, in temporary mining camps, in continuously changing working settings, total improvisation or 'making it up as you go' will not last long; people will get together and arrange a system, no matter how precarious, unjust or violent it may be, to channel things. Total anarchy and unpredictability is bad news for *all* players, whatever their working and living configurations. We will return to this point later.

Where the state has no control, others take over. Some will often monopolise the use of violence and control over people's moves and work. In the Suriname case just described, the 'security' employees of the concession holders most often assume the role of police, even when it concerns persons and issues not related to the firm and mining on the concession. So when there is a fight in the neighbouring settlement, they are the ones that will go there to settle issues, administer justice and restore peace, sometimes even keep perpetrators captive for a short time, or some days, before handing them over to the national police officers who are called from town. The miners also appeal to them to settle quarrels between spouses, or to ask help with disagreements regarding payments between miners.

In other parts of the Amazon, the hosts to the migrant miners may also take on the role of local authority and control the conflict and violence. In Brazil, this is commonly the 'owner of the mine', who usually is the person who first found gold in the location and therefore obtained the right to ownership in the mining culture. Although the 'owner of the mine' eventually may have (partly) formalised his access to the gold, usually his power extends to other arenas of social life in the mine. 'Inside' is a world in itself, where gold-mining defines activities, identities and rules (Larreta 2002: 45). Inside, the negative qualifications of small-scale gold miners that are prevalent 'outside' are absent. Inside is also where the marginality of the territory gives way to the authority of the owner of the mine, who not only safeguards his access to the gold, but also decides on other issues in the *garimpo*, the mining site, such as resolving conflicts between workers.

In quite a similar way, the inhabitants of the Grankriki region in Suriname control the Brazilian migrant miners' access to the mining grounds—not only to mine, but also to live there. They are miners themselves, and work closely together with the Brazilians. The Brazilians refer to the local Maroons as concession owners, although they have never obtained mining concessions from the competent state authorities. The situation here is that the concession was given to a company that does not use its right to explore the area for gold findings and, unlike the firms described above, also did not set up any infrastructure to control the exploration and exploitation of others. As a result, the Maroons of the area have turned their lands meant for planting and hunting into mining 'concessions' too. The families from the villages each have a specific part of the area as their claim, where they mine themselves and give access to migrant miners to mine as well. The migrant miners pay a part of the gold production, usually 10 per cent, to the family on whose 'concession' they work.

The 'concessions' are small and pass from the family elders to the younger men, who then take care of the social and economic organisation. They collect the contributions from the miners, and keep peace and order—for which they are highly respected by the *garimpeiros*. If there are fights, they simply send the troublemakers away. And because access to the gold is what brings the people here together in the first place, being sent away is a serious punishment. In this situation, the local control of the mining area works well.

Mining 'claiming' claims in the margin

Now, we come to the fourth instance where mining-related conflicts and marginal territory interact with each other. This time, we see the connection the other way around: the marginal territories are also in a way created by the mining activity. They are constituted, constructed and built in reaction to the mining operations that go on there. In other words, we have to look at the way in which mining activities are constitutive for territorial organisation and claims, hence the 'situatedness' of mining regions, which in turn lie at the base of conflicts.

The goldfields in the south of Suriname continue at the other side of the border (which is the Lawa River) into French Guiana. Goldfields do not respect national borders. For the Aluku Maroons and Wayana Amerindians as well, the region is one continuous territory where they have lived for many decades, and even centuries. The local population moves, lives and travels in the territory on both sides of the river and national border, to plant their gardens, build their houses and search for hunting and fishing areas. They also freely use both sides of the river to gain direct access to the gold, and to try and control the access of others to it. So when it became too difficult to mine on the French side, where the state has a stronger presence as compared to Suriname, some Aluku moved to the Suriname border of the Lawa River to mine. Although they may not have set foot there for years, now that they had become miners, the territory turned interesting again and they claimed it as tribal territory without further ado. The concession holders would obviously not acknowledge or respect any such claim based on traditional settlement, but the migrant miners, other Aluku and the members of indigenous groups and other Maroons in the region were willing to admit such entitlements. The status of a region and one's right to live, work or mine there hence proves to depend on different parameters for different local groups or actors. 'Legitimately mining' somewhere, or defining an area as 'justifiably accessible' for someone, depends on whom, and with what imposition of power, responds to such claims.

The local inhabitants also found other ways of making a living from the gold rush and, here too, they in a way *created* the nature, 'identity' and (indirect) accessibility of a disputed mining region. Along the Suriname border of the Lawa River, settlements of mechanics and equipment maintenance personnel, merchants, bar owners, sex workers, smugglers and temporarily unemployed miners and cooks have emerged since 2006.

Some have as many as 50 houses, bars and shops, but several other smaller places have also emerged on the Suriname bank of the river. The Aluku and Wayana collect monthly payments from the migrants who have built houses and run their commercial businesses catering for the gold miners here. At these locations, the land that was unused and untouched before gold-mining became significant. It was 'community forest' without particular ownership claims or customary destinations for these locations. But now, due to the recent gold rush in the region, these marginal territories became important and, with that, they also emerged as sources for conflicts between local people, migrants and, from time to time, the state.

In assessing the importance and 'constituting force' of these settlements, we need to bring in the fact that French Guiana and Suriname have very different politics with respect to small-scale gold-mining. In comparison to the Suriname policy of laissez-faire, French Guiana actively tries to eradicate illegal small-scale gold-mining with a large police and military force. However, the French have little success with their policies because ASM is maintained and supplied from Suriname. Additionally, Suriname offers 'sanctuary' for those miners that need to flee from French law enforcement officers, having been caught illegally mining or bringing provisions. The newly occupied marginal territory generates opportunities for different actors here. Miners cross the borders easily and, most of the time, completely unnoticed. Miners find facilities close by and, at the same time, out of reach of the French. In turn, the Surinamese shop and workshop owners obtain a livelihood. This is only possible due to the marginal status of this territory.

Conclusion: Gold conflicts in marginal territories

Reflecting upon the role and importance of physical distance—of being only difficultly governable and being isolated from the key urban centres—in the conflicts surrounding mining, we come across various important features.

What we can say is that a lack of effective state governance and authority characterises the marginal areas where much of the small-scale gold-mining takes place. In our research in the Amazon, we found different effects when it came to the emergence and evolution of conflict.

To begin with, we found that the idea that mining areas beyond the effective range of the state embody a complete mobocracy or ochlocracy, a full collapse of any system, mere lawlessness and complete explosion of violence seems to be untrue. Although possibly unjust, and partly based on physical power, some rule and some routine about how to interact, whom to pay and whom to obey exists. Different players, even in competition, benefit from some predictability, a certain acceptance of power and a certain acknowledgement of the other's position and legitimacy. We must add, however, that the territorial marginality may both contribute to the ineffectiveness of government in the mining areas, as well as be the result of local dynamics undermining and altering the state's definition of the region, and the state's proposal on how it should be administered.

Second, the focus on the marginal status of the mining regions shows that the lack of state control opens opportunities for other actors to usurp authority and embody legal and less-legal power. These actors are often from circles close to the state, and borrow the power of the state to impose themselves locally. They will, however, often need to make concessions because they cannot call in the full force of the state to enforce their will.

In other cases, however, we see that it is not the one closest to the state that prevails. The situation of marginality also opens opportunities for those who have no such proximity to the power structures of the state. These actors will often bring in their creative and innovative vehicles to promote their interests and obtain good results through their well-planned interactions with the other players. Their power or authority is based on other elements, such as geographic or technical knowledge, traditional or customary entitlements, or established rules and expectations.

In all cases, the total or partial absence of the state does not result in chaos or a blunt measure of physical strength. Violence, or the permanent threat of violence, benefits no one. One way or the other, the situation of not being governed will lead to some form of creative, albeit often unfair, and yet to some degree agreed-upon, self-governance.

To summarise, we note that in the marginal areas, the access to mining grounds is taking place in many different ways: sometimes triggering conflicts between actors involved, while in other situations leading to the formation of surprising collaborations. Where conflicts occur, they are intimately related to the state of marginality of the mining areas.

Not rights or judicial force, but an independent dynamic of strengths and understanding interdependencies and interest balances seem to be the drivers that regulate conflicts. We can recommend that government interventions in such constellations always carefully take note of the situation on the ground if they do not want to do more harm than good.

References

Carrillo, F., T. Salman and C. Soruco, 2013. 'Cooperativas de Minería de Pequeña Escala en Bolivia: De Salvavidas de los Pobres a Maquinaria de Manipulación Política.' *Letras Verdes. Revista Latinoamericana de Estudios Socioambientales* 14: 235–54.

Damonte, G.H., 2016. 'The "Blind" State: Government Quest for Formalization and Conflict with Small-Scale Miners in the Peruvian Amazon.' *Antipode* 78(4): 956–76. doi.org/10.1111/anti.12230

Damonte Valencia, G., 2013. 'Formalizing the Unknown. The Stalemate over Formalizing Small-Scale Mining in Madre de Dios.' *The Broker: Connecting worlds of knowledge*, 5 November. Available at www.thebrokeronline.eu/Articles/Formalizing-the-unknown

Foucault, M., 2009. *Security, Territory, Population: Lectures at the Collège de France 1977–1978. Vol. 4*. Basingstoke, New York: Palgrave Macmillan.

Heemskerk, M., C. Duijves and M. Pinas, 2015. 'Interpersonal and Institutional Distrust as Disabling Factors in Natural Resources Management: Small-Scale Gold Miners and the Government in Suriname.' *Society & Natural Resources* 28(2): 133–48. doi.org/10.1 080/08941920.2014.929769

Larreta, E.R., 2002. *'Gold is Illusion': The Garimpeiros of Tapajos Valley in the Brazilian Amazonia*. Stockholm Studies in Social Anthropology. Stockholm: Department of Social Anthropology, Stockholm University.

Ribot, J.C. and N.L. Peluso, 2003. 'A Theory of Access.' *Rural Sociology* 68(2): 153–81. doi.org/10.1111/j.1549-0831.2003.tb00133.x

Salman, T., and M. de Theije, 2017. 'Analysing Conflicts around Small-Scale Gold Mining in the Amazon: The Contribution of a Multi-temporal Model.' *The Extractive Industries and Society* 4(3): 586–94. doi.org/10.1016/j.exis.2017.03.007

Theije, M. de and E. Bal, 2010. 'Flexible Migrants. Brazilian Gold Miners and their Quest for Human Security in Surinam.' In E. Bal, T. Eriksen and O. Salemink (eds), *A World of Insecurity: Anthropological Perspectives On Human Security.* London/Virginia: Sterling/Pluto Press.

Theije, M. de and M. Heemskerk, 2009. 'Moving Frontiers in the Amazon: Brazilian Small-Scale Gold Miners in Suriname.' *European Review of Latin American and Caribbean Studies* 87: 5–25. doi.org/10.18352/erlacs.9600

Theije, M. de, J. Kolen, M. Heemskerk, C. Duijves, M. Sarmiento, A. Urán, I. Lozada, H. Ayala, J. Perea and A. Mathis, 2014. 'Engaging Legal Systems in Small-Scale Gold Mining Conflicts in Three South American Countries.' In M. Bavinck, L. Pellegrini and E. Mostert (eds), *Conflict Over Natural Resources in the Global South—Conceptual Approaches.* Abingdon, UK: CRC Press/Taylor & Francis. doi.org/10.1201/b16498-9

13

Small-scale gold-mining: Opportunities and risks in post-conflict Colombia

Alexandra Urán

In the last 50 years, the guerrilla group Colombian Revolutionary Armed Forces (Fuerzas Armadas Revolucionarias de Colombia, FARC) has made progress in furthering its operations, mostly in areas with a relatively poor government presence. In these areas, on the margins of the official government, this armed group has spread the construction of a political setting in order to develop a set of rules that control rural areas. It is not a mere coincidence that these areas are the same places where small-scale mining (SSM) activities occur in conditions that are out of reach of the official legal system (Toro et al. 2012). The strong presence of the FARC guerrilla and other illegal armed groups in these rural areas have increased procedures for managing and administrating natural resources separately from the official legal system and outside of the government's control, and they have also fabricated an extended conflict in response to the political competition for power (Álvarez 2015). This chapter focuses on the situation of gold-mining to show how the Government of Colombia in the last decade has intensified and reproduced a neo-extractivist model of an industrial and mechanised paradigm for mining: a strategy that has excluded small-scale miners as local agents and decision makers. The main objectives of this chapter are twofold: to evaluate how the radical left in the political scenario may be included and recognised as a political actor in order to change the rules on mining and the administration of mining resources; and to generate a better understanding of the violence and analyses of the socio-environmental conflicts associated with the large-scale exploitation of mining resources. The purpose is to foster an international debate around the post-conflict process in Colombia and, perhaps even more importantly, to underline how including the left army group in the sphere of politics and recognising them as an actor can create opportunities for the SSM in the near future, as well as opportunities for the construction of legal pluralism in the exploitation of mining resources as a way for democratising one of the most complex fields and disputes in Colombia.

Peace process: Background

In 2011, President Santos manifested his intention to the guerrilla group to resume talking about the peace process[1] with the enactment of the Victims and Land Restitution Law (*Ley de Víctimas y Restitución de Tierras*). This law included a list of the issues that the government proposed to address, the representatives that they suggested and the possible places in which meetings for peace talks could be held. After some discussion, the government and the FARC chose Cuba and Norway as the two nations that would mediate the talks: Cuba for being the headquarters of their first meetings in 2010 with other countries and Norway for its experience as a moderator in conflict resolution. The Catholic Church in Colombia also offered their assistance as moderators, as they had participated in humanitarian efforts with the FARC in the last two decades.

The final outline for the agenda consisted of five items:

- **Integral agricultural development policy:** In their agricultural proposal, the FARC placed an emphasis on the creation of a 'designated land for farmers' (*Territorios Campesinos*), which would be collectively owned and managed by individuals chosen by the community.
- **Political participation:** The FARC and the government would jointly identify regions particularly affected by the conflict, in which the national government would commit to creating a total of X Special Provisional Peace Districts, in order for a total of X House

---

1   The negotiations lasted over two years (*Los Tiempos Internacional* 2012).

Representatives to be elected temporarily and for X electoral periods. The topic was also presented that special attention should be given to the parties that arise as a result of the peace talks, such as the FARC-EP party, both in regard to their access to the media, as well as having their programs reach a greater public.

- **Ending the conflict:** The surrender of arms by the FARC, which the FARC say should be accompanied by a restructuring of the military. The surrendering of arms should be done in order for them to return or enter into civilian life through a legal framework for peace—an effort that is a defining element of the suspension of legal processes for subversive agents.

- **Reach a solution to the problem of illegal drugs:** The FARC proposed to prioritise physical eradication as a method for eliminating drugs. In agreement with the cultivators and producers of illegal drugs, it was agreed that the cultivation of plants for ancestral use could not be eliminated.

- **Identify the victims of the armed conflict:** While members of civil society at large are also to be considered victims, special acknowledgement would be given to farming communities— ethnic victims of the attacks made by FARC-EP. Additionally, acknowledgement would be given to the victims of government-inflicted violence, such as the political organisations of leftist and/or trade union groups, which had experienced repression and violence by the government.

It is important to recognise that all the items listed in the agenda for the peace talks are absolutely necessary in order for an agreement to be reached. It is also important to recognise that they may influence the development of mining in Colombia. Simultaneously, each of these items of negotiation may be changed, resolved or negotiated upon, due to decisions made on mining policies. As this is a process of negotiation that involves mining, not only will the efficiency of production improve, but as a result of the resolution of the extensive conflict in Colombia and political polarisation, it will also further develop the democratic process and encourage freedom of speech for different players.

The situation in 2015 urgently demanded the signing of a peace agreement. The government has indicated that only through the solidification of the peace process can economic stability be established in the country. The main question to be asked then is: what principles should the peace agreement secure in order for the mining strategy to be implemented successfully? The discussion that follows is focused specifically around SSM. It will review the case of gold-mining, as in this particular case study different levels of investment resources and players are involved, which will enrich the discussion on the opportunities and risks that the post-conflict era in Colombia brings to the exploitation, use and management of natural resources.

Gold-mining and social involvement

Gold-mining in Colombia is an activity that involves a diverse group of social agents. Unlike in other extraction activities, there is a high demand for the participation and inclusion of different agents in order to regulate and manage this activity. Gold-mining is performed on different scales, from its artisanal form to its most industrial form. On the smallest scale, it is performed without machines and by local communities. The semi-mechanised mining today involves the use of excavator hoes and technology that require high levels of investment by miners. Large-scale or industrialised mining requires high levels of investment in capital and technology and is performed by large companies, the majority of which are international. Each of these scales of mining have different ways of standardising and formalising their processes, different economic impacts, different levels of social involvement, different forms of operation and varying levels of environmental impacts.

SSM is performed mainly in the collective territories, by communities of African descent and some indigenous communities. The government has acknowledged the right of these communities to perform de facto or traditional mining[2] as a form of subsistence, provided that it is not performed intensively. This form of mining does not require an environmental licence. Small-scale mining is also performed as an informal activity, particularly in regions where large-scale mining leaves scrap material that cannot be used with mechanised mining systems. Miners from everywhere in Colombia come to the Chocó region to

---

2 'Traditional Mining is considered to be that which is performed by individuals, groups of people or communities that mine government property without a permit registered in the National Mining Registry and that can verify through commercial and technical documentation that their work has been continuous over the last five (5) years and that the work they perform has been in existence for at least ten (10) years prior to the date in which this law came into effect.' Mining Code, Law 1382 of the 2010 Colombian Congress, Paragraph 1/Article 1.

perform mining manually, which is not very profitable, but becomes a form of subsistence for those who do not have other opportunities in the formal labour market. As this activity is performed using artisanal techniques, in collective territories where de facto mining is performed, or in regions where mining companies have environmental licences and mining rights, it is not a requirement for the miners who work in SSM to have individual mining permits. Recently, however, the use of an identification number, RUCOM,[3] has been implemented in order to sell gold—an implementation that has been met with great resistance.

Following small-scale mining is mid-scale mining, which today is extremely complex. Production units that operate on this scale are identified on the basis of mining contracts. However, the actual geographical distribution of such mining contracts is unknown, and thus, no clear structures exist that can regulate their functioning. Mid-scale mining contracts are frequently established in areas that receive special environmental protection, such as parks and highland areas, and even in the collective territories of ethnic communities where environmental authorities are not present. This is why the effects of its impact on these ecosystems are yet to be determined. While legally, a mineral title and environmental licence is needed (as the activity involves mechanised or semi-mechanised mining), local authorities are not in a position to closely audit the territories for such permits. It is important to note that the process of obtaining a mineral title is a heavy burden for a mid-scale mining company. This is one of the reasons why the mid sector of gold-mining, at best, becomes a form of outsourcing for the mining industry—in regions where a mineral title exists, mid-scale miners sign agreements allowing them to utilise the large-scale or industrial mining areas, provided that the gold is sold exclusively to the big company that owns the permit, who then sells and receives the profits of the mineral.

---

3   Commercial Mining Registration Number (Registro Único de Comercio Minero, RUCOM), established under Article 112 of Law 1450 of the year 2012 of the National Development Plan. Through the RUCOM, the National Mining Agency mandates for controls to be implemented on the sale of minerals and that a list of those that hold mineral titles be published, as well as information on the agents that are authorised to sell minerals. The purpose of this regulation is to generate greater transparency and control around the commercial activity of minerals in Colombia. This regulation was implemented in January 2014.

Such arrangements between large-scale mining companies and mid-scale mining contractors do not account for more than 13 per cent of mining practices. The remaining 87 per cent of mid-scale mining is performed illegally, without the permission of the permit owners or on land where no titles exist (Güiza 2014: 100). In other words, the majority of mid-scale mining operates without any type of monitoring or environmental control,[4] which also contributes to tax evasion, fosters money laundering and the financing of illegal armed groups, guerrillas, paramilitary groups and groups that perform ordinary crimes. This phenomenon has become evident in many parts of the world, as suggested by Di John's (2008) analysis of the causes of what he calls the 'failed states', which, according to him, focus their economies on mining processes themselves and not on the commodification processes of the raw material that promote national industry and generate employment.

Last is large-scale mining, for which legislation, some regulations and specific standards exist, such as mineral titles, environmental licences and management and deforestation plans. Although regulatory guidelines exist, the government's structure for the monitoring, assessment and control of activities related to the management and mitigation of environmental impacts is still unstable and not widespread. The Colombian Government has few officials in the Ministry of the Environment; and, while the government does have regional environmental authorities (CARs),[5] in the majority of cases where the ethnographies were conducted, these officials were found to be influential allies of the mining companies and continued to receive large financial benefits from them. Sometimes, they are even owners or partners of illegal mining contractor groups, which demonstrates the high level of corruption and the resulting lack of impartiality when it comes to performing their duty of monitoring as environmental authorities.

---

4   The estimated average amount of deforestation caused by illegal mining in Colombia between 1990 and 2019 is 310,349 hectares a year, which amounts to an estimated 6,206,000 hectares of destroyed forest land, or 5.4 per cent of Colombia's surface area (See Güiza 2014: 112).
5   Regional Environmental Corporations (Corporaciones Autónomas Regionales, CAR), part of the Ministry of the Environment.

Colombia: A state of competition for mining land rights[6]

The Right of Preference[7] privileges ethnic groups for the use and management of the natural resources in their territories. Although this right is given to the ethnic communities of the Collective Territories of Afro-Colombians (in the case of Afro-Colombians), or to the *resguardos* (a reservation system of communal landholdings) in the case of indigenous groups, evidence shows that in many cases these groups cannot compete with large-scale mining. When they apply for mineral titles, the majority of times they are denied and instead the titles are directed towards the large international mining companies. This is the case of the Greater Community Council of Condoto Río Iró in Condoto in the Department of Chocó. Since 2010, this council has applied for mineral titles for the main gold deposits of the collective territory, for them to be mined by the Afro-Colombian community using industrial methods. In 2013, concessions for the majority of these deposits were given to the international companies—Anglo Gold Ashanti Colombia and Roque de Jesús Homez Robayo International Business and Investments LTDA. The Greater Community Council of Condoto Río Iró, however, were only granted mineral titles that corresponded to less than half of those they applied for—the reason for such rejections being that they did not meet the requirements established by the government, nor did they have the financial or technical resources to perform environmental management. The Right of Preference, therefore, is not given to these communities and, instead, a different law is employed, which is a result of the Constitution

---

6   Concept introduced by Urán (2008) to examine the transformation of the Colombian state, seeking its incorporation into the global market using two complementary strategies: competitiveness and militarisation. The Colombian case is presented as a key example of the Militarised Competitive State, which opens up the path for extended capital accumulation, but also exacerbates the levels of social and economic inequality and social conflict. In this article, it is used to talk about the (mining) land-right competition on the new model of the Colombian state, where unequal policies have been used by the state to protect the interests of the multinational corporations as a strategy to 'clean' the land, removing the 'unproductive' productive sectors—illegal, artisanal and SSM—and to create the idyllic productive state, based on the extraction of mining resources.

7   The Right of Preference is the privilege granted by law to ethnic minorities that allows the government, by way of the Ministry of Mining and Energy, to issue these groups a special exploration and drilling permit in the mining regions of individuals of African descent, in order for them to extract the non-renewable natural resources that are traditionally extracted by these communities. The Mining Code, Law 685 of 2001/Article 133, states, 'In the granting of concessions by the mining authority, priority will be given to the Afro-Colombian community with regard to the concessions for oil field and mineral deposits located in the mining regions of these communities'.

of 1991. This law establishes the government as the owner of the subsoil,[8] and refutes the pre-existing Right of Preference. The new mining code gives privilege and guarantees for multinational companies in terms of production and regulation: the concession or allowances for exploitation of mineral resources was structured as a privileged form of interaction between the state and the private agents, there was an increase on tax exceptions, a weakening of environmental regulations and limited need of consultation with communities, among others characteristics.

Along with the mining boom, these large international companies receive greatly beneficial privileges from the government, which is currently focused on attracting foreign investments. These benefits allow them to mine large mining reserves even in regions that had previously been identified as special-use and environmental management regions, such as collective territories, national parks for biological conservation or research, or lands that had been considered as land reserves apt for agricultural activities. Citing the reason that these lands need to be more productive, today these regions are recognised as mining regions. The government justifies this by claiming that it is important to create a strategy that allows for the promotion of a development model that fosters 'prosperity for all, more employment, less poverty and greater security' (Departamento Nacional de Planeación (DNP) 2011).

More specifically, the Government of Colombia has insisted upon the need to strengthen the industrial mining sector in order to further develop the country, which is already included in the 'National Plan for the Development of Mining and Environmental Policy, Vision Colombia 2019' (UPME 2013). The surge of large-scale mining in Colombia is part of a strategy that seeks to take advantage of the high international demand for minerals: mining in developing countries in order to fuel industry in influential countries or in emerging large economies that produce high value-added products, prioritising the economies of mining countries (*La Razón Pública Journal* 2015). This mode of production based on the extractive industry is idealised by formal institutions and public entities due to its potential economic outcomes and also due to the level of resource mobilisation that it could achieve (Bebbington and

---

8   In the 1991 Political Constitution of 'Title XII of the Economic Framework for Public Finance', Chapter 1, General Provisions, Article 332 states: 'The government is the owner of the subsoils and of non-renewable natural resources, notwithstanding the rights acquired and established in accordance with previous laws'.

Bury 2013). For many academics, this mode of production consists of the 'neo-extraction project', which has been adapted as a financial strategy in Latin America (Acosta 2011).

The mining resources of the country have been conceptualised as commodities that need the expertise and technical capital of multinational companies to become assets to the state; as such, all the resources of the country have become available in legal and political terms for the large-scale extraction performed by private companies. In consequence, local communities and miners are understood as occupiers of those resources, instead of their legitimate historical users or marginal subjects in need of policy intervention (Bebbington 2012). Historically, public entities and official institutions have concentrated their formal efforts to promote and create secure conditions for the large-scale extraction of resources (Palacios 2006). This dynamic of changing the occupational means of production on collective property lands without prior consultation,[9] or without a general agreement from the communities that are directly affected, has caused social action to be taken, such as blocking of main roads and roads used for commercial distribution, confrontations, the closing of highways and even mining sites, and other acts of protest that have affected the operations of some of the companies operating in Colombia. These companies include Gran Colombia Gold, an international company from Canada that has 111 mineral titles in Segovia and Remedios in the Department of Antioquia; and Anglo Gold Ashanti, which has been in the country for eight years and is the largest investor in the gold-mining of La Colosa, located in the highlands of the Department of Tolima and in Chocó, in the region where collective permits are used, as mentioned earlier.

As an example, Gran Colombian Gold has experienced serious confrontations with the public with regard to the validity of the mineral title; these confrontations have continued since Gran Colombian Gold received the permit from the international company Frontino Gold Mines Ltd of the United Kingdom. Gran Colombian Gold received the contract for operations in the region that initially had been granted to the company from England in the first quarter of the nineteenth

---

9   Prior consultation is the fundamental right of the people of ethnic groups when measures (legislative and administrative) are taken, or when projects or activities are going to be conducted within their territories. The purpose of prior consultation is to protect their cultural, social and economic integrity and to ensure their right to participation (Sentencia SU-039 of 1997, cited in Rodríguez 2014).

century, under operational contract number RPP00140. Under these companies, however, the property has been in a lawsuit since 1976, when the Frontino Gold Mines entered Concordato and when mining rights were being demanded by workers who, at the time, were being laid off. These mining rights were demanded as part of the negotiation of their benefit settlements. In the year 2000, the national government placed the mine up for a public bid. Finally, in 2002, in a unilateral decision made by the government, the licence belonging to Frontino Gold Mines was awarded to Gran Colombia Gold. It wasn't until the year 2010 that Gran Colombia Gold could begin operations, as protests arose rapidly. The figures on union-related violence in North of Antioquia during these decades are among the highest in Colombia. Historical reference can be made to events, such as the Massacre of Segovia and Remedios in 1988,[10] and other actions against the public and the community of union workers, which, due to forces of coercion, had to abandon their hopes of obtaining mining concessions. This demonstrates that though the company has been granted the concession of the mineral title by legal means, several different interests exist that arise in violent protest in order to preserve the mining land rights.

In addition to the issue of violence in these mining regions is the problem of illegal mining. According to the data from the Colombian Mining Association (Asociación Colombiana de Minería, ASM), only 13 per cent of the mining activity in the region is legal or formalised. As national media outlets purport, this issue of violence is a direct product of the illegal aspect of mining and not due to a lack of consensus among the different players. This attitude is reflected in the following news remark (*Semana* 2015):

> Gold is part of an illegal empire ruled by criminal organizations that inflict serious problems on the environment, to the region and in the economy. Of the 55 tons of gold that are produced each year in Colombia, only seven belong to large companies, the majority of which are foreign investors that follow regulations. Santiago Ángel, President of the ACM, affirmed that if everything was formalized, the country would receive more than two billion dollars. In other words, this sector would be one of the greatest generators of income in the country.

---

10   The period between 1982 and 1997 in Segovia and Remedios, Antioquia, was an ongoing period of political violence against the public that specifically targeted those who expressed political dissidence: social movements—community associations, unions, public boards, human rights committees—and the Patriotic Union (Centro Nacional de Memoria Histórica 2014: 45).

Paradoxically, the municipalities[11] that report larger productions of gold also report experiencing vulnerable conditions and the presence of armed conflict, as the investigation made by Ibañez with Laverde (2014: 217) revealed:

> Mining municipalities, particularly those that mine silver and gold, demonstrate a lack of institutional structure and experience vulnerable socioeconomic conditions that existed prior to the initiation of mining practices. These municipalities do not have a strong institutional presence, they experience more armed conflict, are more isolated from main centers of production and tend to be more dependent regions. They are also municipalities with less potential for agricultural production.

Although evidence of these vulnerable conditions has been well documented by researchers and experts in the main gold-mining regions, the national government has refused to admit to the correlation between mining and violence. For instance, Rettberg and Ortiz-Riomalo identify 12 mechanisms whereby mining may be nurturing conflict and criminality, which we group in two categories: direct and indirect mechanisms. Direct mechanisms refer to activities whereby illegal armed actors seek direct access to mining-related rents. This may take the form of actually running mining operations or taking part in distribution and trade of mining output. Indirect mechanisms are those whereby mining feeds into and exacerbates existing social conflict, or contributes to funding illegal actors, for example, via protection payments (Rettberg and Ortiz-Riomalo 2014).

In another study conducted between January and December 2011 by Centro de Investigación y Educación Popular (CINEP 2012), researchers registered 274 collective social actions in their database of social struggles related to the extraction of oil, coal and gold, which, over time, has shown an increase since 2005, and consistent growth between 2008 and 2011.

For the government, violence is not an indicator of the need to question the efficiency of mining activities, as the parameters used by the government to measure the competence of mining do not specifically measure the capacity of mining to make efficient social reproduction possible. In other words, the government focuses on showing the positive impacts created by mining, such as a strong source of income, but does not show how it also creates negative social impacts, such as violence. In terms of its capacity to

---

11 Translators note: Municipalities in Colombia are the sub-regions by which departments are divided.

generate financial surpluses and earnings for those that have the privilege of owning mineral titles, mining in Colombia is therefore valued for its economic efficiency, even though society may not be effectively involved in the activity. Additionally, the legitimacy of the mining process and its justification as the present development model, based on the exploitation of natural resources is specifically associated with economic growth and is limited to the meeting of regulations (even though these are flexible and manipulated), and not with their ability to produce in a socially responsible way. This perspective needs to be debated, given the critiques of the efficiency of the neo-extraction model, which have demonstrated that the limitations of the model are connected to the limits of nature itself, as well as to the strategies of social reproduction (Brand 2014).

Mining activities should therefore be thought of as a way of not only creating material reproduction, but also political reproduction. Further, the peace process should be perceived not just as a new political agenda, but also as related to the reproduction of the economic system.

Parties' positions on the issue of mining

The participation of illegal groups in mining has been so radical that the official sources estimated that the extraction of natural resources has become their main source of financing (BBC News 2011). The government has argued that in order to bring in investment to the country, national security and stability are fundamental. Arguments like these are based on figures such as those shown by 2014 Control Risks,[12] which registered a 60 per cent increase in road and oil well blockades as compared to the year before, which caused a decrease of 65 per cent in foreign investment during this period (UNCTAD 2014). In the negotiations in Havana, the topic of agrarian reform and its relationship to mining and energy policies has yet to be properly addressed. The government has argued that there are other issues of greater priority to be addressed, such as the establishment of a ceasefire as well the guidelines for the end of the conflict, which is what negotiations have focused on until now.

---

12 Control Risks is an independent, global risk consultancy specialising in helping organisations manage political, integrity and security risks in complex and hostile environments. They support clients by providing strategic consultancy, expert analysis and in-depth investigations, handling sensitive political issues and providing practical, on-the-ground protection and support.

Even though the FARC guerrillas have been accused of participating in illegal mining, they have publicly established their interest in reforming the mining and energy sectors. They have expressed the need for Farmer Reserve Zones (Zonas de Reservas Campesinas, ZRC) to be established, which consists of identifying lands with an agricultural priority or under agrarian reform, as an initial topic to be included in negotiations. They have also called for the establishment of clear regulations for land restitution. This includes defining who the victims are and what arrangements are needed in order for restitution to occur, as well as defining the processes for the productive inclusion of both the victims and the arrangements for restitution, which is part of the fifth topic of the suggested negotiation agenda.

In order to reach an agreeable compromise that considers both positions— government and opposition—mining policies must be reformed to take land-use regulations into account. These regulations need to be based on an environmental plan that ensures the conservation of the environment and the implementation of a rural development policy that makes agricultural economies viable for farmers. This implies an adequate use of the soil, the organised management of agricultural land boundaries and the institutionalisation (formalisation or legalisation) of mining activities. In this way, under the framework of a rural development policy, mining policies may acknowledge the rights of citizens in territories affected by the mining of natural resources, as well as the right of the application of Mining Code, Law 1448, on the reparation and restitution of land.

It is evident that while the peace talks in Havana have given priority to the idea of putting an end to the conflict through practical actions, such as a ceasefire, other issues are still being left by the wayside and may have to wait longer for resolution. This suggests that signing a peace agreement, which is what is being discussed today and what national and international opinion calls for, is not the same as an actual agreement *for* peace. The hasty signing of an agreement may mean that guidelines regarding the mining of mineral resources may not be established. While it is possible that the signing of a peace agreement may generate a climate of financial security, which will surely provide greater incentive for foreign investment, it is also very possible for it to bring legal difficulties that may negatively affect operations in the mining and energy sectors. Some of these risks may include, for example, ambiguity around environmental licences, as well as confusion regarding prior consultation in the mining regions of indigenous and Afro-Colombian communities related to the topic of victimisation, which is currently happening on a large scale throughout the country.

The peace agreement is key

President Juan Manuel Santos believes that the mining industry is the key to financing 'the closing of the peace process—if a final agreement is signed in Havana' (Bermúdez 2015). However, the social cost, which in this case is a product of the social and environmental impacts of mining, also limits the reproduction of the model based on the exploitation of natural resources. This is why these impacts must be accounted for in the process of analysing the efficiency of mining in Colombia.

SSM is gaining public attention, but the state through its formal institutions is pressuring its actors and using categories such as illegal, criminal or *mineros de hecho*, de facto as synonymous without any precision or empirical knowledge (Álvarez 2015). The historical nature of the diverse forms of extraction has been abandoned for more simple explanations that reduce the activities to legal and illegal ones. Therefore, the intervention of the state is also simplified into the options of formalisation or criminalisation as the two policies controlling the involvement of the official legal system in mining (ibid.).

In light of these issues, the following 10 recommendations should be held in consideration in order to put the mining system back on track, or in order to reach what we can refer to as the Agreement for Peace. Classification of the different types of mining would include: *formal* (which meets all established regulations); *artisanal* (which includes mining performed by indigenous Afro-Colombians and farmers); *informal* (which does not meet some environmental, labour, mining or health regulations, but has the intention to formalise); *illegal* (which does not meet any regulations, but mining is not their vocation); and *criminal* (which finances illegal activity). Other possibilities are the establishment of an office for SSM to convert formalisation into a gradual process; formation of an agency that assists small-scale miners in the formalisation process; hunting down the 'big fish' of criminal mining; and, last but not the least, the development of a social map of mining that allows for information on the mining cadastre (with all the mineral titles identified) to be cross-identified with ethnic group territories and protected environmental areas, which shows where miners of all types and races work (from ancestral to informal miners) and locates criminal miners. Other plausible options include the establishment of the 'consolidation' of mining zones, taking item one from the peace agreement negotiations into account; an agency for intercultural dialogues, in order for these ongoing dialogues with grassroots communities to have an institutional space; or organisation of prior consultation for larger-scale formal mining. Two out of every three mining projects considered to be of 'national interest' by the government currently experience difficulties with prior consultation, which has created the idea that prior consultation is an obstacle for development (Bermúdez 2015).

By carrying out these recommendations, the government will be able to consolidate a more participatory mining process, as current assessments made by the government on mining in Colombia reveal a high degree of improvisation on behalf of the national government, with regard to the use and management of its natural resources, and a low degree of participation and distribution of its benefits. SSM should not be reduced to a premodern set of practices on its way to disappearing, but should evolve as a social expression of discontent with official institutions and social phenomena, which creates a need to conceptualise alternative legal descriptions capable of capturing its complexity and social relevance (Álvarez 2015). These aspects make mining a highly political issue, which should be open to debate and to seeking solutions among different sectors and agents in Colombia.

Signing a peace agreement that is based on an *agreement for peace* will make this process a more open strategy and will serve as a more active exercise of democracy, which could fuel the successful reproduction of mining companies. Demobilising the guerrillas is not a guarantee of the successful process of social progress, as the national government suggests. The peace process should combine demobilisation of the guerrillas with the active participation of all the political forces in the country, including the left and the opposition in general. This would give a new voice to rural sectors, including environmental and human rights defenders, and would therefore increase the pressure on the government with regard to the management of mining and oil exploration and exploitation permits.

The peace talks have already tangibly impacted the mining and energy sector and have the potential to substantially influence reforms for regulations around investing, which could radically restructure the national investment map. Nevertheless, it is important to emphasise that the possibility of greater influence from the political left on economic and development policies, as a result of the political participation of the FARC-EP guerrillas, may broaden the panorama of decision-making and may not negatively affect the interests of a considerable part of Colombia's corporate sector, as suggested by those who oppose such a development. After all, no nation may rebuild itself democratically without opposition.

Conclusion

The role of the state as the promoter of mining through the insertion of multinational companies into the country has generated a collision of interests that had been almost invisible or marginal before (Álvarez 2015). The complexity and diversity of actors, mainly the small-scale and local mining communities, institutions and rules on the mining ground need to be studied more deeply and empirically to achieve an efficient arena of participatory post-conflict Colombia.

The effect of post-conflict implications in Colombia on the mining sector is indisputable. The participation in the political and formal life of a group with ideas of opposition, such as with the FARC, may make for a debate that is broader in its vision on the management of the mining sector. Demobilisation, as well as the political participation of the FARC, may broaden the debate on the mechanisms and strategies that allow a political and economic operation to be carried out successfully, with regard to the use and benefit of mining resources. Political pluralism is a reality that exists in society or, more precisely, in multiple social fields. Consequently, the participation of the more radical left, with a discourse that will most likely be strongly influenced by the idea of nationalising resources, would also be a position that should be heard, as has been done in other countries in the region. While the case of Colombia has remained far from the other political tendencies in Latin America, where governments from the left have reached power (even though it is not as likely that a coalition of forces from the left wins a majority in elections in Colombia), the possibility of a peace agreement with the guerrillas may and should change its governability, based on this first peace agreement between Colombian Government and FARC-EP 2017, which could be reproduced with the National Liberation Army (Ejército de Liberación Nacional, ELN).

References

Acosta, A., 2011. 'Extractivismo y Neoextractivismo: Dos caras de la misma Maldición.' In M. Lang and D. Mokrani (eds), *Más Allá del Desarrollo. Grupo Permanente de Trabajo sobre Alternativas al Desarrollo.* Quito: Fundación Rosa Luxemburg with Abya-Yala.

Álvarez, J.D., 2015. 'Governing Mining Resources in the History of Colombia: Between Official Institutions and Resistance.' *Law and Development Review* 9(11).

BBC News, 2011. 'La Minería Ilegal Financia el Crimen en Colombia'. BBC Mundo, 11 March. Available at www.bbc.com/mundo/movil/noticias/2011/03/110310_colombia_mineria_ilegal_az.shtml

Bebbington, A., 2012. *Social Conflict, Economic Development and the Extractive Industry: Evidence from South America.* London, New York: Routledge Press.

Bebbington, A. and J. Bury, 2013. *Subterranean Struggles: New Dynamics of Mining, Oil and Gas in Latin America.* Austin: University of Texas Press.

Bermúdez, A., 2015. 'Los Coqueteos de Santos II a los Mineros'. *La silla vacía*, 26 abril. Available at lasillavacia.com/historia/los-coqueteos-de-santos-ii-los-mineros-50094

Brand, U., 2014. 'Social-ecological Transformation as Basic Condition for a Realistic Post-2015 Agenda.' *Die Post-2015 Agenda.* Reformoder Transformation. Wien: Österreichische Forschungsstiftung für Internationale Entwicklung.

Centro de Investigación y Educación Popular (CINEP), 2012. 'Minería, Conflictos Sociales y Violación de derechos humanos en Colombia.' Informe Especial 2. Bogotá: CINEP/Programa por la Paz.

Centro Nacional de Memoria Histórica, 2014. *Silenciar la Democracia. Las Masacres de Remedios y Segovia, 1982–1997.* Bogotá: CNMH.

Departamento Nacional de Planeación (DNP), 2011. *National Development Plan 2010–2014.* Bogotá: DNP.

Di John, J., 2008. 'Conceptualizing the Causes and Consequences of Failed States: A Critical Review of the Literature.' *Working Paper Crisis Series* 25: 1–51.

Göbel, B. and A. Ulloa (eds), 2014. *Extractivismo Minero en Colombia y América Latina*. Bogotá: Universidad Nacional de Colombia.

Güiza, L., 2014. 'Informe Colombia.' In C. Heck and J. Tranca (eds), *La Realidad de la Minería Ilegal en los Países Amazónicos de SPDA*. Lima: Sociedad Peruana de Derecho Ambiental.

Ibañez, A. with M. Laverde, 2014. 'Los Municipios Mineros en Colombia: Características e Impactos Sobre el Desarrollo.' In J. Benavides (ed.), *Insumos Para el Desarrollo del Plan de Ordenamiento Minero*. Bogotá: Universidad de los Andes.

La Razón Pública Journal, 2015. 'El Trilema Minero: La Gran Minería Sostenible y Socialmente Responsable es una Falacia.' 29 August. Available at www.razonpublica.com/econom-y-sociedad-temas-29/2349-el-trilema-minero-la-gran-mineria-sostenible-y-socialmente-responsable-es-una-falacia.html

Los Tiempos Internacional, 2012. 'La negociación duró más de dos años.' 17 September. Viewed at www.lostiempos.com/diario/actualidad/internacional/20120917/la-negociacion-duro-mas-de-dos-anos_185766_394085.html (site discontinued)

Palacios, M., 2006. *Between Legitimacy and Violence: A History of Colombia, 1975–2002*. Durham: Duke University Press. doi.org/10.1215/9780822387893

Rettberg, A. and J. F. Ortiz-Riomalo, 2014. 'Conflicto Dorado: Canales y Mecanismo de la Relación entre Minería de Oro, Conflicto Armado y Criminalidad en Colombia.' Informe final. Bogotá: Universidad de Los Andes/ Centro de Estudios sobre seguridad y Drogas–CESED.

Rodríguez, G. A., 2014. *De la Consulta Previa al Consentimiento Libre, Previo e Informado*. Bogotá: Universidad del Rosario Press.

Semana, 2015. 'El Drama del Oro Colombiano.' 31 January. Available at www.semana.com/economia/articulo/mineria-el-drama-del-oro-colombiano/416246-3

Toro, C., J. Fierro, S. Coronado and T. Roa, 2012. *Minería, Territorio y Conflicto en Colombia*. Bogotá: Universidad Nacional de Colombia with Censat Agua Viva.

Unidad de Planeación Minero Energética (UPME), 2013. 'Plan Nacional de Desarrollo: Minero 2010–2018.' Ministerio de Minas y Energía, República de Colombia. Available at www.upme.gov.co/Docs/Plan_Minero/PNDM_2010_2018_dic_31.pdf

United Nations Conference on Trade and Development (UNCTAD), 2014. *World Investment Report, 2014: Overview*. United States and Geneva: United Nations. unctad.org/en/PublicationsLibrary/wir2014 _overview_en.pdf

Urán, A., 2008. *Colombia un Estado Militarizado de Competencia: Las Fallas Estructurales Para Alcanzar la Explotación Sustentable de los Recursos Naturales*. Kassel: UniversitätPress.

14

Muddy rivers and toxic flows: Risks and impacts of artisanal gold-mining in the riverine catchments of Bombana, Southeast Sulawesi (Indonesia)

Sara Beavis and Andrew McWilliam

In 2008, the Bombana District (Kabupaten) of Southeast Sulawesi in eastern Indonesia experienced a dramatic influx of artisanal miners following the chance discovery of alluvial gold in the Tahe Ite and, later, Wumbubangka rivers (located in the sub-districts of Rarowatu and Rarowatu Utara, respectively). As stories of the golden bounty spread, a flood of gold prospectors arrived in the area, many of them opportunistic, temporary migrants from other regions in Indonesia, and began working the alluvial goldfields immediately adjacent to, and within, the river systems. Within months, there was said to be a 'sea of people' (*lautan manusia*) streaming into the area as thousands of hastily constructed prospecting camps sprung up along 15 km of the main channel of the Tahe Ite River and then expanded into other areas.

The initial reception of the artisanal mining was positive. The Governor of Southeast Sulawesi at the time was a strong promoter of mining development in the province, and declared the discovery a sign of the grace and blessing of Allah (*rahmat dan berkah*) (IKAPERMAB-

Yogyakarta 2009). The local district head (*bupati*) issued a government regulation—Surat Keputusan No. 10, 2008—which decreed that the miners would have to pay as much as IDR250,000 (US$24) per person in licensing fees to work the prospective sites (Kamil 2009). By mid-2009, according to news reports, there were approximately 60,000 gold panning licences issued and as much as IDR15 milliar (over US$1.5 million) flowing into regional government coffers (IKAPERMAB-Yogyakarta 2009). At the same time, the local government also issued a series of formal exploration and mining licences (called Kuasa Pertambangan, KP; and, since 2009, Ijin Usaha Pertambangan, IUP) within their district jurisdiction to a growing number of gold exploration companies seeking to secure potentially lucrative leases.

The initial enthusiasm for mining activity, however, was soon tempered by a litany of problems and complaints that quickly emerged from the inadequate oversight and compliance monitoring of artisanal and small-scale mining (ASM). Among the impacts from the dramatic influx of artisanal miners was visibly increased river channel erosion and sediment loads from the finely sifted materials, described as a 'thick porridge' (*bubur kental*), which flowed downstream from the scattered diggings in the riverbed. Damage to roads was readily apparent as monsoon rains turned dirt tracks into impassable bogs, and conflict with local villagers expanded over access and use of customary lands, extensive illegal logging and disruption of river flows downstream extending to neighbouring settlements. Downstream rice farmers in villages utilising the Langkowala River, for example, reported major declines in water supply for irrigation. Finally, a growing number of artisanal prospectors had died in their attempts to excavate gold ore from the narrow tunnels and deep holes (*lubang tikus*) dug into the river channels (Kamil 2009), and there were reports of an increase in a variety of social ills, such as prostitution, drunkenness, muggings and violence on the goldfields.

In response to the growing public outcry, the government implemented a series of quasi-effective crackdowns (police 'sweeping' operations) on mining activities, especially in relation to the proliferating unlicensed artisanal operations (known as *penambang tanpa injin*, PETI). Their attempts to reduce and regulate the influx of people seeking good fortune (*rezeki*) from gold planning had some initial success. But, in the chaotic circumstances, many commentators and reporters of the unprecedented developments in Bombana questioned whether the golden bounty was really a blessing or a disaster for the general community.

Sahrul, representing the advocacy non-government organisation Sagori, which has protested against the destructive impact of mining in the region, argued that mining really only benefited a few and was simply an 'ATM' for the benefit of certain government and security authorities, while local people were marginalised in the process (Antara 2012).

Anatomy of a gold rush: Hidden risks

Integral to artisanal gold-mining in the region has been the widespread and sustained use of elemental mercury for processing the gold. This practice has resulted in extensive, unregulated releases of mercury into the river systems that support valuable downstream land and riverine activities, including irrigation, *tambak* fish ponds and estuarine fishing. Seven years on since the first discoveries of alluvial gold, the active mining population has reduced substantially as a result of declining gold yields and greater government control over licensing and access for authorised mining companies. But even as the intensity of the gold rush has diminished, continuing hydraulic mining of the riverbanks and channels has left a long-lasting deleterious environmental legacy. In previously worked areas, there is an extensive accumulation of unprotected mine tailings and drainage pits, within and adjacent to streams and waterways. Downstream flows in the two affected river catchments still exhibit regular, extreme levels of turbidity and sediment loads—a result of the intensive mining of riverbanks and beds in the upper catchments, in combination with heavy seasonal monsoon rains.

This chapter reflects on the initial results of a targeted sampling program of river water and sediments at key points of the Langkowala and Kasetahi river systems during the year 2015 (Figure 14.1). The sampling was undertaken, in part, as a training exercise with selected staff drawn from regional government environmental agencies and university staff, and as a demonstration of the value of scientific analysis and monitoring to facilitate and support public policy debate and planning. The survey was complemented with a social mapping exercise to better understand human–water interactions and experiences along the river channels. Here, we consider both the immediate impacts and implications of the downstream effects of mining, as well as the longer-term risks and environmental management challenges of the riverine environments for the thousands of residents and artisanal miners who derive livelihoods from their flows.

The research forms part of a three-year Australian Government funded project entitled, *Artisanal and Small Scale Mining for Development* (eastern Indonesia) (2014–2017). It brought together Australian and Indonesian research partners in a project designed to strengthen social and environmental impact assessment methodologies, and to improve policy aspects of mining governance in the region.

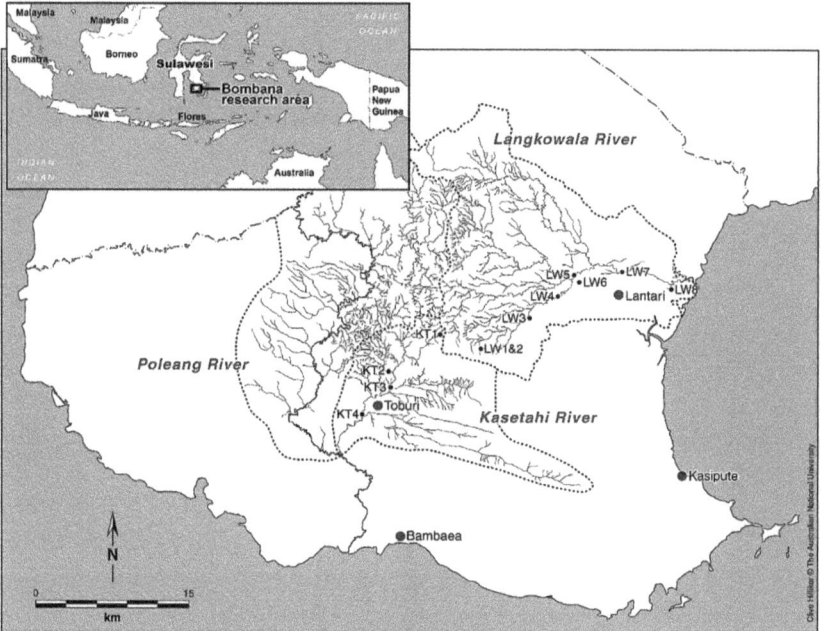

Figure 14.1: Location of study area and sampling sites
Source: Clive Hilliker ©The Australian National University.

The mercury cycle and its pathways

Mercury is widely used in artisanal gold-mining to facilitate gold processing through the chemical bonding of gold and mercury to form an amalgam. The mercury is then removed through the application of heat (either by direct flame or retorting). However, the process is highly toxic and its indiscriminate use is associated with deleterious environmental and public health risks. In Bombana, as in most other small-scale gold-mining sites around the world where the unregulated use of mercury is widespread, local ignorance and disregard of mercury's toxicity and potential harm is commonplace.

By way of overview, mercury is also added to the environment from natural sources (for instance, degassing from soil and water, volcanic eruptions, fire, weathering of mercury-bearing rock), as well as anthropogenic sources (for instance, mining, industry, cremations). It is cycled through the atmosphere and lithosphere, and can be converted to a more soluble and volatile organic form known as methyl mercury by biologically mediated processes within the soil and water, or once ingested by organisms (for example, within the rumen of cattle or the muscle of fish). Mercury is readily taken up by organisms and it bio-accumulates over time; the longer the exposure to mercury, the higher the risk (Lechler et al. 2000; Limbong et al. 2003).

Mercury also bio-magnifies as a step-by-step process along the food chain (Veiga et al. 1999; Limbong et al. 2003). Top-level predators are therefore particularly vulnerable to the accumulated effects of mercury contamination within the environment. For people living and working in a mercury-contaminated setting, such as artisanal small-scale gold-mining, the pathways that mercury can take are various. They include inhalation of vaporised mercury released during torching of the mercury–gold amalgam; ingestion of contaminated water and food; and direct skin contact with contaminated water, sediment, soil and gold amalgam (van Straaten 2000a, 2000b; Cortes-Maramba et al. 2006; Kitula 2006). Importantly, mercury not only passes the blood–brain barrier, but also through breast milk and the placenta. Unborn foetuses and young babies are therefore at significant risk if their mothers have been exposed to mercury. The risk to individuals varies according to the form of mercury (whether inorganic or organic), the pathway of uptake and the length and magnitude of exposure. In the district of Bombana, risks are mainly associated with inhaling vaporised mercury through the process of unprotected burning of amalgam, drinking contaminated water and having direct skin contact with mercury or contaminated water during mining operations.

Mercury sampling in Bombana, 2015

In considering the impact of mercury released into the environment, it is well attested in global literature that one of the most direct and dangerous interactions with mercury is through the burning of amalgam to isolate gold (Bank 2012). As part of our collaborative research activities in Bombana, the project is working with an non-government organisation anti-mercury advocacy group, BaliFokus, to highlight public health risks and effects from direct contact with mercury.[1] As part of their activities, BaliFokus has undertaken mercury vapour analysis at a range of locations in the study area, with some striking results. Levels of airborne mercury were recorded in a variety of locations within the active residential mining areas of Wumbubangka Village (Langkowala catchment) and Rau Rau Village (Kasetahi catchment). These results are shown in Table 14.1. The findings reveal a wide range of concentrations across locations with the highest values at four specific sites that are well above US-determined safe levels of exposure. These sites are mainly local shops (*toko*) that trade in gold and are likely to be places where mercury is volatised through burning. When these activities are undertaken in enclosed spaces with little ventilation, dense mercury vapour may linger in the air and expose the occupants to increased mercury inhalation and ingestion. Occupants in these contexts include women of the household, who may combine amalgam-processing with nurturing care of infants. Sustained exposure to these levels of mercury vapour can lead to severe lung damage, neurological problems, tremors, skin rashes, ataxia, insomnia and kidney abnormalities (McKelvey and Oken 2012).

Table 14.1: Concentrations of airborne mercury in the Kasetahi and Langkowala catchments, Bombana District, South East Sulawesi

Mercury ambient emission sampling, Bombana, March 2015		
Sampling point	Area	Hg (ng/m^3)
Toko klontong/emas Resky (shop)	SP2	3,196.91
Rumah Ibu Niartini (private dwelling)	Watu-Watu	n/a
Toko emas (gold shop)	Wumbu Bangka	287.58
Toko emas Pak Salam (Toko Obat 99) (gold shop)	Wumbu Bangka	54,931.84
Toko Pulsa Ibu Wira/Pak Yasin (shop)	Wumbu Bangka	13,118.07
Toko emas Misan (yard of gold shop)	Wumbu Bangka	303.76
Toko emas Misan (gold shop near site of amalgam burning)	Wumbu Bangka	17,363.23
Restan (near site of amalgam burning)	Wumbu Bangka	6,532.08
Toko emas Pak Andika (gold shop firebox)	Wumbu Bangka	1,363.75
Ruang tengah Pak Andika	Wumbu Bangka	6,246.60
Tempat pembarakan Pak Andika (amalgam burning site)	Wumbu Bangka	27,210.71
Rumah Suci (private dwelling)	Wumbu Bangka	258.81
Rumah Pak Pere (private dwelling)	Rau-Rau	32.69
Rumah Dita (private dwelling)	Rau-Rau	28.07

Note: Hg = mercury; SP2 = identifier for a specific transmigration settlement.

Source: Field survey data provided by BaliFokus, 2015

1   BaliFokus has also been working with local public health service providers (*Depkes*) to raise awareness of environmental mercury applications and their potential human health risks. To date, the Indonesian health system has not developed any effective public health campaigns around mercury toxicity and artisanal gold-mining, despite its widespread distribution across Indonesia and demonstrated deleterious health impacts for people closely involved in the practice (see, for example, Limbong et al. 2003; Global Mercury Project 2006; Serikawa et al. 2011).

In order to determine the concentration of mercury in the river systems affected by mining, samples of water and streambed sediments were collected at four sites along the Kasetahi River and eight sites along the Langkowala River. A background reference site sample was also taken from a stream within the boundaries of the neighbouring Rawa Aopa Watumohai National Park to the north where no mining activities occur. All samples were analysed for dissolved and total mercury.

Findings for dissolved mercury in the river samples were all below detection limits. This indicates that mercury in the river channel is adsorbed into, and forms fine coatings on, suspended particulate matter (for example, clay particles, organic matter and colloidal material) that subsequently settle out as stream velocity slows. Figures 14.2 and 14.3 show the concentrations of total mercury at the sampling sites, with reference to international and/or national water- and sediment-quality guideline values for drinking water, aquatic ecosystem health, aquaculture and recreation (WHO 2011; ANZECC 2000; Canadian Council of Ministers of the Environment 1995). At the time of sampling, it was evident that more active mining was taking place within the Kasetahi catchment than the Langkowala, and this is reflected in the data; for example, the volumes of suspended sediment being transported by stream flow are shown in Figure 14.4. The results reflect the level of disturbance and the consequent mobilisation of materials off-site by the action of flowing water. Two sites on the Kasetahi River (KT2 and KT3) are characterised not only by the highest suspended sediment concentrations, but also of total mercury, with water- and sediment-quality guideline values for multiple uses being well exceeded. By contrast, the lower concentrations of mercury in the Langkowala River (LW1-8) can be attributed to less active mining at the time of sampling, and the dilution effects of tributary inflows. Nevertheless, water- and sediment-quality guidelines have also been exceeded at a number of sites along this river system.

Interestingly, although mercury is being added to the system, there are two sites on the Langkowala River (LW4 and LW7) where mercury is depleted, relative to the top of the catchment above the mining area. These sites are (i) immediately below a small irrigation dam, where heavy material will tend to fall out of suspension and be trapped behind the dam wall, depleting waters immediately downstream of particulates; and (ii) the site of river dredging, where coarse river sand is being removed from the stream for industrial use.

Figure 14.4 illustrates the variable suspended sediment concentrations in the two rivers and highlights the different conditions prevailing during sampling. As noted earlier, at the time there was significant alluvial and hydraulic mining being undertaken in the upper catchment of the Kasetahi, and river flow was visibly turbid. By contrast, in the Langkowala, where such activity was not apparent, the water was less turbid and this is reflected in the low concentrations of suspended sediment.

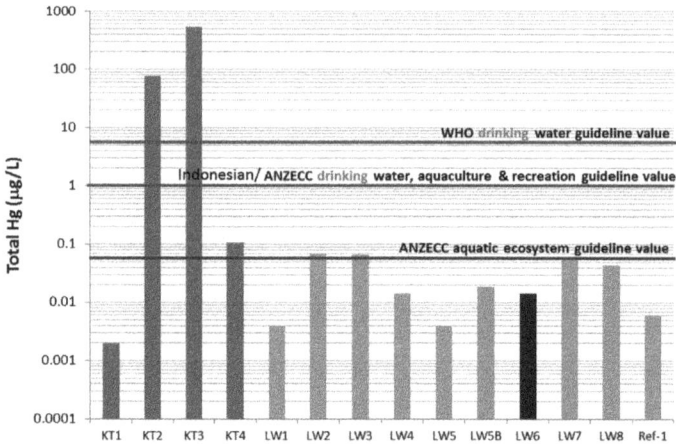

Figure 14.2: Concentrations of total mercury in stream waters of the Kasetahi and Langkowala rivers

Note: KT = Kasetahi River samples; LW = Langkowala River samples; WHO = World Health Organization (2011); ANZECC = Australian and New Zealand Environment Conservation Council (2000); Hg = mercury. Site LW6 refers to groundwater sampled from a village well; site Ref-1 is the reference site outside of the two catchments.

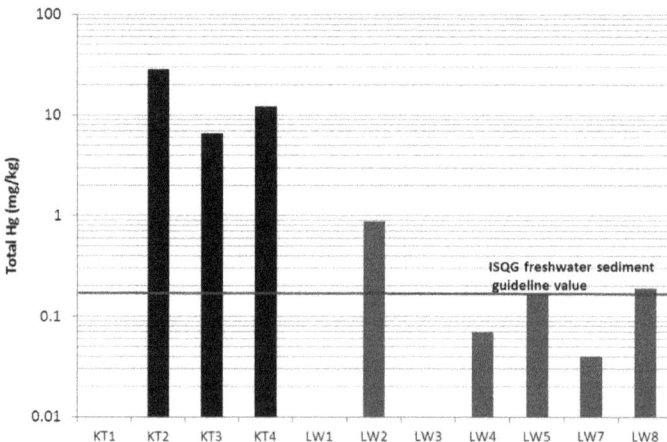

Figure 14.3: Concentrations of total mercury in the bed sediments of the Kasetahi and Langkowala Rivers

Note: KT = Kasetahi River samples; LW = Langkowala River samples; ISQG = International Sediment Quality Guideline; Hg = mercury.

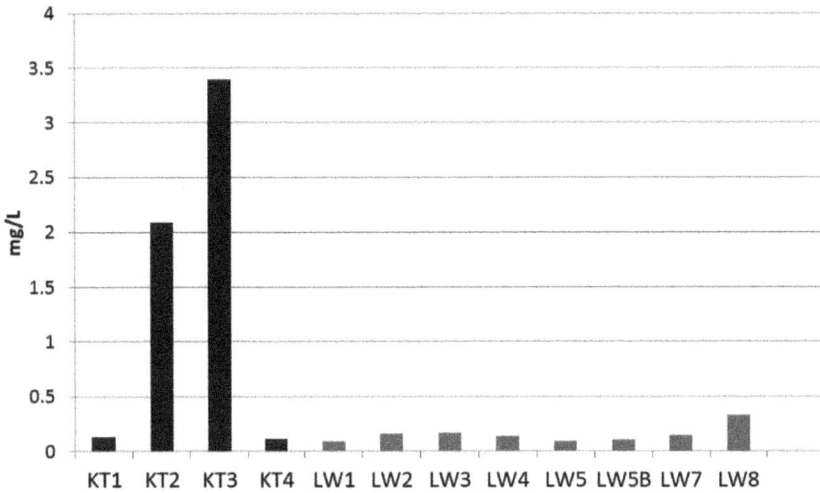

Figure 14.4: Concentrations of suspended sediment in waters of the Kasetahi and Langkowala rivers

Note: KT = Kasetahi River samples; LW = Langkowala River samples.

The monitoring results demonstrate that mercury is being added to the system and that the location and intensity of mining activity are playing key roles in the spatial patterns of contamination. This is consistent with findings in the broader literature (Gray et al. 2000; van Straaten 2000a, 2000b; Fadini and Jardim 2001; Roth et al. 2001). However, the concentrations of mercury in the water and sediments are not extremely high. This finding is likely to be a function of sampling in the middle of the dry season, when surface run-off and stream flow are relatively low. Seasonality is recognised as a salient factor in determining stream pollutant loads; in this case, sampling took place immediately after the fasting month of Ramadhan, when mining had generally ceased for the Idul Fitri/ Lebaran holiday. One of the preliminary results of this study, therefore, suggests that the contaminant concentrations observed in the river channels represent a current, best-case scenario. Higher concentrations of mercury, with associated risks, can be expected during the wet season and particularly in the first flushing events, when materials accumulated from mining over the dry season are mobilised and transported downstream.

Social mapping of human–water interactions

Water and sediment sampling and analysis in the Bombana District has provided an initial, objective snapshot of unregulated mercury contamination and potential risks to human and environmental health in the region. As part of the collaborative research engagement established under the auspices of the project collaborative activities with the University of Halu Oleo and local government staff, field research extended to a social mapping exercise in the two mining-affected river catchments of Bombana. The survey sought to gather information of the nature and scope of human–water interactions in the riverine environments of the Kasetahi or Poleang and Langkowala rivers that are subject to the mercury contamination outlined in the foregoing section. An initial iteration of the social mapping was undertaken in July 2015 and involved visiting a number of locations along the river channels—from the top of the catchments (*hulu*) to the river mouths (*hilir*), observing conditions at these locations and more or less randomly interviewing local residents about their use of the river water, their observations and experiences of river flows since the onset of mining upstream.

Large numbers of residents and temporary or recent migrants live along the key river channels of the Kasetahi and Langkowala rivers, and have in the past drawn on the benefits of the river flows for many daily household activities. Water for household consumption and broader productive livelihood activities has long attracted farmers to the region. This range of interactions includes, for example, fresh water for consumption and domestic use (bathing, washing clothes), as well as places where children might play in the riverbeds and channels.

River water is also used widely for irrigation purposes—principally for irrigated rice (*sawah*) on the fertile sloping mid-catchment and plains of both river valleys, but also for line and net fishing and harvesting freshwater prawns downstream, for stocking fishponds and watering wandering livestock (Bali banteng cattle in particular), as well as a resource for hand-watering vegetable gardens, establishing fruit trees and for washing vehicles. There are established sites for seasonal sand mining from deposition of material in the riverbeds and for use in house construction and building, as well as water for cement mixing and much else.

While the implementation of this social mapping exercise remains at an early stage, the responses of households living along the rivers were highly consistent and generally dissatisfied with the present condition of the rivers. Based on initial discussions and brief interviews with a range of respondents, there was a consensus view that there had been major and long-lasting negative impacts as a direct result of the enthusiastic but poorly regulated mining practices in the upper catchments.

Among the issues raised by respondents living and working in villages and townships along the main river channels was the widespread complaint that the river no longer provided a plentiful supply of potable water. These days, and for a number of years, following the advent of mining in the upper catchment, river quality has dropped dramatically and the flows are often very turbid or muddy (*keruh*), especially following rains. The situation is such that many respondents no longer draw on the river water for their household consumption, only using it when alternatives are not available, and even in these circumstances they will draw water selectively, when flows are visibly clearer. Most rely instead on a combination of nearby alternative spring water sources, limited government-provided reticulated water supply services and the purchase of potable water from mobile traders in the form of large plastic containers known as *galon*. The latter has become a regular feature of daily life in the goldfields area, and one that substantially increases living costs for householders and miners alike.

Another significant deleterious impact from hydraulic mining of riverbanks and beds, particularly in the Langkowala catchment, has been the disruption of river flows to substantial areas. Rice farmers in downstream villages of the Langkowala River, for example, reported major declines in water supply for irrigation, reducing land under production by nearly 500 ha, or the equivalent of 2,500 t of harvested rice (based on average yields of 4–5 t/ha), and considerably higher in terms of opportunity costs over time. Despite protests by farmers and farmer groups about the threat to their livelihoods as a result, little corrective action was taken by the local government. Consequently, many farmers who formerly relied on gravitational river flows have now switched to shallow pump well irrigation methods to ensure adequate supplies of water at critical times. The availability of groundwater resources has helped cushion the damaging impacts of mining, but has resulted in higher investment costs for farm inputs. Any impacts on groundwater resources from this increased use are not known.

A third impact of the dramatic rise in river sediment flows and increased deposition of sediments in the river channels over recent years is the perceived higher incidence of localised flooding. Inundation of maturing cereal crops (especially rice and maize) with mud slurries can result in major crop losses for farmers cultivating crops adjacent to the river channels.

In the course of the social mapping exercise, participants visited a series of locations throughout the two catchments. The responses recorded were remarkably consistent in recognising the much-reduced water quality and heavy sediment loads during periods of active mining. At the mouth of the Langkowala River, a visit was made to a group of fishing households who have for many years fished (*ikan bilang*) and trapped river prawns (*balachan*). At Tanjung Baropa, a series of stilt and platform houses hug the southern bank and provide a staging base for working and managing a series of lift net (*togo*) structures erected in the main river channel.

Discussions with the group revealed a litany of complaints about the changing conditions of the river that they have been relying on for their livelihoods. They highlighted the heavy sediment loads deposited in the river mouth, which they attribute directly to the upstream mining activity. This has caused silting in the river mouth, preventing access to larger trading boats that used to purchase their catch on a regular basis. The traders are now forced to anchor further out and require the fishers to use their own boats to make an additional transfer of their catch. The fishing households have also observed declining yields from their fishing efforts, which they attribute to the turbidity of sediment loads in the river flows, reduced flows at times[2] and the frequent flooding after rains in the upper catchment, which damages their lift nets and reduces their ability to sustain productive yields. In these and other ways, a highly sustainable and low-impact fishery that has flourished for years is now under threat from the disregard and neglect of environmental standards, and the absence of effective artisanal mining procedures.

---

2   Fishing households also blame the upstream fish pond (*tambak*) operators who hold back water in their retention ponds at low tide.

Conclusions

Artisanal mining in the Bombana district of Southeast Sulawesi and its downstream effects are strikingly evident in the quality of water and sediments examined in this study. Through targeted water and sediment sampling at different locations in the river systems, we can see what is being added to the system over time. We can also see how these impacts are modified by their dynamic interaction within the riverine systems. Test results reveal how and where mercury-contaminated particulates are trapped or removed from a stream. The pattern of mining intensity also influences how the contaminant concentration varies over time and space within waterways (for instance, during the Ramadhan fasting month), as does the seasonal pattern of monsoon rainfall, where the onset of the wet season has a flushing effect on river sediments and the mercury load that they carry. Identifying hotspots of contamination provides insight into the risks of exposure, for both people and aquatic ecosystems.

Preliminary results drawn from this study suggest that the risks of mercury suspended in river water downstream of mining activity occur at a whole-of-river scale. The transport of mercury within the water column, mainly adsorbed into particulate matter, creates risks for biota (especially catfish, freshwater shrimp, etc.) to take up mercury along the river course. The long-term direct and cascading effects of mercury ingestion within these catchments remain unclear without further research. The results of the sampling survey thus illustrate the need for active and longer-term monitoring of river and sediment quality as an important environmental and public health initiative by local government. Continuing efforts of the research partners to promote and support policy initiatives directed to these ends form an integral part of the action research objectives of the present project.

Finally, the complementary social mapping survey into human–water interactions along the mining-affected river systems has highlighted a series of adverse environmental and ecological impacts from the massively increased sediment flows due to unregulated hydraulic mining of riverbeds and banks. In particular, damage and destruction to the river amenity and its associated high-value, downstream benefits for local communities and rural livelihoods is a major and unfortunate legacy of artisanal mining. It is by no means evident that the short-term and limited economic benefits of the Bombana gold rush offset the long-term environmental costs that are borne by the majority of residents.

References

Antara, 2012. 'Tambang: LSM minta Pemda Bombana moratorium izin tambang.' 17 February. Viewed at www.bisnis-kti.com/index.php/2012/02/tambang-lsm-minta-pemda-bombana-moratorium-izin-tambang/ (site discontinued)

Australian and New Zealand Environment Conservation Council (ANZECC), 2000. *Australian and New Zealand Guidelines for Fresh and Marine Water Quality—Volume 1: The Guidelines*. Canberra, ACT: Australian and New Zealand Environment and Conservation Council, and Agriculture and Resource Management Council of Australia and New Zealand.

Bank, M.S. (ed.), 2012. *Mercury in the Environment: Pattern and Process*. Berkeley: University of California Press. doi.org/10.1525/california/9780520271630.001.0001

Canadian Council of Ministers of the Environment, 1995. *Protocol for the Derivation of Canadian Sediment Quality Guidelines for the Protection of Aquatic Life*. CCME EPC-98-E. Ottawa: Environment Canada, Guidelines Division, Technical Secretariat of the CCME Task Group on Water Quality Guidelines.

Cortes-Maramba, N., J.P. Reyes, A.T. Francisco-Rivera, H. Akagi, R. Sunio and L.C. Panganipan, 2006. 'Health and Environmental Assessment of Mercury Exposure in a Gold Mining Community in Western Mindanao, Philippines.' *Journal of Environmental Management* 81: 126–34. doi.org/10.1016/j.jenvman.2006.01.019

Fadini, P.S. and W.F. Jardim, 2001. 'Is the Negro River Basin (Amazon) Impacted by Naturally Occurring Mercury?' *Science of the Total Environment* 275: 71–82. doi.org/10.1016/S0048-9697(00)00855-X

Global Mercury Project, 2006. 'Environmental and Health Assessment Report: Removal of Barriers to the Introduction of Cleaner Artisanal Gold Mining Extraction Technologies. Brazil, Indonesia, Laos, Sudan, Tanzania Zimbabwe.' Global Environment Facility, United Nations Development Programme, United Nations Industrial Development Organization.

Gray, J.E., P.M. Theodoraka, E.A. Bailey and R.R. Turner, 2000. 'Distribution, Speciation and Transport of Mercury in Stream Sediment, Stream Water and Fish Collected near Abandoned Mines in South Western Alaska, USA.' *Science of the Total Environment* 260: 21–33. doi.org/10.1016/S0048-9697(00)00539-8

IKAPERMAB-Yogyakarta, 2009. 'Bombana dpt Lailatur Qadr.' 15 September. Available at ikapermab-yogyakarta.blogspot.com.au/2009/09/bombana-dpt-lailatur-qadr.html

Kamil, S.Y., 2009. 'Tambang Emas Bombana: Berkah Atau Ancaman?' *Sagori Hijau Indonesia*, 23 July. Viewed at sarekathijauindonesia.org/2009/07/107.html (site discontinued)

Kitula A.G.N., 2006. 'The Environmental and Socio-economic Impacts of Mining on Local Livelihoods in Tanzania: A Case Study of Geita District.' *Journal of Cleaner Production* 14: 405–14. doi.org/10.1016/j.jclepro.2004.01.012

Lechler, P.J., J.R. Miller, L.D. Lacerda, D. Vinson, J.C. Bonzongo, W.B. Lyons and J.J. Warwick, 2000. 'Elevated Mercury Concentrations in Soils, Sediments, Water and Fish of the Madeira River Basin, Brazilian Amazon: A Function of Natural Enrichments?' *Science of the Total Environment* 260: 87–96. doi.org/10.1016/S0048-9697(00)00543-X

Limbong, D., J. Kumampung, J. Rimper, T. Arai and N. Miyazaki, 2003. 'Emissions and Environmental Implications of Mercury from Artisanal Gold Mining in North Sulawesi.' *Science of the Total Environment* 302: 227–36. doi.org/10.1016/S0048-9697(02)00397-2

McKelvey, W. and E. Oken, 2012. 'Mercury and Public Health: An Assessment of Human Exposure.' In M.S. Bank (ed.), *Mercury in the Environment: Pattern and Process*. Berkeley and Los Angeles: University of California Press. doi.org/10.1525/california/9780520271630.003.0013

Roth, D.A., H.E. Taylor, J. Domagalski, P. Dileanis, D.B. Peart, R.C. Antweiler and C.N. Alpers, 2001. 'Distribution of Inorganic Mercury in Sacramento River Water and Suspended Colloidal Sediment Material.' *Archives of Environmental Contamination and Toxicology* 40(2): 161–72. doi.org/10.1007/s002440010159

Serikawa, Y., T. Inoue, T. Kawakami, B. Cyio, I. Nur and E. Elvince, 2011. 'Emission and Dispersion of Gaseous Mercury from Artisanal Small-Scale Gold Mining Plants in the Poboyu Area of Palu City, Central Sulawesi, Indonesia.' Paper presented at the 10th International Conference on Mercury as a Global Pollutant, Halifax, Nova Scotia, Canada, July.

van Straaten, P., 2000a. 'Human Exposure to Mercury due to Small Scale Gold Mining in Northern Tanzania.' *Science of the Total Environment* 259: 45–53. doi.org/10.1016/S0048-9697(00)00548-9

van Straaten, P., 2000b. 'Mercury Contamination Associated with Small Scale Gold Mining in Tanzania and Zimbabawe.' *Science of the Total Environment* 259: 105–13. doi.org/10.1016/S0048-9697(00)00553-2

Veiga, M.M., J. Hinton and C. Lilley, 1999. 'A Comprehensive Review with Special Emphasis on Bioaccumulation and Bioindicators.' Proceedings of National Institute for Minamata Disease, Japan, pp. 19–39.

World Health Organisation (WHO), 2011. *Guidelines for Drinking-Water Quality*. 4th edition. Geneva: WHO. Available at apps.who.int/iris/bitstream/10665/44584/1/9789241548151_eng.pdf

15

Artisanal and small-scale mining governance: The 'emerging issue' of 'unregulated mining' in Lao PDR

Daniele Moretti and Nicholas Garrett

'Illegality' and 'formalisation' are key topics in the global literature on artisanal and small-scale mining (ASM) (see, for example, Fisher 2008; Spiegel and Veiga 2009; Spiegel 2011; Lahiri-Dutt et al. 2014). A recent body of research argues that, in many developing countries, national governments and donor agencies have deliberately designed and implemented mining regulations and policies with the exclusive interests of large-scale mining (LSM) in mind, thus accepting the parallel outcome of illegality of ASM (see, among others, Baker et al. 2007: 13–14; Spiegel 2011: 189, 201; Spiegel and Hoeung 2011: 2; Hatcher 2012; Holden et al. 2011, cited in Spiegel 2014: 301, 307; Fisher 2008; Hilson 2009; Hilson and McQuilken 2014: 112).[1] Yet, it has been pointed out that

---

1   The term 'illegal ASM' indicates ASM activities that are neither licensed nor compliant with regulations, taxes and other parts of the formal economy. By contrast, 'informal ASM' is a broader category that includes situations where ASM activities are 'unregulated by the institutions of society in a legal and social environment in which similar activities are regulated' (Castells and Portes 1989: 12). In some cases, this may be because no regulatory frameworks exist. In others, relevant licensing agencies, regulatory agents and taxation officials may be non-existent, unreachable or set unreasonable barriers to compliance, such as indefinite delays to administrative processes or ambiguous procedures that prevent the legal compliance and formalisation of ASM. For example, it may be that authorities do not make available the registration documents ASM operators are legally required to have. Conventionally, formalised ASM is taken as a necessary prerequisite of legitimacy, but this is not always the case. It may be that some ASM operators in a given locality meet all government laws and

broader analyses of the many reasons behind the growth of illegal mining and its more negative impacts are needed to gain a fuller picture, and better guide future policy development and implementation. Among others, this includes attention to the wider 'complexities' that characterise specific ASM governance regimes (Spiegel 2011: 202).

This chapter draws on fieldwork by Nicholas Garrett on the operation of the regulatory environment, desktop research and on joint fieldwork by the authors in 2014. The research was carried out as part of a consultancy for a mining company that sought to develop group standards and an engagement strategy for ASM in various countries, including Lao PDR. The authors visited Vientiane and current and former ASM sites and ASM-impacted areas in Vientiane and Xieng Khouang provinces. The visits offered a basis for direct observations supplemented by interviews with 40 informants, including local community leaders, gold traders, local- and national-level government and technical officials, representatives of international donor agencies and staff from two LSM companies. A significant constraining factor in our research is that we were unable to directly interview active gold miners as we normally do in our research work. Health and safety concerns prevented us from entering one site. The seasonal and transient nature of ASM is also a well-known impediment to ASM research (Noetstaller et al. 2004: 16; Eftimie et al. 2012: 32–3, 44), and our fieldwork in Lao PDR coincided with low times in the seasonal pattern of local ASM activities. Some of the areas we visited had already undergone partial rehabilitation, which limited opportunities for direct observations and interactions. The short time we had available in the field further constrained our capacity to directly engage and interview former and active ASM actors. This is especially significant in a context like Lao PDR, where miners are often reluctant to identify themselves as such, to take part in research and to disclose information on their activities due to illegality and tax issues (Baker et al. 2007: 6–8; see also Noetstaller et al. 2004: 16; Eftimie et al. 2012: 32–3). This was compounded by our association with an LSM company when undertaking field research on ASM in October 2014. We draw more on the viewpoints of government agencies, LSM operators and ASM-impacted communities than those of still-active miners, but we have ensured that our global experience

---

regulations, for example, but that these exclude certain standards deemed critical by the international community. Conversely, in circumstances where miners are operating 'informally' because they cannot reasonably comply with government regulations, or there is no applicable legal framework, informal ASM operators are not necessarily illegitimate.

in interviewing ASM operators has also shaped our analysis. There is comparatively little up-to-date research on ASM governance in Lao PDR, and we hope to fill this gap with an overview of evolving challenges. This chapter also reports recent developments in ASM governance, suggesting how these could be made more effective in future.

The next section offers an overview of the history and present context of ASM in Lao PDR, including its key subsectors and their technical and organisational characteristics. Central to this is a discussion of some key (nationwide) 'push' and 'pull' factors that have driven the sector's expansion and mechanisation over the past 10 years. Following that, we consider several 'complexities' that so far have limited the effectiveness of the country's ASM governance system, inhibiting attempts to promote better practices and ensure accountability. Recent expansion and increased mechanisation of artisanal mining have led to greater concerns over negative health and safety, environmental and social impacts. These concerns are behind recent media reports and perceptions by the government that 'unregulated' or 'illegal' ASM is an 'emerging issue' that needs to be tackled through new regulations (Vaenkeo 2014a). They have also largely motivated recent interventions to improve sectoral policies and management systems. The concluding section of the chapter reviews these interventions, which have ranged from training and awareness programs among miners and local communities, to capacity building in relevant government agencies and current efforts to draft ASM regulations. While each of these individual interventions has achieved some limited results in its own way, ASM governance to date has remained largely 'ad hoc and ineffective' (Baker et al. 2007: 18; see also BGR 2014). Moving forwards, it is argued that the Government of Lao PDR (GoL) should consider adopting a more 'strategic' approach to ASM. Such an approach would draw on international good practice and take into account current capacities to implement. It would adopt an integrated suite of interventions tailored to the particular characteristics of different ASM subsectors that is cognisant of specific national development priorities. While designing and implementing such a strategy is fraught with difficulties, several guides and toolkits are now available to help governments seeking to better manage ASM activities (such as Paget et al. 2015; see also United Nations Environment Programme 2015; Eftimie et al. 2012; Hinton and Hollestelle 2012). To be effective, such a strategy ought to build on up-to-date and country-specific sectoral research that also considers ASM's relationship to other economic sectors and the wider regional economy, national industry strategies and rural development plans (Hilson and McQuilken 2014; Paget et al. 2015), as well as a review of existing capacity gaps within the government. Further, it must be effectively implemented by well-resourced departments with adequate capacity, working in coordination with one another (Paget et al. 2015). For this to be achieved, the strategy would have to be developed and implemented with the support of international donors and in iterative consultation with relevant sectoral stakeholders (including artisanal miners, local communities, civil society, government, LSM and other supply chain actors). Global experience shows that such cooperation is a key factor of success in promoting better practices and achieving a better balance between positive and negative impacts within the sector (Paget et al. 2015; see also Aubynn 2009; Spiegel and Hoeung 2011; Spiegel 2014).

ASM in Lao PDR: History, present context and main drivers of expansion

ASM is a still-debated term used to describe many different forms of extractive activities, from artisanal and unlicensed to more mechanised and licensed medium-scale operations (Lahiri-Dutt et al. 2014: 7). In Lao PDR, ASM has a long history that predates French colonisation. Presently, it involves a long-standing mixture of local kin groups, private and state-owned domestic enterprises, and partnerships of national and regional companies (Larsen 2010: 6–7). In more recent times, however, it has come to include a range of independent foreign operators, sometimes operating in small groups and sometimes in more organised, commercial forms. It has also been increasingly shaped by inputs of financing, technology and know-how from neighbouring countries. Lao PDR ASM is therefore not a sheltered national sector but one that is increasingly integrated with the wider regional artisanal and small-scale mining and economy.

Based on the literature and our own observations, artisanal and small-scale mining in the Lao PDR can be divided into two main categories. The first involves small-scale mining undertaken without modernised production systems. These operations extract a mix of mineral and non-mineral resources, including gold, and comprise mines run by the government and privately owned mines with investment from local and/or foreign investors from countries including China, Thailand, Vietnam and Korea (Shingu 2006: 1, 7; Larsen 2010: 6–7; International Council on Mining and Metals (ICMM) 2011).

The second category includes manual and partly mechanised artisanal mining, primarily of gold, though other precious metals and stones, base metals like tin and construction materials are also mined and quarried (Larsen 2010: 6–7; ICMM 2011: 17; Latsaphao 2014; Vaenkeo 2014a, 2014b). Our informants referred to this ASM type as 'handicraft mining' (*kun kut hatagam*) or, in the case of partly mechanised artisanal mining, as 'semi-handicraft mining' (*kut kun kung hatagam*).

There are no precise estimates, but the large numbers of part-time ASM miners are thought to be equivalent to between 15,000 and 50,000 full-time miners, with 35,000–46,000 being the figure used in the most recent sectoral projections (Larsen 2010: 7, 16, 20, 23, 30; ICMM 2011: 17). Non-mechanised artisanal mining is known to take place in at least 11 out of 17 provinces, and is estimated to involve around 11 per cent of all ASM workers, while partly mechanised alluvial and hard-rock ASM has been reported in 9 out of 17 provinces, and involves around 89 per cent of the ASM workforce. However, both geographical distribution and participation levels are systematically underestimated. This is due to various reasons, including the fact that not all provinces have been covered in recent surveys, the itinerant and subsistence nature of ASM, lack of monitoring capacity by GoL and the miners' reluctance to reveal themselves and talk about their activities for reasons considered above (Baker et al. 2007: 4–7).

While mining laws allow for and require permits for it (detailed in the next section), non-mechanised alluvial and hard-rock mining is often carried out informally by whole families, including women (who make up between 50 per cent and 80 per cent of the country's ASM operators) and children. At times, individuals or groups of unrelated miners also participate in this type of activity, especially in the case of non-mechanised hard-rock mining (Earth Systems Lao (ESL) 2003: i, 3; Shingu 2006: 8; Baker et al. 2007: 2; Larsen 2010: 7; ICMM 2011: 11, 17; Eftimie et al. 2012: 6–7, 88, 93; Lahiri-Dutt et al. 2014: 16–18). For the most part, these miners are not migrants but locals who mine nearby areas of alluvial sediments revealed by lower water levels in the dry season (between January and June). Alternatively, they mine hard-rock deposits uncovered by past and present larger mining companies or small outside operators with greater mining know-how and financing. This type of ASM is undertaken mainly during downtimes in the agricultural work cycle, as a seasonal subsistence activity to supplement rice and vegetables cultivation, rearing of livestock, fishing and gathering from forests, which remain the primary sources of livelihood, alongside other cash-earning activities. Depending on location, such cash-earning activities can include limited cash cropping, sale of livestock, tourism and textile and other industries (ESL 2003; Shingu 2006: 8–9; Baker et al. 2007: 2, 5). Increasingly, however, this type of ASM is also undertaken by small groups of outside operators or more organised commercial ventures from other parts of the Lao PDR and neighbouring countries. Most alluvial artisanal and small-scale gold-mining (ASGM) operators in Lao PDR mine by hand, using simple tools like bowls, buckets, chisels, pickaxes, hoes, floats and water pumps to move water for sluicing and open sluice boards with sack linings to pan and sieve the ore (ESL 2003: i, 19–21; Shingu 2006: 8; Baker et al. 2007: 2; Lahiri-Dutt et al. 2014: 16–18). In some areas, such as Borikhamxay Province, men reportedly dig deep vertical shafts and mine gold gravel to be panned in the river (Shingu 2006: 9). In Vientiane Province, Xieng Khouang Province and many other parts of the Lao PDR, operators also increasingly mine primary deposits, mainly using simple tools adapted from agriculture and construction, such as hoes, shovels and bowls to dig pits, and rudimentary, unsupported tunnels to extract and wash ores, which are then processed by the miners themselves or by larger traders using mechanised crushing, sluices, copper plates, soaking tanks and chemicals like mercury and cyanide (Baker et al. 2007: 5; Latsaphao 2014; Vaenkeo 2014a, 2014b).

Mechanised and semi-mechanised alluvial and hard-rock operations require official permits and close reporting and supervision. Yet, they too are often undertaken without formal licensing or regulation and with little monitoring and environmental or health and safety awareness. In some cases, such unlicensed or unregulated operations are directly run by foreign companies or small groups of miners from abroad. More frequently, they tend to be run by local operators (and, until relatively recently, even by the army), but often with the assistance of foreign investors from neighbouring countries, who may provide them with financing, know-how or technological inputs, as well as commercialising the minerals that they produce (Baker et al. 2007: 5, 9; Larsen 2010: 6–7; Lahiri-Dutt et al. 2014: 14). Again, here too increasing numbers of small or more organised groups of foreign operators have also directly entered this sector in more recent times, primarily from neighbouring countries like Vietnam and China. Processing of ores is minimal and technologies simple, often including only barges fitted with excavators, conveyors and sluices or old hand drills, compressors and tracks, but often with more sophisticated adits, pits and tunnels for hard-rock mining. Where gold flakes are coarse, drying and blowing by mouth are commonly used to separate them. Where particles are fine, however, or in the case of hard-rock mining, mercury amalgamation and, more recently, cyanide have also been increasingly employed (Larsen 2010: 7; Latsaphao 2014; Vaenkeo 2014a). The number of employees found in these more mechanised artisanal and small-scale operations range from just 10 to 150 each, with the majority having around 20 workers. Many of these workers are from nearby villages but, by contrast with mostly alluvial, non-mechanised operations, a greater number of migrants often participate in this type of venture, which therefore tends to be less seasonal, more enduring and not as strictly subsistence related (Shingu 2006: 7; Baker et al. 2007: 5).

If ASM has been a widespread activity in rural communities for some time, a combination of familiar push-and-pull factors contributed to the considerable expansion of participation in the sector over the past two decades, which reportedly reached a peak in around 2008–09, as well as its increased mechanisation (Shingu 2006; Baker et al. 2007: 2, 5, 8, 10–13; Larsen 2010: 7; ICMM 2011: 5; Eftimie et al. 2012: 90; Lahiri-Dutt et al. 2014: 8–9; Latsaphao 2014; Vaenkeo 2014a, 2014b).[2] As is the case globally (ICMM 2010: 5; Eftimie et al. 2012: 3), ASM participation in Lao PDR tends to be linked to poverty and lack of alternatives that could provide better income opportunities. The sector has thus emerged as an important source of cash and livelihood diversification, with much informal alluvial mining practised by farmers to supplement other livelihood strategies (ESL 2003: 19; Shingu 2006: 7; Insouvanh 2015).[3] To appreciate the financial attractiveness of ASM as a complementary

---

2   It is important to note, however, that the sector (especially in its fully or partly mechanised component) has reportedly shrunk somewhat over the past two years, partly as a result of a 2012 nationwide GoL moratorium on all new ASM ventures, and inspections and closures of existing mines with poor environmental records and likely also partly because of decreased gold prices since 2012–13.

3   This factor alone may be somewhat weaker in explaining the rise of more mechanised ventures and operations that, as discussed above, tend to be more commercial, meaning not strictly subsistence-related, more often financed by outside operators and more likely to involve migrant labour (Baker et al. 2007: 5).

livelihood strategy, it is sufficient to note that Lao PDR alluvial panners can reportedly extract up to 31 grams of gold per annum, which could be equivalent to an additional annual income of over US$641.[4] Moreover, mechanised and semi-mechanised operations, especially of primary deposits, are likely to yield even greater returns than this (Larsen 2010: 7). For example, our informants suggested that manual hard-rock gold miners in Xieng Khouang Province could reportedly earn up to between 6 and 40 per cent of the average local annual income in a single day of mining during the peak phase of two minor, highly localised rushes.

Over the past two decades, institutions like the Asian Development Bank and the World Bank have also encouraged GoL to adopt structural reforms and promote foreign investment as part of a transition towards a liberal market economy meant to promote national and local economic development (Boungnaphalom 2010; Larsen 2010: 4, 13–14; ICMM 2011: 17; Hatcher 2012: 5–8; Lahiri-Dutt et al. 2014: 2–3). While it has brought benefits and opportunities to Lao PDR and its people, this process also engendered new forms of poverty and inequality linked to changes in demographics, the economy, occupational opportunities and land use, and thus also to opportunities for livelihood diversification towards a more cash-based economy and sources of cash other than agriculture (Rigg 2005; ICMM 2011; Lahiri-Dutt et al. 2014: 3–5). Therefore, the recent rise in ASM activities across the country should be understood, not only through the lens of poverty and lack of alternative opportunities per se, but also in the context of the effects of recent structural reforms (Spiegel 2009: 39, 2011: 192; Hilson and McQuilken 2014: 111–13).

---

4   This is based on recent estimates that local miners receive no more than half of the international price of gold when they sell to local traders at or near the mines (ICMM 2011: 17), and on the average annual gold price of US$1,282 in 2014, when our research was conducted (as obtained through 'Historical Charts and Data for Gold, 2014' at www.kitco.com). Not much is known about the downstream supply chain of Lao ASGM operators, though a reported pattern involves the miners selling gold to itinerant gold buyers who visit their villages every week during the mining season. These buyers then further process it for sale in the regional markets or sell it directly to jewellers (ESL 2003: ii, 22; Insouvanh 2015). This pattern was confirmed by our gold trading and jewellery industry informants, who estimated that around 15–20 per cent of all gold traded in Vientiane may originate from ASM, a figure that was said to have declined in the last couple of years, as a result of the 2012 government moratorium on alluvial ASGM. According to them, gold is traded through intermediaries, often in transactions involving only small quantities of up to 15 grams at a time. The destinations vary, and include regional and Vientiane-based markets and jewellers, with some of it also thought to be smuggled to neighbouring countries, like China and Thailand. Although some of the ASM gold is then exported in the form of jewellery, it is also commonly purchased for savings and marriage-related payments.

But if participation in ASM offers an alternative source of livelihood in the face of wider structural changes, it also causes significant environmental degradation, particularly in the case of rush situations and more mechanised ASM or medium-scale mining activities. This can, in turn, engender further vulnerability, thus promoting further participation in ASM to make up for ASM's own impacts on agricultural livelihoods (Insouvanh 2015; Lahiri-Dutt et al. 2014: 3, 9–15, 24–6). In line with these facts, our research indicated that in Vientiane and Xieng Khouang provinces, participation in ASM (and especially manual and semi-mechanised artisanal mining) tended to be linked to situations where local communities were experiencing livelihood stresses due to a combination of population growth (such as through immigration), land shortages, increased food prices, limited employment and alternative cash earning opportunities, and liquidity issues in between times when most crops become available for consumption and sale.

Another factor at play in the rise of ASM may relate to issues of ownership, state legitimacy and political agency. While Lao PDR mining laws place minerals under centralised and unified management by the state (Article 3, Mining Law 1997; Article 4, Mining Law 2012), local actors can hold alternative understandings of mineral ownership. For instance, in Xieng Khouang Province, we interviewed local community members about a recent gold rush. The informants were not themselves miners, but they belonged to communities located near the ASM sites. Contrary to national laws, they believed that minerals belonged to, and could therefore be freely worked by, all Lao PDR citizens (and not just local residents). It has further been argued that many Lao PDR villagers see the state as a distant abstraction of which they are not fully part and that they do not necessarily feel wholly included in the benefits generated by recent economic changes (Mansfield 2000: 2; Lahiri-Dutt et al. 2014: 3). In this light, it has been suggested that engaging in illegal ASM and reserving the right to sell extracted resources to illegal buyers paying the highest price is not just down to an economic logic, but is also a means for some villagers to exert political agency by asserting control over their livelihoods, and over land, in a rapidly changing context driven by distant outside forces (Lahiri-Dutt et al. 2014: 4, 23–6).

In addition to these drivers, there are a number of pull factors at play in the expansion of ASM in Lao PDR. Like any other business, ASM can be undertaken only where the costs of extracting a resource are lower than the income made by selling it (ICMM 2010: 9). Participation levels will also be influenced by how favourably the rewards obtained through ASM compare with those afforded by other economic opportunities available to ASM operators (Spiegel 2014: 301, 307); by the degree to which ASM's profits are necessary to mitigate the risks and/or the cessation of other activities; and by the extent to which ASM can be undertaken alongside alternative forms of livelihood. It is therefore unsurprising that in Lao PDR, even when undertaken as a means of livelihood diversification, participation in ASM is always guided by a combination of lack of alternative opportunities *and* the 'favourableness' of mining itself (ESL 2003: 19–20; Baker et al. 2007: 4). As such, a commonly reported explanation for the recent growth in ASM activities is the rise of mineral prices over recent years (Lahiri-Dutt et al. 2014: 11; cf. Spiegel 2014: 301, 307). This, however, can be contrasted with the 2012–13 gold price crash and the fact that the gold price has since remained low have (along with the aforementioned 2012 moratorium) reportedly led to an overall reduction of activity.

In many parts of the country, including Vientiane and Xieng Khouang provinces, a complementary and partly related push factor has been the increased availability of financing, technology, inputs and outside mining know-how from nearby countries like China, Vietnam and Thailand in recent times (Baker et al. 2007: 4; ICMM 2011: 5). This has facilitated the expansion of ASM, particularly in its more mechanised and 'rush' forms. In many cases, this can take the form of outsiders (mainly from neighbouring countries but also from other parts of Lao PDR) coming into an area to mine, with locals initially supplying services and mining labour (Spiegel 2014: 305; Insouvanh 2015). In so doing, locals acquire skills that they later use to mine independently, with outsiders still transporting, processing and trading the mined ores. In some areas at least, the gold buyers themselves also supply miners with mercury and other chemicals (ESL 2003: 22). In the Lao PDR context, the start of more mechanised small-scale, medium-scale and large-scale mining operations (Baker et al. 2007: 6–7), as well as other forms of large-scale resource development, like hydropower, have also sometimes encouraged greater involvement in manual and mechanised ASM by both locals and outsiders. This is possibly linked to a combination of changes in the local socio-economic and environmental contexts (for example, greater monetisation of the economy, rising prices, land impacts, easier and cheaper exploitation of local deposits, etc.). Greater participation of this kind also increases the 'favourableness' of mining; for instance, through the transmission of know-how, the creation of infrastructure that reduces access and transportation costs, or the unearthing of deposits or disposal of tailings that can be more easily and rewardingly worked through ASM techniques. In certain cases, the appearance of more mechanised mining activities in the given areas can even encourage locals to develop greater levels of organisation, including at the village level. It can motivate these units to shift away from alluvial mining and towards semi-mechanised mining of primary deposits, which typically yields greater returns, but also requires more local coordination to raise funds and/or outside investment or seed money (ibid.: 8).

Current ASM governance 'complexities' and the 'emerging issue' of 'unregulated mining' in Lao PDR

The aforementioned expansion of artisanal and small-scale mining has been mirrored by growing preoccupation in media reports and within the government at all levels about the increase of 'unregulated or illegal ASM' activities and their negative impacts. These concerns are exemplified by the issuing of two nationwide moratoria on new ASGM activities and consequent rounds of inspections in 2005 (Baker et al. 2007) and 2012, as well as ad hoc provincial and district-level moratoria on specific forms of ASM at various times. Most recently, they led to the National Assembly's characterisation of 'unregulated or illegal ASM' as 'an emerging issue' to be addressed by GoL through improved regulation (Vaenkeo 2014a).

The current Lao PDR legislative framework does (primarily through the amended Mineral Law 2012) provide for the licensing of ASM, but it also poses several restrictions and requirements to operate legally (cf. Spiegel 2009: 42, 2014: 301–3; Spiegel and Hoeung 2011: 3). For example, ASM operators are legal only when:

- They are Lao PDR citizens, resident in the local community.
- They acquire appropriate mining licences and mine within the boundaries of the licence or permit area, in accordance with the techniques and modes of organisation and management appropriate to that kind of licence (for example, no machines or machines under five horsepower and only up to 10 labourers).

- They mine only 'alluvial gold, alluvial tin and tailing'—where alluvial gold is defined as 'mechanical weathered mineral, which have [sic] flown and accumulated in some location naturally'—whose exploitation 'is not appropriate for industrial mining' (revised Mining Law 2012).
- They pay all required taxes.[5]
- And/or they are Lao PDR entities with the required permits who follow the more stringent requirements for 'small-scale mining' and 'business artisanal mining', such as reporting regularly to the relevant energy and mines sector authorities and following strict laws on health and safety, environmental and social protection and sustainable development, including minimising and compensating for environmental and social impacts, and maintaining roads and other infrastructure.

Equally, small-scale processors and gold traders only process and trade legally when:

- They acquire the appropriate licences.
- They do not deal in misappropriated or otherwise illegally mined gold.
- They pay all required taxes.

In practice, this means that many, if not most, ASM operators and gold transporters, processors and traders are de facto working illegally because, among others, they do not hold the required permits and pay required taxes; some are not Lao PDR citizens (or Lao PDR citizens who are local residents); they mine primary deposits rather than alluvial deposits, as defined by the law; and they lack environmental planning and rehabilitation. Moreover, GoL issued two moratoria on all new ASGM activities in 2005 and 2012, the second of which was still effective at the time of fieldwork, with related inspections of existing operations and their closure in case of noncompliance.

Under the current governance framework, inspections and closure processes of non-compliant operations are undertaken by ad hoc bodies involving a multitude of agencies, from local representatives of the departments of mines, planning, investment, natural resources and environment, finance, and public security to local administrative authority representatives. Similarly, ASM permits are administered at the local rather than central level, with business ASM permits being the

---

5   See also the 'Draft Presidential Decree of Lao PDR President on the Tax Rate of Natural Resources/Royalty'.

competency of provincial-level authorities under the technical supervision of provincial-level offices of the Department of Mines (DoM), whereas those for non-business (or 'handicraft') artisanal mining are issued by district-level administrative authorities. While this governance structure clearly involves a degree of decentralisation, this is not necessarily a significant devolution of power insomuch as a process intended to build decentralised capacity so that the central GoL has a better view of the activities, issues and challenges on the ground across the country. Nevertheless, it means that Lao PDR law requires that different agencies and administrative authorities at different levels should be involved in ASM-related issues and also, crucially, that each of them has specific powers and competencies. In turn, this 'complexity' (Spiegel 2011: 191, 193) feeds a potential for local variation in ASM strategies and their implementation. For example, some provincial and district governments have taken a much more favourable attitude to ASM, even asking the central government for permission to resume releasing permits after national moratoria to regain lost revenue from this source (Baker et al. 2007: 5). By contrast, others have shown more opposition to ASM, not only due to concerns about environmental, health and safety and social issues, but also because they regarded it as a threat to key local industries, like agriculture or tourism. In some cases, this may have resulted in misalignments of interests vis-à-vis ASM between local (provincial and district) and central GoL authorities because the former derive a more direct potential benefit from fees for issuing ASM permits (ibid.). It can also limit the state's capacity to ensure accountability in the sector and open spaces where illegal activities can come to thrive. For example, even after the 2012 moratorium, local authorities continued to issue permits for ASM operations extracting construction materials, including gravel from riverbeds. According to anecdotal evidence, many of these operations have then proceeded to illegally mine alluvial gold alongside gravel, at times using chemicals in such a way that caused environmental pollution and health risks for nearby communities and raised public concern (Southivongnorath 2014; Vaenkeo 2014a, 2014b). These challenges are compounded by other widely acknowledged regulatory 'complexities' (Spiegel 2011) that leave Lao PDR with what has been called an 'ad hoc and ineffective system of [ASM] governance' (Baker et al. 2007: 18). As a result, the implementation of national laws and the regulatory framework on ASM has long been, and continues to be, 'weak or wholly lacking' across the country (ibid.; see also BGR 2014). Among others, these issues include a lack of communication between different authorities (and levels of government) involved in mining activities. There are general capacity issues at all levels in DoM, Department of Geology (DoG) and other agencies charged with regulating ASM (such as limited human and financial resources and lack of technical expertise in ministries and departments), which severely limits capacity to both assist and regulate the sector. This is compounded by a lack of detailed ASM regulations accompanying mining legislation (Baker et al. 2007; Insouvanh 2015; see also Spiegel 2009: 42; 2011: 202; Spiegel and Veiga 2009).

Others have argued that GoL has been reluctant to intervene in artisanal and small-scale mining because it views ASM as an activity that requires significant (and largely lacking) resources to be properly managed, with potential risks (such as creating resentment among miners), and for comparatively little benefit in return, particularly from the standpoint of central authorities that gain fewer benefits from fees and taxation (Baker et al. 2007: 4, 8–9, 19; Spiegel 2011: 193). By and large, governments at all levels have tended to accept the presence of micro-level informal ASM activities, especially in the case of subsistence-oriented artisanal mining by local actors. Indeed, this mining has been allowed in spite of the nationwide and local moratoria, as long as it did not cause significant social and environmental issues (see also Baker et al. 2007: 7).

In more recent years, however, the central and local governments' 'tolerance level' of ASM has been more frequently and severely tested. Recent expansion and increased mechanisation of artisanal and small-scale mining mean that significant numbers of new operators have entered the sector, often with limited capacity and incentives to respect good practices. This has resulted in increasingly significant negative impacts in many parts of the country, including land degradation and water pollution that affected agricultural livelihoods and the unsafe use of chemicals like mercury and cyanide for recovery and processing within or near inhabited areas and vulnerable environments, with consequent negative effects like exposure to fumes, dispersal in the environment, bioaccumulation in the food chain and loss of wildlife, fisheries and even livestock (ESL 2003: i–ii, 20–1, 24; Shingu 2006: 7–9; Baker et al. 2007: 2, 4–5, 8–13; Larsen 2010: 7; ICMM 2011: 5; Eftimie et al. 2012: 90; Lahiri-Dutt et al. 2014: 8–9; Latsaphao 2014; Vaenkeo 2014a, 2014b). Many ASM operations do not properly plan for, map and reinforce tunnelling or ensure that they are properly ventilated; cause land erosions and collapse; make little or no use of any personal safety equipment; and make inappropriate and dangerous use of chemicals (ESL 2003: 24). As a result, ASM across the country has also been associated with health and safety risks for the miners themselves and for non-miners living in the proximity of mining and processing areas, including fatalities (for example, 20 known fatalities in one of the districts we visited during a minor, highly localised gold rush). In turn, these environmental, health and safety issues have led to concern among affected local communities and, at times, conflict between operators and surrounding communities—a pattern that is arguably not reported as often as ASM–LSM conflicts but that is found in many contexts globally (see, for example, Hinton and Hollestelle 2012: 69; Eslava 2014; Insouvanh 2015; Soemarwoto 2015). It is largely as a result of these steadily increasing impacts, coupled with the aforementioned 'complexities' and limitations of the current ASM governance system, that ASM has come to be spoken of more and more in terms of 'unregulated or illegal mining' and as an 'emerging issue' in need of proper regulation and management.[6]

Looking ahead: Towards a more strategic approach to ASM

Over the past few years, GoL has taken some steps with an array of partners to improve existing knowledge and management of artisanal and small-scale mining. Overall, these have not constituted elements of an integrated strategy, but isolated interventions with so far limited impacts.

Between 2003 and 2007, the Global Mercury Project (GMP) worked with the government and other stakeholder groups (for example, local communities, non-government organisations and United Nations agencies) to:

- Build ASM-related capacity within the then DoG and DoM and local government agencies.
- Introduce technology and training to miners to reduce mercury use and loss to the environment.
- Develop new legislation specific to ASM activities, such as a new national mercury code.
- Produce educational materials in Lao.

---

6   Our interviews suggest that, at least among some central-level authorities, there is also a growing recognition that many of the resources found in Lao PDR are of a scale unsuitable for large mining companies, which therefore will become commercially viable only if exploited by properly regulated artisanal and small-scale mining.

- Collect data on the extent and magnitude of small-scale and intermediate gold-mining activities in Lao PDR and the Mekong Basin (ESL 2003; Shingu 2006: 7–9; Baker et al. 2007: 2, 7, 12–19; Spiegel and Hoeung 2011: 8).

More recently, the World Bank undertook a survey of (primarily) river-based tin ASM, which yielded fresh data on artisanal and small-scale mining, particularly in relation to its gender profile and gender-related sustainability issues (Eftimie et al. 2012: 87).

The private sector has also contributed to this process, albeit in a limited way. For example, LSM companies have provided some assistance to build capacity within the DoM. They have also conducted awareness training for ASM-affected communities and assisted government agencies in safely disposing of dangerous chemicals and rehabilitating areas impacted by ASM activities. However, this has been limited to their specific project areas within the country.

In the past couple of years, the DoG, DoM and other GoL authorities collaborated with a number of international partners like the World Bank, BGR, the Finnish and Thai governments, and Japan Oil, Gas and Metals National Corporation to develop new regulatory frameworks and further build in-house capacity to more effectively govern the mining sector (see, for instance, Korkiakoski et al. 2012; BGR 2014; Insouvanh 2015). As part of this, and following pressure from the National Assembly due to the aforementioned growing concerns about the negative impacts of some unregulated ASM activities, GoL is drafting new ASM regulations with technical assistance from international donors. At the time of fieldwork, the regulations were still in draft form and undergoing consultation. Many of the existing requirements and restrictions appear to be carried over in the draft stage of the regulations. However, potential departures from previous legislation were being provisionally considered for inclusion in the new regulations, which may eventually come to embody a potential shift towards greater levels of formalisation of ASM, from alluvial and hard-rock mining to individual and collective modes of organisation. These included allowing for the exploitation of both placer and hard-rock deposits by artisanal mining permit holders (though only under specific conditions, and where the latter can be extracted and processed wholly without the use of machines); enabling artisanal miners to register as 'collectives'; and the eventual creation of a fund that will provide training to ASM operators on mine safety, health and environmental protection and rehabilitation. At such an early stage, it is unclear whether the regulations will become law any time soon, if they will be redrafted prior to that following ongoing consultation, and if they will be effectively implemented. Alongside the other interventions outlined above, however, they may eventually constitute a significant step towards reforming ASM governance in Lao PDR and tackling some of the 'complexities' that currently limit its effectiveness.

These individual interventions have all achieved limited results in their own right. Overall, however, the existing ASM regulatory framework continues to be implemented in a largely ad hoc and reactive fashion, remaining on the whole permissive[7] while lacking the capacity to provide effective assistance, steadily improve practices and enforce standards. In its efforts to tackle what is increasingly perceived to be 'the emerging issue' of the negative impacts of 'illegal and unregulated mining' across the country, GoL should continue to build on these recent interventions by taking a more *strategic* approach to ASM. This means developing measures to ensure accountability while addressing the main drivers behind the expansion of ASM and its more negative impacts, and particularly those 'push factors' that are more likely to be modified through the use of appropriate instruments.

An effective ASM management strategy would concern all operations throughout their life cycle, including activities as far upstream as mining and as far downstream as traders, and secondary processing operations that are artisanal or small scale (Paget et al. 2015). It would also be integrated with industry strategies and wider national–rural development plans (Hilson and McQuilken 2014). While developing and implementing such a strategy is certainly no easy task (Spiegel 2011: 197; 2014: 303, 305, 308), a recent draft guide produced by RCS Global for the Intergovernmental Forum on Mining, Minerals, Metals and Sustainable Development (IGF) (Paget et al. 2015) provides more details for governments on how a strategic approach can be achieved in practice.[8]

---

7   For example, even in the context of an ongoing moratorium, low-level ASM activities are often tolerated, especially if undertaken as a local livelihood diversification strategy. For reasons highlighted in the paper, such as a lack of monitoring and enforcement resources, even more commercial ASM, or ASM activities undertaken by larger numbers of local or migrant actors, do take place despite the moratorium.

8   The IGF is a global forum for dialogue between member-country governments, mining companies and industry associations about practical issues related to the sustainable management and development of the mining sector. It includes governments from the Asia-Pacific, such as India, Mongolia, Papua New Guinea and the Philippines, but not Lao PDR (IGF 2014).

The starting point of the guide is that there is no 'best practice' in ASM management, but rather sets of 'good practices' that are continually evolving and whose effectiveness is context dependent. Instead of recommending a cookie-cutter set of reforms to be applied universally, it thus reviews a number of instruments that can be employed by governments to better manage ASM. Among others, such interventions include ASM zones; livelihood diversification schemes; education and training; the provision of services; supply chain initiatives; capacity building within relevant government agencies; and better interaction between national and local level planning and monitoring, including as it applies to different ASM types (for instance, alluvial versus hard rock, mechanised versus manual) and different minerals and metals (Spiegel 2011: 191–3, 202; Spiegel and Hoeung 2011: 3, 5, 7, 10–11; Eslava 2014; Paget et al. 2015: 9, 12–15, 41, 61–4, 69–70, 78–92, 96–9). For each of these tools, the guide highlights advantages and disadvantages, including potential design and implementation challenges and the kind of contexts where they may prove most or least effective.

It also suggests how each tool can be incorporated into different overall approaches that best suit the existing conditions of particular countries or sections of a country's artisanal and small-scale mining, as well as the resources available for their management. For instance, such approaches may emphasise the promotion of 'better ASM practices' through training and incentives rather than forced compliance, particularly where governments lack monitoring or enforcement capacity. In other contexts, they may instead involve full compulsory licensing, monitoring, enforcement and sanctions for non-compliance, as well as effective incentives, either in relation to all artisanal and small-scale mining or to particular sections of it (such as particular kinds of operations, like mechanised or commercial; particular types of metals; or operations within specific areas of a country). The guide further suggests how governments can switch between different approaches to reflect changing conditions and ensure continued improvement in artisanal and small-scale mining's profile. An effective ASM management strategy for the Lao PDR would therefore need to be grounded in up-to-date research on the country's artisanal and small-scale mining, including its positive and negative impacts, supply chain characteristics and relationship to other economic sectors and regional economies. It should also be guided by an analysis of existing capacity gaps within government (Paget et al. 2015). As there is a current dearth of up-to-date data on both fronts, new ASM research should be a priority for the Lao PDR.

A strategic approach to ASM should then be guided by a vision about clear standards to which GoL wishes ASM to operate, keeping in sight what can realistically be achieved given artisanal and small-scale mining's current profile and available resources, as revealed by the aforementioned research and gap analysis. The guiding vision and wider strategy should be developed by an ASM taskforce or coordination cell, including representatives from all government departments relevant to managing ASM. It should also be informed by consultations with other relevant stakeholders so as to maximise transparency, legitimacy and buy-in, and to ensure that it features effective incentives. The vision should, of course, already be incorporated in the final ASM regulations that were being developed at the time of our research.

The strategy would then need to be effectively implemented by departments working in coordination with one another through the ASM taskforce or coordination cell, which could be set up either as a separate government organ or hosted by a single department or several departments (Paget et al. 2015). Crucially, relevant departments would also need to be well resourced and have adequate capacity. In turn, this can be achieved only if such a strategy is developed and implemented with the support of international donors and in iterative consultation with a multitude of relevant stakeholders, including, among others, representatives from each of the ASM mining subsectors; representatives from each of the connected mineral and metals sectors and the ancillary industries that support the ASM mining subsectors; the communities that surround ASM mining areas; civil society and LSM operations. As past experience has shown, such meaningful, collaborative participation is key to mobilising the goodwill and resources, without which it would be impossible to develop a governance system capable of enhancing the positive contributions of legitimate forms of ASM, whilst also ensuring accountability and mitigating the sector's more negative impacts (Paget et al. 2015; see also Aubynn 2009; Spiegel and Hoeung 2011: 8–11; Spiegel 2014: 303, 307).

References

Aubynn, A., 2009. 'Sustainable Solution or Marriage of Convenience? The Coexistence of Large-Scale Mining and Artisanal and Small-Scale Mining on the Abosso Goldfields Concession in Western Ghana.' *Resources Policy* 34: 64–70. doi.org/10.1016/j.resourpol.2008.04.002

Baker, A., H. Wotruba, E. Aucoin, K. Figueiredo and E. Bougnaphalom, 2007. 'Lao PDR Summary Report.' United Nations Industrial Development Organization, Global Mercury Project. Available at iwlearn.net/iw-projects/1223/reports/lao-pdr/lao-pdr-summary-report/view?searchterm=baker+lao

BGR (German Federal Institute for Geosciences and Natural Resources), 2014. 'Lao PDR—Support for a Sustainable Development of the Mining Sector.' Available at www.bgr.bund.de/EN/Themen/Zusam menarbeit/TechnZusammenarb/Projekte/Laufend/Asien/1071_2009-2294-8_Laos_Bergbauberatung_en.html

Boungnaphalom, E., 2010. 'Mineral Development in Lao PDR.' Paper presented at the first United Nations Development Programme International Conference on Mining: Staking a Claim for Cambodia, Phnom Penh, 26–27 May. Viewed at www.un.org.kh/undp/images/stories/special-pages/mining-conference-2010/docs/10.Eravanh%20Boungnaphalom_Minerals%20Development%20in%20Laos.pdf (site discontinued)

Castells, M. and A. Portes, 1989. 'World Underneath: The Origins, Dynamics, and Effects of the Informal Economy.' In A. Portes, M. Castells, and L.A. Benton (eds), *The Informal Economy: Studies in Advanced and Less Developed Countries*. Baltimore: Johns Hopkins University Press.

Earth Systems Lao (ESL), 2003. 'Luang Prabang Artisanal Gold Mining and Sociological Survey.' Final Report for United Nations Industrial Development Organization, 'Removal of Barriers to the Introduction of Cleaner Artisanal Gold Mining and Extraction Technologies'. Vientiane: Earth Systems Lao. Available at iwlearn.net/iw-projects/1223/reports/lao-pdr/luang-prabang-artisanal-gold-mining-and-sociological-survey-lao-pdr/view

Eftimie, A., K. Heller, J. Strongman, J. Hinton, K. Lahiri-Dutt, N. Mutemeri, C. Insouvanh, M. Godet Sambo and S. Wagner, 2012. *Gender Dimensions of Artisanal and Small-Scale Mining: A Rapid Assessment Tool*. Washington, DC: World Bank Group's Oil, Gas and Mining Unit. Available at siteresources.worldbank.org/INTEXTINDWOM/Resources/Gender_and_ASM_Toolkit.pdf

Eslava, N., 2014. 'Latin America Notes: Solutions to ASM Mining Conflicts, a Need for Stakeholder Participation.' RCS Global. Viewed at www.rcsglobal.com/latin-america-notes-solutions-to-asm-mining-conflicts-a-need-for-stakeholder-participation/#more-1699 (site discontinued)

Fisher, E., 2008. 'Artisanal Gold Mining at the Margins of Mineral Resource Governance: A Case from Tanzania.' *Development Southern Africa* 25(2): 199–213. doi.org/10.1080/03768350802090592

Hatcher, P., 2012. 'New Approaches to Building Markets in Asia.' Working Paper No. 41. Singapore: Lee Kuan Yew School of Public Policy, National University of Singapore.

Hilson, G., 2009. 'Small-Scale Mining, Poverty and Economic Development in Sub-Saharan Africa: An Overview.' *Resources Policy* 34(1–2): 1–5. doi.org/10.1016/j.resourpol.2008.12.001

Hilson, G. and J. McQuilken, 2014. 'Four Decades of Support for Artisanal and Small-Scale Mining in Sub-Saharan Africa: A Critical Review.' *The Extractive Industries and Society* 1: 104–18. doi.org/10.1016/j.exis.2014.01.002

Hinton, J. and M.R. Hollestelle, 2012. 'Methodological Toolkit for Baseline Assessments and Response Strategies to Artisanal and Small-Scale Mining in Protected Areas and Critical Ecosystems.' WWF and Estelle Levin Ltd. Viewed at www.profor.info/sites/profor.info/files/docs/Methodological%20Toolkit.pdf (site discontinued)

Insouvanh, C., 2015. 'Lao PDR Khmu Ethnic Group Women in Artisanal Gold Mining.' ARC Linkage Project 'Going for Gold' Case Study, ASMAsiaPacific. Viewed at asmasiapacific.org/wp-content/uploads/2015/10/Lao-PDR-Case-Study-Final.pdf (site discontinued)

Intergovernmental Forum on Mining, Minerals, Metals and Sustainable Development (IGF), 2014. 'A Voluntary Partnership for Global Dialogue on Sustainable Mining and Development.' Presentation by Eng. Paul M. Masenja, CEO Tanzanian Minerals Agency. Viewed at www.globaldialogue.info/IGF%202013%20Flyer%20(updated).pdf (site discontinued)

International Council on Mining and Metals (ICMM), 2010. *Working Together—How Large-scale Miners can Engage with Artisanal and Small-scale Miners.* International Council on Mining and Metals (ICMM), Communities and Small-Scale Mining (CASM) and International Finance Corporation (IFC) Oil, Gas and Mining Sustainable Community Development Fund.

International Council on Mining and Metals (ICMM), 2011. 'Utilizing Mining and Mineral Resources to Foster the Sustainable Development of Lao PDR.' Mining: Partnerships for Development Report. London: ICMM. Viewed at www.icmm.com/page/59737/utilizing-mining-and-mineral-resources-to-foster-the-sustainable-development-of-the-lao-pdr (site discontinued)

Korkiakoski, E., P. Schmidt-Thome and J. Laukkanen, 2012. 'Cooperating Country Report of Finland.' Coordinating Committee for Geoscience Programmes in East and Southeast Asia, 48th CCOP Annual Session, 4–8 November, Langkawi, Malaysia. Available at www.ccop.or.th/48as.59sc/48as4.19_Finland.2012.pdf

Lahiri-Dutt, K., K. Alexander, and C. Insouvanh, 2014. 'Informal Mining in Livelihood Diversification: Mineral Dependence and Rural Communities in Lao PDR.' *South East Asia Research* 22(1): 103–22. doi.org/10.5367/sear.2014.0194

Larsen, M. 2010. 'Lao PDR Development Report 2010: Natural Resource Management for Sustainable Development—Hydropwer and Mining.' Washington, DC: World Bank.

Latsaphao, K., 2014. 'Illegal Gold Mining in Xieng Khuang Ends.' *Vientiane Times*, 29 September.

Mansfield, S., 2000. *Lao Hill Tribes: Traditions and Patterns of Existence.* Oxford: Oxford University Press.

Noetstaller, R., M. Heemskerk, H. Felix and D. Bernd, 2004. 'Program for Improvements to the Profiling of Artisanal and Small-Scale Mining Activities in Africa and the Implementation of Baseline Surveys.' Washington, DC: Communities and Small-Scale Mining, World Bank.

Paget, D., N. Garrett and N. Eslava, 2015. *Guidance for Governments on Managing Artisanal and Small-Scale Mining.* Consultation Draft. London: IGF and RCS Global.

Rigg, J., 2005. *Living with Transition in Laos: Market Integration in Southeast Asia.* London: Routledge.

Shingu, K., 2006. 'Final Report for Mining, Infrastructure and Environment: Sector Plan for Sustainable Development of the Mining Sector in Lao PDR.' Washington, DC: World Bank. Available at siteresources.worldbank.org/INTLAOPRD/Resources/FR_4_Mining Infrastructure_.pdf

Soemarwoto, R., 2015. 'Informal Gold Mining in Bayah Beach, Banten, Indonesia.' ARC Linkage Project 'Going for Gold' Case Study, ASMAsiaPacific. Viewed at asmasiapacific.org/wp-content/uploads/2015/09/Bayah-Case-Study-Indonesia.pdf (site discontinued)

Southivongnorath, S., 2014. 'MONRE: 10 Projects to be Cancelled.' *Vientiane Times*, 24 September.

Spiegel, S., 2009. 'Resource Policies and Small-Scale Gold Mining in Zimbabwe.' *Resources Policy* 34(1–2): 39–44. doi.org/10.1016/j.resourpol.2008.05.004

Spiegel, S., 2011. 'Governance Institutions, Resource Rights Regimes, and the Informal Mining Sector: Regulatory Complexities in Indonesia.' *World Development* 40(1): 189–205. doi.org/10.1016/j.worlddev.2011.05.015

Spiegel, S., 2014. 'Rural Place-Making, Globalization and the Extractive Sector: Insights from Gold Mining Areas in Kratie and Ratanakiri, Cambodia.' *Journal of Rural Studies* 36: 300–10. doi.org/10.1016/j.jrurstud.2014.09.007

Spiegel, S. and S. Hoeung, 2011. 'Artisanal and Small-Scale Mining (ASM): Policy Options for Cambodians.' Policy Brief. Cambodia: United Nations Development Programme.

Spiegel, S. and M.M. Veiga, 2009. 'Artisanal and Small-Scale Mining as an Extralegal Economy: De Soto and the Redefinition of "Formalization".' *Resources Policy* 34(1–2): 51–6. doi.org/10.1016/j.resourpol.2008.02.001

United Nations Environment Programme, 2015. 'Guidance Document: Developing a National Action Plan to Reduce, and Where Feasible, Eliminate Mercury Use in Artisanal and Small-Scale Gold Mining.' Available at wedocs.unep.org/bitstream/handle/20.500.11822/11371/National_Action_Plan_draft_guidance_v12.pdf?sequence=1&isAllowed=y

Vaenkeo, S., 2014a. 'Government Takes Action on Illegal Mining.' *Vientiane Times*, 28 July.

Vaenkeo, S., 2014b. 'Chinese Firm Escapes After Illegally Mining for Gold.' *Vientiane Times*, 29 July.

16

Reassembling informal gold-mining for development and sustainability? Opportunities and limits to formalisation in India, Indonesia and Laos

Keith Barney

In the past two decades, research activity and policy development have intensified on the issue of formalising artisanal and small-scale mining (ASM), or informal mining. Numerous experts and influential institutions, including the World Bank, now view formalisation and legal registration as primary policy responses to the socio-economic, environmental and human health-related challenges posed by illegal or informal mining (see, for instance, Siegel and Veiga 2009; Maconachie and Hilson 2011; World Bank 2013). A number of African countries have made substantial progress towards the formalisation of artisanal and small-scale gold-mining (ASGM) through establishing legal rights for miners, with Ghana implementing initial provisions as early as 1989 (Maconachie and Hilson 2011). In the Asia-Pacific, Mongolia has arguably emerged at the forefront of formalisation through the provision of small-scale and community mining licences (Purevjav 2011); indeed, the Mongolian Government now views ASGM as important for maintaining national economic stability in the country's post–commodity boom era (*Financial Times* 2014). Bougainville Island passed its notable *Mining Act 2015*, which contains legal provisions for small-scale mining that would involve local authorities in regulating this activity (O'Faircheallaigh et al. 2016). There has been more partial policy support for community mining licences and permits in countries including the Philippines (Verbrugge 2014a), Indonesia (Spiegel 2012) and Cambodia (Cuddy and Seangly 2015; Spiegel 2016). Yet, overall across the Global South, progress with connecting the insights of applied research and advocacy into national policy frameworks and moving formalisation to the forefront of mineral governance agendas has been uneven and halting at best (Hilson and Gatsinzi 2014). In some country contexts, such as Zimbabwe, established policies supporting ASGM have been subject to rollback and renewed political contestation (Spiegel 2009). Many other mineral-rich countries, such as India and Lao PDR, remain ambivalent towards formalisation as a route for addressing the social and environmental externalities and promoting rural economic livelihoods (Moretti and Garrett, this volume).

This chapter draws upon empirical fieldwork conducted in three locations in India, Indonesia and Lao PDR between 2013 and 2016, as part of a multicountry research project aimed at understanding the role of new technologies in informal mining, and the relationships between ASGM, farmers and agrarian transitions in Asia. I examine key variations in ASGM practices and governance contexts as a basis for better understanding the potential role of legal formalisation. The cases illustrate how there is no singular ASM gold economy and, thus, there are distinctive potentials and challenges for formalisation initiatives across different country contexts and regions. To better understand this complexity, I locate the historically mediated process whereby different, grounded socio-natural ASGM 'assemblages' (as involving combinations of gold-bearing tracts; mining–agrarian populations, technologies and practices; market arrangements; civil society groups and transnational actors; and state institutional governance relations) become organised and relationally territorialised in particular places (see Collier and Ong 2005; Li 2007; Ouma et al. 2013; Vandergeest et al. 2015). Such assemblages of informal mining, the regimes of extraction and the distributions of surplus and rents, take form through multi-scaled state–society relations, and are mediated through diverging agrarian transitions and distinct historical trajectories of resource governance (Peluso 2016).

The complexity and diversity of informal arrangements in the three mining sites contribute to the primary arguments put forth in this chapter. First, I argue that formalisation interventions would necessarily need to be conceived in relation to the scale and organisation of the informal mining assemblage in question—from the micro-artisanal end to the medium-scale mechanised end of the spectrum. Second, technical templates for formalising informal mining, as a way of engaging with the sector's social and environmental challenges, are unlikely to gain traction. ASGM initiatives could instead seek to connect with civil society and be responsive to local collective actions, thereby developing informed, grounded understandings of local livelihood contexts in ongoing mining–agrarian transformations, and of the political and class-based struggles occurring in ASGM communities. Third, in medium-scale ASGM contexts, there can be significant trade-offs between ecosystems and agrarian sustainability and local health concerns on the one hand, and income generation from ASGM operations on the other. Regularisation and formalisation initiatives in Asia would face difficult choices between these priorities; such shades of grey could heighten political risks to external intervention. Such environmental politics are also reflected within local communities themselves, in terms of their debates and contestations on the appropriate pace, scale and location of informal mining.

The cases themselves also present novel insights into the political and governance arrangements for informal mining in each country context. In Jharkhand state in India, I argue that formalisation as basic regularisation would represent a necessary first step for supporting miners' livelihoods, lifting the burden of restrictive state control and promoting the basic right of the poor to access and extract minerals. In practice, the political ecology of informal mining in Jharkhand is framed through distinctive federal, state and local–*adivasi*[1] resource politics and contested social struggles (Shah 2013). In Central Kalimantan, Indonesia, there is potential to progress with ongoing state formalisation reforms, in a context where mechanised 'medium-scale' gold dredging, on both land and rivers, can represent a profitable rural enterprise. Here, as well, the burden of petty exploitation visited by local police upon the poorest and most marginal miners relegates some ASGM activity to the suppressed 'illegal' sphere, with pernicious results for the poor (see also Spiegel 2012). In Central Kalimantan, formalisation reforms require careful consideration, as the more capitalised end of the informal mining sector involves a distinctive

---

1   An umbrella term used to denote a heterogeneous group of tribes and indigenous group of people in India.

extractive assemblage, operating in connection with local authorities, which can be associated with clearly negative impacts on common pool resources, particularly for fisheries and aquatic habitats.

In Lao PDR, the last decade of neoliberal-inspired mining reforms has focused policy formulation on the large-scale corporate mining sector, neglecting the long tradition of local people's mining (cf. Maconachie and Hilson 2011). Agrarian displacement due to state-backed, large-scale land acquisitions for resource development has been identified as an important 'push' factor for local farmers to seek cash income though involvement in informal gold-mining (Insouvanh 2015). Despite a centrally issued moratorium on new mining concessions, in Lao's southern Xekong Province, fieldwork documented medium-scale dredging and sluicing by cross-border Vietnamese (and Lao) 'backhoe' miners on former streamside rice paddy fields. Here, as well, a more mechanised, medium-scale range of informal mining and its transnational connections present governance challenges with few easy solutions.

This chapter thus develops a critical interpretation of the formalisation literature as related to ASGM, and outlines the significant heterogeneity of informal mining practices and governance arrangements across field sites in India, Indonesia and Laos. The study concludes with reflections on the potential pathways and limits to formalisation beyond a technical policy 'fix'. It explores how applied participatory research programs might seek to more rigorously understand ASGM assemblages and to identify practical policy options that are responsive to collective actions and social mobilisations, involving both informal gold miners as well as broader agrarian communities.

Conditions of informality in artisanal and small-scale gold-mining

The terminology and concept of 'informal mining' is useful for moving beyond state-centric discourses of illegitimacy and illegality, and for recognising the sector's crucial role in numerous Global South rural development contexts. This chapter does not approach formalisation through a step-wise template, or an easily replicable 'formalisation fix' (Dwyer 2015; Spiegel 2016). Instead, we can understand ASGM in relation to multi-scaled mineral regulatory regimes, contested political ecologies and situated mining practices and territorialisation, all of which are enmeshed in contexts of ongoing struggles for livelihood, rights and recognition in agrarian–mining communities (Tschakert 2009a, 2009b, 2016; Singo and Levin 2016). Following Lahiri-Dutt (2004), I view ASGM not as an anachronistic, preindustrial activity, nor as primarily a subsector of modern corporate mining. I also avoid locating ASGM primarily through the lens of state corruption and uncontrolled extraction, or as necessarily tied to negative social outcomes—for example, in discourses of informal mining as 'fast money' just as quickly spent by miners on alcohol, drugs, gambling and prostitution. While there can be exploitative social relations in ASGM, such peasant mining can also be understood as part of a vibrant, vast, historically significant 'informal economy' operating largely outside of modern private and public institutions. In Hart's (n.d.) interpretation, the informal sector is that segment of the economy that is 'irregular, unpredictable, unstable, even invisible'. As Hilson (2013, 2014) notes, it is also important to understand how the informal mining economy can be actively reproduced by state and donor policies, and to understand the environmental challenges posed by ASGM, not as a technical environmental management issue, but rather as a 'development problem with an environmental dimension'. This move places miners and local mining-dependent communities, their political and economic concerns and their livelihood struggles as the starting points for analysis.

Critical research on ASGM and informality

The work of Hernando De Soto (2000) serves as the starting point for research on ASM formalisation. De Soto's framework lends support to a 'legal system that transforms assets into capital' (Hall 2004: 402), and builds upon notions of the (potential) entrepreneurialism of the poor. Siegel and Veiga (2009) situate their analysis of informal mining within this frame, in which the creation of formal, legal property rights is conceived as a key (albeit not exclusive) basis for effective poverty reduction and improved regulation. Formalisation is viewed as a process with three basic elements: a right to land title, a right to minerals and a right to mine. A series of supporting reforms would accompany this, including legalised trading rights; the clarification of overlapping or muddled legal frameworks; and the streamlining of bureaucratic obstacles that can limit smallholder formalisation in practice. Siegel and Veiga (2009) also note that formalisation is at times avoided by informal miners, as it can create new bureaucratic costs and overheads, and limit flexibility in accessing new mining sites. Along these lines, Barreto (2011) advocates for establishing clear legal, spatial planning and zoning frameworks for ASM, followed by targeted policy packages for legalised miners, including support for technological upgradation, new techniques and mining methods, improved labour and environmental management standards, advice for business development and access to formal credit.

Tempering the neoliberal-inflected views of De Soto, Buxton (2013: 6) points to a paradox of informality, in that it can 'both increase resilience by providing an economic livelihood activity and increase vulnerability as it removes the protections and opportunities provided by the government'. Buxton outlines informality, livelihood vulnerability and marginalisation as three core and recursive social processes characteristic of ASM. Recognising the numerous barriers that serve to block effective reforms in informal mining, Buxton argues for new types of structured 'knowledge networks', as institutional vehicles for participatory research and engagement (see also Barreto 2011). Given the huge array of informal mining practices across different countries, formalisation could have intended or unintended exclusionary consequences—for instance, through high licensing fees, and creating new hierarchies between legalised and capitalised small-scale operators and more actively criminalised miners. Thus, critical ASM observers caution that formalisation does not represent a singular solution to the negative manifestations of informal mining (see Langston et al. 2015). Formalisation programs could, however, tap local knowledge and invite the participation of miner organisations to build more pragmatic, responsive and effective regulatory frameworks (Barreto 2011; United Nations Environment Programme (UNEP) 2012).

Hilson (2013) and Hilson and Gatsinzi (2014) further emphasise that the informal ASM economy is not a phenomenon that is external to the power and territorial authority of the state. They identify how the conditions and incentives for informality are continually reproduced, not only due to bureaucratic complexity and high transaction costs of legal compliance, but other factors as well (De Soto 2000). Also at issue are inappropriate or poorly coordinated policy interventions into the informal sector, the favouring of corporate mining interests in policy frameworks and persistent state underinvestment in sustainable agrarian livelihoods. These authors reiterate that the informal sector is not necessarily founded in small-scale miners' intentional avoidance of taxes and government regulation, as indeed it is very often the selective and exploitative involvement of state actors, such as the police and military, that can reinforce the underground, patronage-based 'shadow state' (Reno 1999) character of the informal mining economy (Peluso 2018).

Of critical importance is how locally mined gold is transformed into a global commodity. Fold et al. (2014) elaborate on the complex geographical co-production of formal and informal gold value chains. A key implication of their analysis is that any formalisation program would need to be based upon a clear understanding of the socio-spatial organisation of both formal and informal gold value chains, examining how and where these merge into a single marketing chain, and tracing through how certain gold commodity markets can lock in exploitative labour arrangements and poor environmental practices. Along these lines, significant efforts have been made in recent years to connect legalised and fair trade artisanal gold production to buyers in the Global North. Here, Singo and Levin (2016) argue that a 'failure to formalise' stymies efforts to restructure how artisanal gold miners are connected with global gold markets:

> If miners don't formalise, they can't professionalise. Without professionalisation, miners are not able to accommodate the due diligence requirements of 'responsible' buyers. This exacerbates buyer disengagement.

Such double (indeed triple) binds with producing and marketing 'sustainable' ASM gold are characteristic of the political economy of informal mining.

In his recent work, Boris Verbrugge has significantly extended our understanding of the political economy of informal mining through a sophisticated historical–geographical examination of the state and capital interests that support ASGM activity in the southern Philippines. Verbrugge (2014a, 2014b) argued that potential formalisation policies need to be more closely considered in relation to the internal and external power relations that structure overall labour–capital dynamics of ASM. He characterises complex governance regimes of ASGM in Mindanao as comprised of multi-tiered labour structures, linked in various permutations to the 'capital interests' of local state officials. In the Philippine uplands, such politically sanctioned informal mining can take on highly exploitative and hazardous forms, including forms of debt bondage. This is occurring in a context of elite consolidation and upland state formation, as well as a national policy shift away from large-scale mining (Verbrugge 2015a).

For Verbrugge (2014a), tracing the connections between informal mining and elite interests in Mindanao requires detailed political economic analysis, based on a grounded understanding of the context of mineral governance, the politics of decentralisation and the dynamics of local politics in southern Philippines.

Verbrugge (2015b) also offers new insights into the political relations and forces of capital that structure emergent class hierarchies between a poverty-driven labour force and informal ASM entrepreneurs. Conceptually, this move focuses attention on 'the diverse origins and implications of informality in ASM … the vested interests in the (partial) persistence of informality, in the form of a labour force that is bereft of formal legal recognition and vulnerable to exploitation' (ibid.: 1042). Verbrugge thus provides a detailed and grounded examination of power, marginality and labour exploitation in ASGM in the Philippines, examining the reproduction of domination through changing regimes of mining governance. In this way, Verbrugge and Besmanos (2016) draw attention not just to the rights of mineral resource tenure claimants, but also to the labour conditions of highly exploited workers.

While work from scholars such as Verbrugge identify the complex terrain of political–economic power relations in ASGM, critical observers such as Labonne (2014) voice a strong pessimism regarding the actual potential for formalisation and other pro-poor state policy interventions to introduce positive changes. In a blunt challenge, Labonne (ibid.: 123) argued that ASM formalisation is more likely to emerge from 'the success of governments in combating poverty in the non-mining rural communities'. While caution on the prospects for formalisation is surely warranted, agrarian studies scholars have also argued that contemporary agricultural labour markets are often insufficient for reducing rural poverty (Otsuka et al. 2010). In many contexts, poverty and underdevelopment, as well as new livelihood opportunities, are created through rural people's interactive engagement in both agrarian and mineral-based activities, and thus Labonne's analysis risks an undue compartmentalisation between these livelihood portfolios. Simply refocusing policy interventions upon improving agrarian livelihoods and facilitating urban–industrial transitions is unlikely to be an adequate response to the challenges of ASGM in many contexts.

This review of the core and current debates on ASM and formalisation adds significant complexity to governance interventions founded upon either top-down regulatory enforcement or a neoliberal, market-friendly approach based upon the simple legalisation of property rights as a path to a more sustainable and beneficial form of rural people's mining. I next turn to briefly reviewing variations in ASGM practices and organisational forms as a basis for discussing the prospects and limits of formalisation in the three field cases.

The diversity of ASGM assemblages and practices

A wide heterogeneity of sites, actors, technologies and practices, institutional–capital arrangements and agrarian–environmental relations characterise informal and small-scale gold-mining. Veiga (n.d.) usefully distinguishes between three broad clusters that are often grouped together as ASGM: a) micro-scale, informal, artisanal; b) small-scale, illegal, semi-mechanised; and c) medium-scale, illegal, mechanised. On the micro-artisanal side, ASGM can involve relatively independent peasant miners working marginally productive alluvial gold tracts, experiencing high levels of vulnerability and holding subordinate livelihood positions within the local agrarian economy, but who also find support through non-market access to gold and other market or subsistence-based ecological resources. Towards the more medium-scale end of the spectrum, informal gold-mining moves into more complex organisational and hierarchical labour arrangements, working deeper and more valuable alluvial, hard rock or riverine deposits with more significant levels of technology, investment and debt relations and operations linked to powerful business interests and political patrons (Verbrugge 2014a; Verbrugge and Besmanos 2016). This form of mining can even involve transnational capital connections. Across this range, informal gold-mining is also characterised by changing socio-technological practices, multi-scaled governance systems, varying degrees of miners' geographical and livelihood mobility and complex socio-environmental transformations (Verbrugge et al. 2015; Ferring et al. 2016).

This diversity of ASGM actors, practices, institutional arrangements and environmental outcomes has significant implications for program and policy development on formalisation. Indeed, Ferring et al. (2016) question the extent to which this diversity is actually reflected in current policy discourses on ASGM in Ghana, a key informal gold-producing country.[2] Insufficient recognition of the heterogeneity of informal mining could limit the scope for formalisation initiatives to particularly favourable sites, rendering policy interventions ineffective on a broader scale. Even under optimal conditions, the formalisation of small- and medium-scale mining would often involve complex reallocations of resource access and property rights, as ASGM typically monopolises and even eliminates most future land uses (Putzel et al. 2014). There is also no guarantee of improved environmental performance through increased legal regulation (MacDonald et al. 2013).

Where does this understanding of the political economy of persistent informality, and of highly complex and diverse ASGM governance regimes and assemblages, broadly lead us in terms of policy approaches to formalisation? Most basically and directly, these factors would pose significant challenges to the effectiveness of expert-led, sector-constrained, technical-based formalisation policy reforms. Insights can be drawn from Michael Dwyer's (2015) perspective on the formalisation of land tenure and ownership in Cambodia through donor-supported land titling programs as a 'formalisation fix'. Dwyer usefully highlights a need to extend beyond technical-driven policy interventions, and towards a grounding of formalisation policies within distinct, spatialised and place-based historical–geographical contexts and socio-political struggles (see also Bridge 2002; Peluso et al. 2012; Spiegel 2016). A close understanding of the local political ecologies and political economies of informal mining might better reveal the democratic possibilities, as well as some hard realities of formalisation as a legal policy approach.

I draw upon the above literature to examine local ASGM assemblages in three field sites: in India, Indonesia and Laos. I direct analysis into the realm of state–capital interests, and hierarchical labour regimes in ASGM that reproduce vulnerability and establish the conditions for perpetuated (Spiegel 2012) or persistent (Verbrugge 2015b) informality. I locate these variations in relation to the material and environmental realities of mining activity; to agrarian–ecological class relations (Akram-Lodhi et al. 2010); and to local socio-political struggles for rights to livelihood by both local miners and resource-dependent communities.

2   Drawing upon Mitchell (2002), Ferring et al. (2016: 3) identify a managerialist impulse in new donor and policy frameworks aimed at regulating and formalising small-scale mining, which can be based upon unwarranted simplifications: 'Actually existing complexity is discursively erased in policy and other official representations so that a particular type of expert and intervention can rule'.

Territorial assemblages of informal gold-mining

Artisanal-peasant gold-mining along the Subarnarekha River, Jharkhand state, eastern India

Indigenous (*adivasi*) or tribal people have a long history of shallow alluvial gold panning and mining along the lower Subarnarekha River (the 'thread of gold') in India's eastern Jharkhand state. The local mining taking place in the Subarnarekha watershed is typically unregulated, irregular and based upon seasonally ephemeral gold resources. Lahiri-Dutt (2004) refers to such peasant mining as an 'elusive, unquantifiable and uncertain section of the mineral economy' (ibid.: 123), while the autonomous and informal structure of such activity renders it 'conceptually, methodologically and theoretically difficult to define in terms of its precise nature, size and significance' (ibid.: 126).

Deb et al. (2008: 195) identify the majority of miners and practices in these locations in India as informal or 'non-legal' (as opposed to illegal), as mining is carried out 'beyond the purview of the law'. Gold is classified as a 'major' mineral under India's *Mines and Minerals Act 1957*, and legally any gold-mining occurring more than 6 inches beneath the surface is prohibited. Thus, the federal Indian Government allocates gold-mining permits to large-scale miners only (Lahiri-Dutt 2004). In practice, as Deb et al. (2008: 207) observe, the '[l]ack of available official or unofficial data related to the practice of artisanal mining in India makes the whole issue very nebulous'. In Jharkhand, I suggest that any effort to support artisanal miners' livelihoods and to improve practices would face a number of challenges, in terms of the low intensity, spatially dispersed and relatively marginal economic value of the resources currently extracted, and the rather poor prospects for Indian state policy reform. Indeed, ASGM is the target for periodic crackdowns by the provincial and district constabulary, while the state promotes investments into larger-scale, capitalised gold-mining in the Subarnarekha watershed (Bose 2015).[3]

---

3   In Seraikela-Kharsawan district, Geological Survey of India has been drilling since May–June 2011, where there are potential reserves of gold (Bose 2015).

Fieldwork in Jharkhand's Seraikela-Kharsawan District in February 2014 highlighted two miner communities. The first involved women micro-scale miners, working in small groups but panning on an individual basis or with children assisting. Village women were panning for alluvial gold in low-lying agricultural lands along the banks of the Chandil Reservoir, using basic hoes and wooden pans (*donga*), scraping the soil surface and panning for minute particles of gold. The local historical context is important to their livelihood situation. The Chandil Reservoir was created in 1983 through the World Bank–funded Subarnarekha Multipurpose Project (SMP), involving a series of barrages and canals targeting electricity generation and irrigation. The major Chandil and Icha dam components of the SMP involved the flooding of some 30,000 hectares of local farms and forest lands; the under-compensated displacement and resulting impoverishment of approximately 68,000 *adivasi* people in 160 villages (with 38 settlement sites submerged); and the imprisonment and violent deaths of a number of protestors and community leaders by local security forces in the late 1970s and early 1980s (Probe Alert 1991). Notably, the Chandil Dam, as well as other reservoir projects, also submerged productive sites of traditional gold extraction by local *adivasi* communities (Lahiri-Dutt 2004; Bose 2015). During our fieldwork, we also encountered evidence of colonial-era gold-mining infrastructure (old elevated roadways, stone-reinforced tunnels) that testifies to this history of mineral extraction in the area.

While ethnic Santhal[4] men plough agricultural fields or undertake evening-time reservoir fishing, women pan for gold along the water's edge, or travel upstream to pan at the Subarnarekha's confluence. Artisanal gold-mining is an important means of supplementing agricultural, wage labour and subsistence livelihoods. The ecological impacts from their activities would be negligible, as they do not use mechanised digging equipment or sluice boxes, or mercury for gold amalgamation. Their low capital-intensive gold-extraction activities are not a matter of concern to local officials, which is likely why such *adivasi* women are able to continue to access and exploit this local resource.

One elderly female village informant pithily identified a number of intersecting livelihood issues:

4   One of the largest *adivasi* groups in India, with an estimated population of some 2.4 million in Jharkhand state.

In one week I can get one lakh rupees, or nothing.[5] Our old farming land is now the home for fish … [But] we don't get a fair price [from the gold traders], and it's not enough to keep our stomachs filled. (Interview, 3 February 2014)

For these women, the painstaking work of daily panning for gold represents an important contribution to household livelihoods, following their historical displacement due to a major infrastructure project. This relationship between population displacement due to large-scale projects and subsequent 'informal' access to remaining niche common property resources has been a recurrent theme in the literature on development-induced displacement (see, for instance, Kibreab 2000). In Jharkhand, women's engagement with micro-artisanal mining is moreover based upon an explicitly gendered access to localised gold resources, which is at a small enough scale not to concern local officials, who can act to block miners' access to waterways for mining in other nearby locations (Bose 2015).

Another group of gold miners we interviewed, who were also displaced from the original SMP, involved a community of approximately 50 male miners. The men were working in two groups, digging pits of up to 4 m deep along the dry season bed of the Subarnarekha River, about 8 km downstream of the main barrage, and some 15 km upstream from Jamshedpur city. As the agricultural season ends, local ethnic Mundari (*adivasi*) men bicycle daily from villages up to 20 km away to undertake gold extraction as an alternative to wage labour in nearby brick kilns, or cutting sawn wood for sale from local forests. This form of mining involves more specialisation in labour activities and the use of diesel hydraulic pumps to increase extractive capacity.[6] During our visit, two different work teams were subdivided into task groups, removing rocks and boulders, scooping water and carrying sand from the streambed for panning. Gold is sold to a single trader at the local Chandil market, at a reported price of INR2,300 (US$37) per gram (representing a 22 per cent discount on the India spot price for gold on that day).[7] Miners reported that daily profits were shared equally amongst work team members. At an upper estimate of group production of approximately 8 grams of gold per

---

5   INR10,000, or approximately US$150. The reported value may have been more of a rhetorical flourish. Tarun Bose (2015: 10) reports that women from nearby Moisara village could earn INR300–400 (US$4.80–6.40) per week on average from a similar style of gold-mining.

6   Purchase price of diesel pumps at INR15,000/US$242 per pump; averaging 20 litres of fuel per day, costing INR1,000, or US$16.

7   To compare, India gold spot price on 4 February 2014 was INR2,969/gram: www.goldpriceindia. com/gold-price-february-2014.php.

day, the gross yield of a work team of 25 members might represent some INR18,400 (US$294). Subtracting daily fuel costs of US$16 provides net revenues of US$278, or US$11 per worker, if shared completely evenly (purchase and maintenance cost of pumps, compensation for the owner of the machinery or investor, and any other fees also need to be subtracted). In interviews, individual wages were also reported to lie between INR50 and 250 per day (US$0.80–4.00), so their average daily income is perhaps within these lower and upper bounds, depending on the organisational structure of the work teams and rates of gold recovery from the riverbed.[8] Local miners indicated that if they were discovered finding larger gold pieces, or if they ventured further into higher horsepower pumping machinery, it would attract the attention of local officials (field interviews, 4 February 2015). While their mining activities are quite visible, these miners generally aim to operate outside of the focus of attention of local officials as much as possible.

The broader issue in these *adivasi* lands in Jharkhand, identified by researchers such as Alpa Shah (2013), is that of a stalled agrarian transition. Historic Jharkhand witnessed neither an agricultural consolidation by a landlord class, nor a capitalist transition to commercial agrarian commodity production. To be sure, in Jharkhand, forests have been logged and liquidated, major mineral resources are controlled by private or state-owned corporations, including Coal India, and large-scale state development projects, such as the SMP, have produced significant enclosures and dispossessions. Yet, Shah (2013) argues that the majority of the region's rural population has been situated outside of the primary circuits of capital, assembling a multifocal livelihood through low-productivity family-based agriculture, and non–market based access to increasingly marginal natural resources, including gold and forest products. In response to this rural crisis, there is widespread engagement in regional labour migration under highly precarious circumstances, while demographic pressures contribute to increasing levels of rural landlessness and pauperisation (ibid.). An ongoing Maoist insurgency in Jharkhand (including various splinter groups in competition with each other) leverages upon conditions of poverty and rural discontent, which in turn can invite heavy state and police pressure, including on peasant gold extraction activities that extend beyond an acceptable scale.

---

8   There can be constraints with securing accurate information on daily income from ASGM, as miners have a broad interest in not drawing attention to their activities.

While clearly important for the communities involved, the small and micro-scale subsistence-supporting nature of peasant gold extraction activity, an ongoing political situation involving disaffected Maoists, and the relative degrees of suppression of gold-mining by local state actors, would make this a challenging arena for policy intervention. Deb et al. (2008: 200) advocate for local governments (*Panchayats*) to play a supporting role in artisanal mining under their jurisdiction, and propose a review of state regulatory frameworks to remove ambiguity around the legality of ASGM sites. [9] Here, promoting a basic right of the poor to mine (rights over land, rights to trade and rights to basic working conditions) could potentially assist local gold artisanal panners and miners to secure a foothold in the formal economy, and possibly alleviate some of local conditions of extreme poverty. This would entail a reorientation of an often coercive relationship between local state actors and the rural poor, and would require support for the miners to advocate on their own behalf through collective organisations. However, in Jharkhand at least, there is little indication that state authorities are contemplating such reforms.

Medium-scale informal gold-mining in Central Kalimantan, Indonesia

Central Kalimantan has been represented in some of the largest and most contentious ASGM mining booms in the post–New Order period in Indonesia. Spiegel (2012) and Agrawal (2007) have examined a high-profile 'gold-rush' site at Galangan in Katingan District, Central Kalimantan, in detail. This heavily degraded area of about 50 km² once attracted up to 10,000 miners, using pumps and dredges to work through the mineralised rainforest soils. Miners were drawn from local Dayak communities, from elsewhere in Kalimantan, as well as from Java and Madura. Spiegel (2012: 201) discussed the highly volatile and, at times, violent Galangan–Hampalit gold-rush site in relation to the ambiguous outcomes of administrative decentralisation, the diversity of participants in ASM mining, as well as the 'political and economic interests in perpetuating informality'. International Crisis Group (ICG 2001) also documents Galangan–Hampalit as a particularly contested site of struggles over natural resources. Indeed, a violent encounter between

---

9   Similar principles of involving local communities in ASM governance have been applied in the Bougainville *Mining Act 2015* (see O'Faircheallaigh et al. 2016). I thank Matthew Allen for pointing me towards this.

Dayak and Madurese miners provided the ignition trigger for the 2001 outbreak of ethnic violence between these two communities in Central Kalimantan.

In the forests and rivers of Katingan Regency and Gunung Mas (gold mountain) district, new 'rush' sites can still attract thousands of miners. Typical in-land mining methods involve the use of dredging pumps, sluice boxes and carpets, with a recovery rate for particles of gold reported as approximately 30 per cent. In a site visited in Gunung Mas, dredging pumps can work through 5–10 m^3 of soil per day, eventually yielding approximately 5–10 grams of gold per day per team of four to seven miners, providing a gross revenue of some US\$153–307 per day. Some miners, working on an individual basis (using smaller but less costly pumps that use less fuel), might mine and sell their 'pay dirt' to other processors for onward gold recovery. Miners indicate that a primary calculus governing their activities is the ratio between diesel consumption and grams of gold yielded. Baseline labour costs are discounted, and the eventual net earnings of workers depend upon the value of gold their unit can extract from the rainforest soils or river bottoms.

Major rivers in Central Kalimantan, such as the Kahayan and the Katingan, and associated freshwater ponds that represent their former courses, have also been mined through floating dredging platforms, resulting in significant aquatic ecosystem impacts (see United Nations Institute for Training and Research (UNITAR) 2016). Despite its widely acknowledged environmental impacts, ASM is now a primary income-earning activity in many communities along the major river systems of Central Kalimantan. Gold-mining is considered as more attractive than working in oil palm concessions (which pays as low as IDR30,000 per day for an entry-level worker, or US\$2.50), harvesting rattan, cultivating bananas or tapping rubber. Indeed, sharp declines in rubber and rattan prices in the past years have ushered more rural people into mining (fieldwork interviews, November 2014 and September 2016).

During a visit to the Galangan–Hampalit 'moonscape' (Spiegel 2012: 194) in November 2014, we encountered approximately half a dozen hardscrabble miners—ethnic Dayak, Javanese and Banjarese—working through the tailings long after the end of the main gold rush, in a bleak and heavily degraded landscape. For one miner, working autonomously without the protection of a 'boss', payments of up to IDR100,000 per month were still required to the local police. The man used an 'Alcon' brand of diesel pump, using 5 litres of fuel per day, and sold 30 kg of pay dirt to a local processor. A nearby Banjarese couple living in a rudimentary shack was in an even more precarious situation, estranged from their children due to their poverty, moving itinerantly around the Hampalit landscape in an attempt to avoid payments to the police. For these most marginal of miners, working the tailings from a previous gold rush offers a residual livelihood foothold. Lifting the burden of petty extortion by police forces (see also Spiegel 2012: 195) would seem to offer a first step for some improvement to these miners' desperate predicament.

More capitalised informal mining operations using pump and sluice box systems (for in-land mining), which can be placed upon mobile dredging platforms (for riverine mining), are the primary modes of informal gold-mining present in these districts. Some district miner groups have gained livelihood security from formalisation systems through the combination of Wilayah Pertambangan Rakyat (WPR, Community Mining Areas) and the Izin Usaha Pertambangan Rakyat (IPR, Community Mining Licences). Formalised miners also require an environmental permit from the provincial and district environmental agency (Badan Lingkungan Hidup) in the form of an environmental impact assessment document, as well as environmental management and monitoring plans. However, our local informants indicated that formalised community mining licence arrangements required significant paperwork,[10] while local authorities applied extremely high surcharges for the permits. Limited to 25 hectares, community mining licensing also limit mobility and flexibility in developing new mine sites, and thus would seem more readily suitable to fixed hard-rock deposits, rather than for more mobile and ephemeral alluvial gold-mining. For these reasons, the vast majority of alluvial mining groups in Central Kalimantan have avoided formalisation; indeed, some miners interviewed in September 2016 had never heard of the possibility of a formal licence.

In this legal grey zone, mining 'bosses' facilitate protection from police harassment for unit owners and their workers, as well as provide equipment, financing, (illegal) access to state-subsidised diesel fuel ('solar' in Bahasa) and gold-trading services in exchange for a share of unit profits. Local police also benefit from ASGM through controlling access to 'solar'.[11] State-subsidised diesel (retailed at a discount of IDR5,050 per

---

10   See also Harvard Kennedy School (2011: 93) on the 'high cost of formality' in Indonesia.
11   'Solar' is the Bahasa Indonesian term for 'diesel'.

litre, or US$0.39) should be utilised only for personal and not business use. District police oversee quite a public and visible system whereby, in exchange for a fee, drivers of modified vehicles with expanded fuel tanks load up on 'solar' from petrol stations for onward delivery to gold miners or oil palm plantations, where it can be sold for IDR15,000–20,000 per litre (US$1.15–1.54), depending upon the distance.

In Gunung Mas district, in association with a Dayak small-scale mining association (ASPERA-KT), we visited informal in-land mining sites occurring on individually titled, but formerly customary *adat*, forest land. The unit operations were run by small groups of local ethnic Dayak men, involving up to 7–15 workers per large dredging unit, with reported yields as high as 100 grams of gold per 200-litre drum of diesel fuel, which lasts for five days of dredging.[12] From this gross yield, 50 per cent would be awarded to the land and unit owner—in this case, who held a land title issued by the National Land Agency, under the Ministry of Agrarian and Spatial Planning.

There is a recurring rhetoric that informal miners are engaged in drugs, and that their 'hedonic culture' contributes to a range of other social ills (interviews, Palankaraya, September 2016). However, local discussions in Central Kalimantan with both Dayak and migrant miners identified the use of mining income to pay for daily living costs, purchasing land or a vehicle, building a family house and forming a household, or for their children's schooling. Some miners did report that drugs could be used in order to maintain their energy during times when rich gold strikes were identified, and work continued through the night. A number of parents of young miners also indicated that drugs were an issue in ASGM (interviews, Katingan district, September 2016). However, it is not clear that gold miners are making a highly disproportionate contribution to the province-wide social ills of drugs, alcohol, gambling and prostitution.

---

12   On 27 November 2014, the reported Palankaraya gold shop purchase price was reported at IDR480,000 (US$39.41) per gram, a 6.2 per cent discount on the international spot price for that day. Using 200 litres of diesel (worth perhaps US$197), and with 50 per cent of revenues allocated to the investor/landowner, a unit team of 7–15 workers might then generate US$125–267 per worker for a five-day 100 gram (3.5 ounce) yield of gold. This might then represent US$25–53 per worker per day. Food costs need to be deducted; also, for the investor, capital depreciation for diesel pumps (US$200/pump), sluices (US$175 in materials) and carpets (US$26/m²). Note that this rate of return would represent a better-case scenario. Other interviews indicated estimates of worker's pay at US$11–19 per day. All these calculations depend on the site, yield and recovery of the deposit, which are highly variable. Of course, some days can yield no gold at all for the miners.

What is apparent, as others have noted, is that 'informal' mining in Indonesia is internally variegated, characterised by unequal power relations, connected to particular state institutions and actors, and operating through various relationships with formal state agencies (Peluso 2016, 2018). There is a notable comradely ethic amongst workers, and after the share held by the unit owner and landowner, the revenues from gold extraction are shared equally amongst a unit team. In Central Kalimantan, despite periodic provincial crackdowns and 'patrols' (see, for instance, Lingga 2016), it is possible to discuss the formalisation of local mining with local political figures, signalling a relatively more open and permissive political environment for local ASGM activity, as compared to Jharkhand or Laos.

The strongly ecological character of the mining–agrarian transition (Akram-Lodhi and Kay 2010) in Central Kalimantan is also apparent. In Murung Raya and Katingan regencies, the logging boom is long past, oil palm has not delivered benefits for wage labourers, while larger coal-mining companies have often brought conflicts over land. With draining of peat lands and the associated catastrophic El Nino dry season fires (McCarthy 2013), local observers point to a downward spiral of accelerating resource degradation and livelihood vulnerability. In turn, classic dilemmas have emerged between ASGM and common pool resources, especially involving waterways and fisheries. Informal mining does come at a heavy cost: land is permanently removed from productive use, fishing is no longer possible in many heavily degraded waterways and ponds due to dredging, and mercury contamination is very likely widespread, although the extent of contamination of environments and aquatic food chains is poorly documented (Agrawal 2007). Many villagers in fact voice opposition to gold-mining activities in their lands and river systems, even as mining activities are represented in most rural households along the provinces' major gold-bearing waterways, such as the Katingan River (fieldwork interviews, September 2016).

While many villagers have taken part in the successive Central Kalimantan resource booms, the majority of profits from large-scale resource development have been transferred out of the province. The transition has been too rapid and transformative for local communities to successfully engage with and adapt to the changes. Some community leaders seek a revitalised Dayak customary tradition that could be welded to modern economic and technological transformation, including local involvement in formalised and regulated gold extraction (fieldnotes,

26 November 2014). Yet, the challenges of relying upon a positional indigenous Dayak identity and cultural traditions as a bulwark against unsustainable extraction are also evident (cf. Li 2000), with the widespread involvement and varied motivations and practices of miners; at times significant internal tensions within communities over regulating mining; accelerating mining-linked ecosystem degradation; and the significant numbers of migrants drawn from elsewhere in Indonesia into Central Kalimantan's new gold-mining rushes. Recently, the provincial government has promised 1,500 community mining licences as a step towards greater formalisation. However, the current permitting process for community mining areas and licences in Central Kalimantan is complex, costly and poorly organised, the procedures are not adhered to and formal permits are only used by a few dozen miners as a means of providing partial protection from police patrols (non-government organisation interviews in Palankaraya, 4–5 September 2016).

Artisanal and transnational-mechanised informal mining in southern Lao PDR

In the previous decade there was significant interest in research on ASGM in Lao PDR, and a series of publications were published on informal mining through the United Nations–backed Global Mercury Project (Boungnaphalom 2003). The last 15 years of World Bank–influenced minerals and mining policy reforms in Laos have facilitated the entrance of large-scale mining capital into the country (Hatcher 2015). Two national flagship gold and copper ventures at Xepon and Phu Bia were initially led by Australian mining companies (Oz Minerals and PanAust), but have since been purchased by Chinese firms (China MinMetals Group and Guangdong Rising Asset Management). Copper, lignite coal, potash, gold and tin are key mineral assets for Laos, and together metals and minerals accounted for 40 per cent of the country's export earnings by value in 2012 (United States Geological Survey 2014). As with many other national contexts, in Laos a focus on large-scale mining capital has shifted policy attention away from support for ASM livelihoods. The last decade has also witnessed the entrance of medium-scale regional investors and mining work teams (*Vientiane Times* 2014a), undertaking unregulated in-land and river mining activities with backhoe excavators, pump dredges, sluices and/or floating dredging platforms. This medium-scale mechanised mining, drawing in workers, operators and investors from China and Vietnam, occurs through some level of local government permission, and the operations have often caused a significant amount of environmental degradation (International Union for Conservation of Nature Laos 2005; *Vientiane Times* 2010a, 2010b, 2010c, 2013; Land Information Working Group 2012).

Lao PDR's updated 2011 Law on Minerals maintains legal protection for informal mining; however, these rights are within poorly defined limits of mechanisation. In practice, a range of ASM and more mechanised informal gold-mining is occurring across the country (with periodic booms in other minerals that can be accessed through small-scale techniques, such as alabaster). In 2012, the Government of Laos (GoL) announced a moratorium on new concessions for mining, as a result of the uncoordinated allocations of mining concessions by all levels of government and the resultant stream of negative socio-environmental impacts, as well as losses in national revenue to the Treasury. The 2012 ban on new mining concessions led to uncertainty around how local officials should regulate ASM, and has facilitated (selective) enforcement against informal mining. Reports from the *Vientiane Times* (2014a) indicate that the Laos National Assembly:

> asked the government to formulate regulations to govern small-scale mining and mineral extraction projects, including those carried out by locals, after learning that the operations were causing significant environmental damage.

The Ministry of Planning and Investment subsequently undertook a nation-wide inspection and survey of mining concession agreements and operations (Schoenweger et al. 2012; *Vientiane Times* 2015). The ban on new mining concessions has been extended into late 2016 (*Vientiane Times* 2016); however, the small- to medium-scale mining operations have continued to come under policy pressure since then (*Vientiane Times* 2014b; Radio Free Asia 2016). There have been reports of arrests of local Lao miners tunnelling for hard-rock gold deposits in central Xieng Khouang Province in recent years, in part related to safety concerns (*Vientiane Times* 2014c), and there were reports of a number of deaths due to tunnel collapses.

The research of Chansouk Insouvanh (2015) in Xaysoumboun Province, central Laos, highlights how one community decided to sell communal land to an ersatz 'ecotourism' venture, rather than experience what was likely to be uncompensated losses of communal land due to a state-backed rubber plantation project. Such scenarios are not uncommon in Laos and are part of the creation of new agrarian class relations in the context of new global investment into natural resource extraction and 'resource frontier' state–society relations (Barney 2009). As the 'ecotourism' venture developed, it morphed into a gold-mining operation, and began to displace local peasant traditions of river gold-panning. While villagers were historically engaged in low-technology gold-panning, they also started to access the gold-bearing soil excavated by the company workers, and an orientation towards this informal gold-mining activity turned into a new means to earn cash income. As with the other cases described above, this example highlights the different social, economic and capital structures of 'informal' gold-mining, and the different capital interests (Verbrugge 2014a) in play, which are often linked in different ways back to state actors.

Fieldwork conducted in 2016 in southern Laos' Xekong and Attapeu provinces highlighted the involvement of joint Lao–Vietnamese mining operations in the remote Annamite uplands along the Vietnamese border. In the upper Xekaman River watershed near the town of Dak Cheung, Lao–Vietnamese mining operations were leasing and digging up stream-side paddy fields, using backhoe excavators and *thakeng* sluices. One firm engaged in this activity was known locally as Ong Pheuang, reported as a subsidiary of the Vietnamese conglomerate Hoàng Anh Gia Lai (HAGL), a major corporate investor in southern Laos with interests in plantations, hydropower, infrastructure and mining, and with stated mining interests in Xekong Province. A retired district official indicated that Ong Pheuang had purchased their mining licence from a previous Chinese investor named Jiang Kham company, and then sold it onwards to another Vietnamese firm called Hui That. Another Chinese investor had been engaged in similar gold-mining operations in the district, which were halted after the firm did not pay taxes to the district authorities (fieldwork interviews, Dak Cheung, 12 March 2016).

The lack of transparency and public information on these medium-scale mining arrangements in remote areas of rural Laos, and their generally quasi-legal status, makes this a difficult and nebulous area to research and understand. Fieldwork highlighted that villagers are largely left to their own devices on negotiations when outsiders arrive and request access to community land for mining. Local land brokers might also serve as intermediaries between village leaders and gold-mining firms in these transactions. Villagers can be faced with unresponsive district governors when the mining firms renege on their (verbal) agreements with community leaders. District authorities are in turn constrained in controlling mining operations when they are subsidiaries of large and politically connected external corporate investors in Laos, such as HAGL. There was no indication of any formal regulation of these operations by the Ministry of Natural Resources and Environment. While recently the Lao Government announced a crackdown in provincial Attapeu on unregulated mining (Radio Free Asia 2016), there has been little mention of enforcement against transborder medium-scale mining in other provinces, such as in nearby Xekong.

Marginalised ethnic Katu villagers in the upper Xekaman watershed have little previous experience with such mining operations; indeed, we encountered a series of villages where backhoe mining had caused extensive and likely irreparable damage to wet rice paddy land, which also spilled over to affect nearby landowners who did not lease their land. Compensation by the companies to landowners for a mining 'lease' was reported at a meagre LAK10,000–15,000 (US$1.21–1.82) per square metre for productive stream-side wet rice paddy land. These local agreements purportedly require the mining firms to retill and prepare the soil for a resumption of wet rice paddy cultivation at the end of their operations. In reality, the soil structure can be so thoroughly degraded after intensive sluice mining that the resumption of agriculture is rendered impossible. Villagers reported that staff from mining operations, such as Ong Pheuang, also engage in opportunistic illegal logging during their periods of work in the village locations. Such intensive mining operations also remove any future potential for villagers' own small-scale panning for gold in their local streams and rivers, a livelihood option previously shared by men and women (Baird and Shoemaker 2008).

In the deep south of Laos' Attapeu Province, artisanal (non-mechanised) miners are active in upland locations, such as in the Dong Amphan National Protected Area. One local miner reported yields for dry season stream-side panning for gold at 2–3 'houn' of gold per day (worth US$33.20–49.80), or 1 'baht' per 20 days (US$666).[13] For this yield, his cash income over two days in March 2016 reached LKP170,000 (US$20.59), which was a much better option compared to working as

---

13   According to this Lao measurement system, there are 10 'houn' in 1 'saleung', and 4 'saleung' in 1 'baht' of gold. Field interviews in 2016 indicated that 1 Lao 'baht' of gold represented a value of approximately LKP5.5 million (US$666), representing about 15 grams of gold. Using world gold prices as of March 2016, 1 'houn' of gold might then represent approximately US$16.60.

a labourer in nearby HAGL rubber plantations (at LKP40,000 per day; US$4.84). Working as a smallholder gold panner had the added benefit of flexibility, and maintaining his independence and autonomy outside of the disciplinary surveillance of plantation company supervisors (fieldwork interviews, 16 March 2016).

Similar to the other contexts studied in this chapter, Laos demonstrates a highly complex and dynamic landscape, involving both artisanal and more mechanised forms of gold extraction. A nuanced understanding of the actors, provincial and cross-border capital connections, livelihoods and environmental impacts of ASM is required for considering effective policy interventions. In Laos, one might imagine strengthening legal rights for certain forms of informal mining, in combination with better regulation and oversight for other patterns of mining that are producing more negative social and environmental outcomes. For example, a politically empowered 'mobile informal mining monitoring unit', sponsored by the GoL and donor partners could quite quickly develop an accurate and updated assessment of all the different actors and types of informal mining occurring in rural and upland Laos, and their 'capital interests', as a basis for devising more effective policy interventions that are also responsive to local people's circumstances and resource-based livelihoods.

Reassembling informal gold-mining for development and sustainability?

In India, Indonesia and Laos, reforms to the regulatory framework for small-scale mining have been at best uneven and halting, while the vast majority of policy focus has been on large-scale mining. There is currently little policy learning on informal mining across Asian countries, although much greater sharing of best-practice approaches could be promoted. This study has provided insights into the potential pathways and limits to formalisation beyond a technical policy 'fix'. By viewing informal gold-mining as a rational livelihood opportunity that many different people are engaged in, we can move beyond discussions of 'short-term greed' and allegations of the ASM's various 'social ills'. While I do not seek to underplay the significance of social and environmental concerns, for the most part I have found that informal gold miners are everyday people and family members, making a living under less-than-ideal circumstances. Quite simply, gold-mining can offer a better livelihood than is available from various alternatives, even though there are risks, and the financial benefits are unequally distributed amongst the participants. Much of this ASGM activity does indeed come at a direct cost for agricultural lands, forests and waterways, and problems such as mercury contamination can become very serious public health issues. Yet, small-scale gold miners are hardly alone in trading off daily income in exchange for compromises on longer-term issues of health and ecological sustainability.

In Jharkhand, India, promoting the basic right of artisanal operators to mine (involving rights over land, rights to extract and trade gold and rights to safe working conditions) could assist with alleviating conditions of extreme poverty. This would entail a reorientation of an (at times) coercive relationship between state actors and the rural poor, and would require supporting the ability of miners to advocate on their own behalf through collective organisations. In comparison, in Central Kalimantan, informal mining occurs on a greater scale, and is far more mechanised and capitalised. Here, there is arguably a stronger potential for formalisation, although the relationships between informal mining operations, financiers, state patronage and power, struggles over territory and natural resources and new agrarian class formation needs to be placed within the frame of analysis. In Laos, formalising ASM policies and regulations are hampered by a non-transparent regime of extraction, where capitalised informal mining continues with the apparent backing from state actors. Domestic and cross-border medium-scale investors and backhoe miners are certainly producing negative impacts for marginalised upland ethnic minority communities, even though artisanal gold-mining can offer, for some, an alternative from commodified labour relations in Laos' corporate resource economy.

The diversity of sites, actors, practices, capital connections and agrarian–ecological outcomes encompassed by 'artisanal and small-scale' (and 'medium-scale') gold-mining in India, Indonesia and Laos makes this a highly complex and indeed challenging area of resource studies. Considering ASGM as variegated 'assemblages' of informal mining can help to understand how different components of ASM and scales of mechanisation and extraction are organised, institutionalised and territorialised in different places. We need better information on the key logics through which informal mining is situated in relation to local institutions, labour structures, agrarian relations, systems of enclosure and displacements over land and political power and patronage, as well as how informal mining production connects into global value chains.

Participatory initiatives could start by identifying policy options that are responsive to collective political actions and social mobilisations, from both miners and broader local communities.

Understanding the material, social and political–economic basis for ASGM, miners' livelihood motivations and the variegated 'conditions of informality' that characterise local people's mining is therefore a crucial precondition for effective policy intervention. However, I also suggest that we should not be overly optimistic for fully formalising and regularising the vast and undocumented (perhaps undocumentable) ASGM in Asia, thereby 'solving' associated social–environmental challenges. Some of the ASGM occurring across the region is simply too dispersed, small in scale and of marginal economic value to attract significant project or state-based support. For this 'micro'-artisanal gold-mining, limited formalisation in the form of lifting the burden of illegality, and rolling back systems of petty official rent-seeking, could be the most immediate 'pro-poor' options available. Moving towards the medium-scale of the informal mining spectrum, involving use of machinery and more valuable gold deposits, zoning and use rights acceptable to local communities could be devolved for local benefit with the support of state governments, while also avoiding costly and highly bureaucratic formalisation procedures. State or donor support packages could also target more capitalised and mechanised (and, often, more environmentally problematic) mining activities that fall outside of the artisanal or small-scale range, with a focus on dialogue, establishing and enforcing community zoning regulations, working towards professionalisation, promoting workplace safety and improved environmental standards, and providing access to credit and technology, with the aim of shifting this activity into the fully formal sector (Maconachie and Hilson 2011).

Divorced from local politics and a responsiveness to both miners' and local communities' collective actions, formalisation will remain on the margins of policy development. Given elevated gold prices, in this scenario, informal gold-mining is likely to simply continue to churn through South and Southeast Asia's forest lands, rice fields and rivers. As an alternative, international and donor projects can engage in ASM in a politically informed manner, starting with supporting local collective actions, local institutions and social movements, to build a grounded, responsive and practical framework. Similar challenges have long been apparent in many other resource sectors in Asia (for instance, forests and fisheries), even if the environmental and social challenges posed by informal mining can at times be especially stark. Building a formalisation framework that takes a broader view of regulation that includes community norms and standards, and that links community-based resource management with local governance initiatives, has provided innovative alternatives to failed top-down coercive legal enforcement in natural resource management in Asia (e.g. Tyler 2006). In certain contexts of highly intensive, medium-scale mechanised gold-mining, such as that occurring in Central Kalimantan, this would mean accepting that there will likely be less-than-ideal trade-offs between local ecologies, resource access and mineral-based income streams and livelihoods, and that progress will be challenging and incremental. Downwardly accountable and locally responsive state institutions and an appreciation of the historical logics that support ASM within the vast 'informal economy' of many developing Global South countries are conditions for these interventions. The challenges of building such effective state institutions, for reassembling more locally beneficial and sustainable mining practices, are where much of the conundrum of ASM lies.

References

Agrawal, S., 2007. *Community Awareness on Hazards of Exposure to Mercury and Supply of Equipment for Mercury-cleaner Gold Processing Technologies in Galangan, Central Kalimantan, Indonesia.* Palankaraya: Yayasan Tambuhak Sinta.

Akram-Lodhi, A.H., and C. Kay, 2010. 'Surveying the Agrarian Question (Part 2): Current Debates and Beyond.' *Journal of Peasant Studies* 37(2): 255–84. doi.org/10.1080/03066151003594906

Baird, I. and B. Shoemaker, 2008. *People, Livelihoods, and Development in the Xekong River Basin, Laos.* Bangkok: White Lotus Press.

Barney, K., 2009. 'Laos and the Making of a "Relational" Resource Frontier.' *Geographical Journal* 175(2): 146–59. doi.org/10.1111/j.1475-4959.2009.00323.x

Barreto, L., 2011. *Analysis for Stakeholders on Formalization in the Artisanal and Small-Scale Gold Mining Sector based on Experiences in Latin America, Africa and Asia.* Alliance for Responsible Mining. Available at commdev.org/wp-content/uploads/2015/06/Analysis-stakeholders-formalization-artisanal-and-small-scale-gold-mining-sector-based-experiences-Latin-America-Africa-Asia.pdf

Bose, T., 2015. *Gold Hunt in the Subarnarekha*. Viewed at asmasiapacific. org/wp-content/uploads/2015/09/Eastern-India-Case-Study.pdf (site discontinued)

Boungnaphalom, E., 2003. 'Information about the Project Sites in Lao PDR.' Report to the Global Mercury Project. Vienna: United Nations Industrial Development Organization.

Bridge, G., 2002. 'Grounding Globalization: The Prospects and Perils of Linking Economic Processes of Globalization to Environmental Outcomes.' *Economic Geography* 78(3): 361–86. doi. org/10.2307/4140814

Buxton, A., 2013. *Responding to the Challenge of Artisanal and Small Scale Mining: How Can Knowledge Networks Help?* London: International Institute for Environment and Development. Available at pubs.iied. org/16532IIED/

Collier, S.J. and A. Ong (eds), 2005. *Global Assemblages: Technology, Politics, and Ethics as Anthropological Problems*. Malden, MA: Blackwell.

Cuddy, A. and P. Seangly, 2015. 'Going Straight into the "Forest of Gold."' *Phnom Penh Post*, 1 August.

Deb, M., G. Tiwari, and K. Lahiri-Dutt, 2008. 'Artisanal and Small-Scale Mining in India: Selected Studies and an Overview of the Issues.' *International Journal of Mining, Reclamation and Environment* 22(3): 194–209. doi.org/10.1080/17480930701679574

De Soto, H., 2000. *The Mystery of Capital*. New York: Basic Books.

Dwyer, M., 2015. 'The Formalization Fix? Land Titling, Land Concessions and the Politics of Spatial Transparency in Cambodia.' *Journal of Peasant Studies* 42(5): 903–28. doi.org/10.1080/03066150.2014.99 4510

Ferring, D., H. Hausermann, and E. Effah, 2016. 'Site Specific: Heterogeneity of Small-Scale Gold Mining in Ghana.' *The Extractive Industries and Society* 3: 171–84. doi.org/10.1016/j.exis.2015.11.014

Financial Times, 2014. 'Mongolia Tames 'Ninja' Gold Miners to Support Currency.' 5 March. Available at blogs.ft.com/beyond-brics/2014/03/05/mongolia-tames-ninja-gold-miners-to-support-currency/#axzz2v7TeU9Hs

Fold, N., J. Jonsson, and P. Yankson, 2014. 'Buying Into Formalization? State Institutions and Interlocked Markets in African Small Scale Gold Mining.' *Futures* 62: 128–39. doi.org/10.1016/j.futures.2013.09.002

Hall, D., 2004. 'Smallholders and the Spread of Capitalism in Rural Southeast Asia.' *Asia Pacific Viewpoint* 45(3): 401–14. doi.org/10.1111/j.1467-8373.2004.00248.x

Hart, K., n.d. 'Informal Economy.' Available at thememorybank.co.uk/papers/informal-economy/

Harvard Kennedy School, 2011. *From Reformasi to Institutional Transformation: A Strategic Assessment of Indonesia's Prospects for Growth, Equity and Democratic Governance*. Cambridge, MA: Harvard Kennedy School Indonesia Program.

Hatcher, P., 2015. 'Neoliberal Modes of Participation in Frontier Settings: Mining, Multilateral Meddling, and Politics in Laos.' *Globalizations* 12(3): 322–46. doi.org/10.1080/14747731.2015.1016305

Hilson, G., 2013. '"Creating" Rural Informality: The Case of Artisanal Gold Mining in Sub-Saharan Africa.' *SAIS Review of International Affairs* 33(1): 51–64. doi.org/10.1353/sais.2013.0014

Hilson, G., 2014. 'The Minamata Convention: Exactly What Are We Doing?' Field Diary & Development Anecdotes. 10 November. Available at asmfielddiary.blogspot.co.id/2014/11/the-minamata-convention-exactly-what.html?spref=tw

Hilson, G. and A. Gatsinzi, 2014. 'A Rocky Road Ahead? Critical Reflections on the Future of Small-Scale Mining in Sub-Saharan Africa.' *Futures* 62: 1–9. doi.org/10.1016/j.futures.2014.05.006

Insouvanh, C., 2015. *Khmu Ethnic Group Women in Artisanal Gold Mining*. Paper presented at the Between the Plough the Pick: Informal Mining in the Contemporary World Conference, Canberra, The Australian National University, 5–6 November.

International Crisis Group (ICG), 2001. *Communal Violence in Indonesia: Lessons from Kalimantan*. Report No. 19. Jakarta/Brussels: ICG Asia.

International Union for Conservation of Nature Laos, 2005. *Gold Mining in Attapeu*. Vientiane: Mekong Wetlands Biodiversity Conservation and Sustainable Use Programme.

Kibreab, G., 2000. 'Common Property Resources and Resettlement.' In M. Cernea and C. McDowell (eds), *Risks and Reconstruction: Experiences of Resettlers and Refugees*. Washington, DC: World Bank.

Labonne, B., 2014. 'Who is Afraid of Artisanal and Small-Scale Mining (ASM)?' *The Extractive Industries and Society* 1: 121–3. doi.org/10.1016/j.exis.2014.03.002

Lahiri-Dutt, K., 2004. 'Informality in Mineral Resource Management in Asia: Raising Questions Relating to Community Economies and Sustainable Development.' *Natural Resources Forum* 28: 123–32. doi.org/10.1111/j.1477-8947.2004.00079.x

Land Issues Working Group, 2012. 'Mining in Nonghet, Chameun Village.' Lao Land Issues. Available at www.laolandissues.org/wp-content/uploads/2012/01/Case-Xieng-Khouang-Gold-Mining-Chameun-village-Nonghet.pdf

Langston, J., M. Lubis, J. Sayer, C. Margules, A. Boedhihartono and P. Dirks, 2015. 'Comparative Development Benefits from Small and Large Scale Mines in North Sulawesi, Indonesia.' *The Extractive Industries and Society* 2(3): 434–44. doi.org/10.1016/j.exis.2015.02.007

Li, T.M., 2000. 'Articulating Indigenous Identity in Indonesia: Resource Politics and the Tribal Slot.' *Comparative Studies in Society and History* 42(1): 149–79. doi.org/10.1017/S0010417500002632

Li, T.M., 2007. 'Practices of Assemblage and Community Forest Management.' *Economy and Society* 36(2): 263–93. doi.org/10.1080/03085140701254308

Lingga, M., 2016. 'Polres Katingan Amankan 5 Penambang Emas Ilegal.' *Tribrata News*, 28 August. Available at www.tribratanews.com/polres-katingan-amankan-5-penambang-emas-ilegal/

MacDonald, K., M. Lund, M. Blanchette and C. Mccullough, 2013. 'Regulation of Artisanal Small Scale Gold Mining (ASGM) in Ghana and Indonesia as Currently Implemented Fails to Adequately Protect Aquatic Ecosystems.' Proceedings of International Mine Water Association Symposium. Xuzhou, China: IMWA.

Maconachie, R. and G. Hilson, 2011. 'Safeguarding Livelihoods or Exacerbating Poverty? Artisanal Mining and Formalization in West Africa.' *Natural Resources Forum* 35: 293–303. doi.org/10.1111/j.1477-8947.2011.01407.x

McCarthy, J., 2013. 'Tenure and Transformation in Central Kalimantan: After the "Million Hectare" Project.' In Anton Lucas and Carol Warren (eds), *Land for the People: The State and Agrarian Conflict in Indonesia.* Athens: Ohio University Press.

Mitchell, T. 2002. *Rule of Experts: Egypt, Techno-Politics, Modernity.* Berkeley: University of California Press.

O'Faircheallaigh, C., A. Regan, D. Kikira and S. Kenema, 2016. *Small-Scale Mining in Bougainville: Impacts and Policy Responses: Interim report on Research Findings.* Griffith University. Available at ssgm.bellschool. anu.edu.au/sites/default/files/news/related-documents/2016-07/interimresearchfindings_ssm_bougainville_260516.pdf

Otsuka, K., J. Estudillo, and T. Yamano, 2010. 'The Role of Labour Markets and Human Capital in Poverty Reduction: Evidence from Asia and Africa.' *Asian Journal of Agriculture and Development* 7(1): 23–40.

Ouma, S., M. Boeckler and P. Lindner, 2013. 'Extending the Margins of Marketization: Frontier Regions and the Making of Agro-Export Markets in Northern Ghana.' *Geoforum* 48: 225–35. doi.org/10.1016/j.geoforum.2012.01.011

Peluso, N.L., 2016. 'The Plantation and the Mine: Agrarian Transformation and the Remaking of Land and Smallholders in Indonesia.' In J. McCarthy and K. Robinson (eds), *Land and Development in Indonesia: Searching for the People's Sovereignty.* Singapore: ISEAS, Yusof Ishak Institute.

Peluso, N.L., 2018. 'Entangled Territories in Small-Scale Gold Frontiers: Labor Practices, Property, and Secrets in Indonesian Gold Country.' *World Development* 101: 400–16. doi.org/10.1016/j.worlddev.2016.11.003

Peluso, N.L., A. Kelly and K. Woods, 2012. *Context in Land Matters: The Effects of History on Land Formalisations.* Bogor: Centre for International Forestry Research.

Probe Alert, 1991. *Subarnarekha Project in India: Uproots Tribal People, Transforms River Basin.* September. Toronto: Probe International. Available at probeinternational.org/library/wp-content/uploads/2011/04/September-1991-Probe-Alert.pdf

Purevjav, B., 2011. 'Artisanal and Small-Scale Mining: Gender and Sustainable Livelihoods in Mongolia.' In K. Lahiri-Dutt (ed.), *Gendering the Field: Towards Sustainable Livelihoods for Mining Communities.* Canberra: ANU E Press.

Putzel, L., A. Kelly, P. Cerutti and Y. Artati, 2014. *Formalization of Natural Resource Access and Trade: Insights from Land Tenure, Mining, Fisheries, and Non-Timber Forest Products.* Bogor: Centre for International Forestry Research.

Radio Free Asia, 2016. 'Laos Bans New Mining Projects in a Polluted Province.' 27 June. Available at www.rfa.org/english/news/laos/projects-06272016135647.html

Reno, W., 1999. *Warlord Politics and African States.* Boulder, CO: Lynne Rienner.

Schoenweger, O., A. Heinemann, M. Epprecht, J. Lu and P. Thalongsengchanh, 2012. *Concessions and Leases in the Lao PDR: Taking Stock of Land Investments.* Centre for Development and Environment, University of Bern. Bern and Vientiane: Geographica Bernensia.

Shah, A., 2013. 'The Agrarian Question in a Maoist Guerrilla Zone: Land, Labour and Capital in the Forests and Hills of Jharkhand India.' *Journal of Agrarian Change* 13(3): 424–50. doi.org/10.1111/joac.12027

Siegel, S. and M. Veiga, 2009. 'Artisanal and Small-Scale Mining as an Extra-Legal Economy: De Soto and the Redefinition of "Formalization".' *Resources Policy* 34: 51–6. doi.org/10.1016/j.resourpol.2008.02.001

Singo, P. and E. Levin, 2016. 'What Mongolia's Artisanal Miners Are Teaching Us: The Link between Human Rights and Artisanal and Small-Scale Mining (ASM) Formalisation.' *OECD Insights*, 16 March. Available at oecdinsights.org/2016/03/16/mongolias-artisanal-miners/

Spiegel, S., 2009. 'Resource Policies and Small-Scale Gold Mining in Zimbabwe.' *Resources Policy* 34: 39–44. doi.org/10.1016/j.resourpol.2008.05.004

Spiegel, S., 2012. 'Governance Institutions, Resource Rights Regimes, and the Informal Mining Sector: Regulatory Complexities in Indonesia.' *World Development* 40(1): 189–205. doi.org/10.1016/j.worlddev.2011.05.015

Spiegel, S., 2016. 'Land and "Space" for Regulating Artisanal Mining in Cambodia: Visualizing an Environmental Governance Conundrum in Contested Territory.' *Land Use Policy* 54: 559–73. doi.org/10.1016/j.landusepol.2016.03.015

Tschakert, P., 2009a. 'Digging Deep for Justice: A Radical Re-imagination of the Artisanal Gold Mining Sector in Ghana.' *Antipode* 41(4): 706–40. doi.org/10.1111/j.1467-8330.2009.00695.x

Tschakert, P., 2009b. 'Recognizing and Nurturing Artisanal Mining as a Viable Livelihood.' *Resources Policy* 34: 24–31. doi.org/10.1016/j.resourpol.2008.05.007

Tschakert, P., 2016. 'Shifting Discourses of Vilification and the Taming of Unruly Mining Landscapes in Ghana.' *World Development* 86: 123–32. doi.org/10.1016/j.worlddev.2016.05.008

Tyler, S. (ed.), 2006. *Communities, Livelihoods and Natural Resources: Action Research and Policy Change in Asia*. Ottawa: International Development Research Centre. doi.org/10.3362/9781780440101

United Nations Environment Programme (UNEP), 2012. *Analysis of Formalization Approaches in the Artisanal and Small-Scale Gold Mining Sector: Mongolia Case Study.* UNEP, June. Available at wedocs. unep.org/bitstream/handle/20.500.11822/11630/Case_Studies_ Mongolia_June_2012.pdf?sequence=1&isAllowed=y

United Nations Institute for Training and Research (UNITAR), 2016. *Satellite Mapping of Artisanal and Small Scale Gold Mining in Central Kalimantan, Indonesia.* UNITAR, United Nations Environment Programme (UNEP). Available at www.unitar.org/unosat/map/2368

United States Geological Survey, 2014. *2012 Minerals Yearbook: Laos.* November. US Department of the Interior, US Geological Survey.

Vandergeest, P., S. Ponte, and S. Bush, 2015. 'Assembling Sustainable Territories: Space, Subjects, Objects, and Expertise in Seafood Certification.' *Environment and Planning A* 47: 1–19. doi.org/ 10.1177/0308518X15599297

Veiga, M. (n.d.). *Artisanal Mining: Perspectives from the Field.* Viewed at globaldialogue.info/Nov1_ASM_2013/IGF%202013%20-%20 Artisanal%20Mining_%20Perspectives%20From%20the%20 Fiels%20-%20Marcello%20Veiga.pdf (site discontinued)

Verbrugge, B., 2014a. 'Capital Interests: A Historical Analysis of the Transformation of Small-Scale Gold Mining in Compostela Valley province, Southern Philippines.' *The Extractive Industries and Society* 1: 86–95. doi.org/10.1016/j.exis.2014.01.004

Verbrugge, B., 2014b. 'Artisanal and Small-Scale Mining: Protecting Those "Doing the Dirty Work".' IIED Briefing Note, October 2014. Available at pubs.iied.org/17262IIED

Verbrugge, B., 2015a. 'Undermining the State? Informal Mining and Trajectories of State Formation in Eastern Mindanao, Philippines.' *Critical Asian Studies* 47(2): 177–99. doi.org/10.1080/14672715.20 15.997973

Verbrugge, B., 2015b. 'The Economic Logic of Persistent Informality: Artisanal and Small-Scale Mining in the Southern Philippines.' *Development and Change* 46(5): 1023–46. doi.org/10.1111/ dech.12189

Verbrugge, B. and B. Besmanos, 2016. 'Formalizing Artisanal and Small-Scale Mining: Whither the Workforce?' *Resources Policy* 47: 134–41. doi.org/10.1016/j.resourpol.2016.01.008

Verbrugge, B., J. Cuvelier, and S. Van Bockstael, 2015. 'Min(d)ing the Land: The Relationship Between Artisanal and Small-Scale Mining and Surface Land Arrangements in the Southern Philippines, Eastern DRC and Liberia.' *Journal of Rural Studies* 37: 50–60. doi.org/10.1016/j.jrurstud.2014.11.007

Vientiane Times, 2010a. 'Attapeu Villagers Troubled by Muddy Waters.' 17 May.

Vientiane Times, 2010b. 'Gold Diggers Detained in Vientiane Province.' 15 February.

Vientiane Times, 2010c. 'Illegal Gold Mine Poisons Water Sources.' 26 March.

Vientiane Times, 2013. 'Unregulated Mining Poses Environmental Threat.' 16 September.

Vientiane Times, 2014a. 'Chinese Firm Escapes After Illegally Mining for Gold.' 29 July.

Vientiane Times, 2014b. 'Govt Tackles Illegal Gold Mining in Provinces.' 10 April.

Vientiane Times, 2014c. 'Illegal Gold Mining in Xieng Khouang Ends.' 26 September.

Vientiane Times, 2015. 'Many Miners Found to be in Violation of Agreements.' 12 March.

Vientiane Times, 2016. 'PM Announces Continued Suspension of Mining Concessions.' 20 October.

World Bank, 2013. *Artisanal and Small-Scale Mining*. 21 November. Available at www.worldbank.org/en/topic/extractiveindustries/brief/artisanal-and-small-scale-mining

Postscript

Kuntala Lahiri-Dutt

The contributors to this book have covered wide ground, with many of them charting new paths and providing new dimensions through which to think about informal mining in the contemporary world. The chapters have literally taken the readers from one corner of the globe to another, from Southeast Asia and mainland Asia, to the heart of Africa and Latin America. They have not only followed the historical trajectory of informal, artisanal and small-scale mining at continental scales, but also at national and local scales. In the process, contributors have examined new dimensions of artisanal and small-scale mining (ASM), the idea of informality and the broader political–economic processes within which it takes place, and dealt with the difficult questions that such mining raises. In the process, the contributors have plotted the future direction of research into the broader thematic topic.

Clearly, three important aspects emerge from the chapters as crucial for scholarly understanding. The first is to link the growth of informal mining with political–economic processes. Hidden within these processes are the causes of the sudden mushrooming of informal mining in locations and contexts that have no history of such activities. Dealing with this question would involve understanding the broader context: rural stagnation; structural adjustment programs that leave the rural poor to fend for themselves; unhelpful states pursuing self-interest that outweighs obligation to their citizens; and the global reach of the market in which international commodity prices touch the lives of the poor living in the most remote areas of the world. Even though contemporary rushes for minerals remind us of those of the past, examining the finer details reveals them as somewhat different, precisely because of these factors that constitute the background within which contemporary extractions are occurring. Again, concealed within these broader political and economic processes driven by global forces lie particularities that are essentially local in nature; for example, the specific trigger factor(s) that lead to a sudden enhancement of mining activities in an area.

The second aspect is the need to understand mining organisations and the oftentimes complex and exploitative labour processes that define informal, artisanal and small-scale mining. The lone entrepreneurial miner is largely fictional, and modes of mining production exhibit a vast range, from family units operating solo or in groups, to reasonably capital-intensive operations managed by contractors on behalf of owners of land or capital equipment. The contractors act as intermediaries, hiring labour locally and ensuring that they are paid the lowest wages possible. The precarity of labour, rather than legal attributes, is one of the manifestations of the informality of ASM. The cheap labour, lying at the bottom of the production process, makes this kind of mining possible and productive. Of this labour, we now know a large proportion are women, sometimes acting as independent wage earners, but most often acting as part of the family unit of labour, and producing minerals on piece-rate basis for up to 12 hours per day. Whether these meagre incomes add to their empowerment—that elusive goal for feminist development planners— by providing a way out of poverty and exploitation remains a pressing concern. The back-breaking work that women and men perform in such mining leads to innumerable and serious health issues amongst the workers. Again, the question as to how best to deal with them poses a challenge that remains intractable.

The third aspect is the management of mineral resources and communities' rights. 'Governance' of ASM is a challenge in situations of inadequate knowledge, especially since institutions have typically aimed to facilitate the extraction of minerals by companies and corporations. There remain gaps in knowledge, lacunae in laws and poor recognition of more equitable systems of governance in legal frameworks. These appear as primary obstacles in helping those toiling away in the mines. More importantly, there are undeniably close links between informal mineral extraction and political power. In this context, governance needs to be creative and innovative, taking into consideration the 'assemblage' of factors that create the local context.

Several almost intractable difficulties inherent to informal, artisanal and small-scale mining become apparent in a close reading of the chapters. I will note only three that are important for consideration of the chapters in this book. The first is geographical. Whilst we at once become aware of the in situ, place-based, local nature of this kind of mining, we also become acutely aware of how some of these local transformations are integrally linked to global processes and systems. For example, high commodity process in international markets, and the ease of communication through which a local place becomes global, allowing inhabitants to participate in global market processes and supply chains.

The second inherent contradiction is analytical. In a situation in which absolute ownership of mineral resources by the state is accepted, how does one envisage the judicious, equitable and sustainable management of these resources? Is it even possible to relieve ASM of its perceived illegality? Is it possible even to contemplate co-management of resources that have so far been treated as open and lootable commodities? In other words, and to put it simply, is it possible to integrate contemporary notions of justice and equity, and contemporary processes of participation and inclusion of the marginalised, into the management of mineral resources?

The third problematic is linked to the two noted above, and concerns the elephant in the room—the sustainability of environmental integrity. This concern has somehow taken a backseat in discussions of informal, artisanal and small-scale mining in recent years, as researchers have tried to point to its historical roots, livelihood concerns and to the important economic contributions it makes. The major issue has been mercury pollution, with global research tending to focus on the damages caused by mercury use by artisanal gold miners. However, one could include deforestation, deteriorating water quality, changing river regimes, damages to riverbeds and lowering of air quality through dust generation and open-air coking of artisanally mined coal. Indeed, the harm is not limited to the local area. Whilst such mining provides a crucial livelihood for those who are involved in it, the effects are commonly borne by those often living in areas beyond the mining tracts, and by those who do not receive any benefit whatsoever from it. Therefore, questions of environmental justice and equity assume an important dimension. If a political economy lens allows us to draw attention to the extractivism inherent in states' neoliberal policies, then the need for the protection of environmental integrity for future generations encourages us to consider the political and economic questions that rise at the local scale. However, the widespread nature of ASM implies that there is a growing need to move beyond the micro scale to the macro, and from the present into the future. No extractivism is sustainable, whether practised by states or individuals, irrespective of the pressing, compelling factors that they might be subject to.

The task for future researchers is to broaden the definition of mining, but at the same time link ASM with rural development and labour processes. We now know that stagnating rural economies and oppressive poverty are the primary persuasion pushing many rural folks into mining to 'try their luck', but what we do not know is how many of them—if any— can indeed strike it rich. In other words, how do mining individuals, families and communities handle and perceive money and wealth, and in what ways do they build assets, and of what kind? How do gender relations change once families and communities begin to diversify their livelihoods? This book has provided more questions than answers, and the list could be endless. Indeed, can we ask if it is possible to think of such mining within the framework of a moral economy? We know how large-scale mineral extraction undermines ordinary people's livelihoods and benefits the companies and corporations besides the states. In this context, can one theorise informal mining as an expression of resistance?

These are some questions amongst many for the future generations of scholars to answer. I will end this book with these questions, hoping that they will be debated by upcoming researchers.

www.ingramcontent.com/pod-product-compliance
Lightning Source LLC
Chambersburg PA
CBHW042319210326
41599CB00048B/7146